Applied Nanotechnology and Nanoscience International Conference (ANNIC 2017)

Journal of Physics: Conference Series Volume 987

Rome, Italy
18 - 20 October 2017

Editor:

Yian Tai

ISBN: 978-1-5108-6123-7
ISSN: 1742-6588

Printed from e-media with permission by:

Curran Associates, Inc.
57 Morehouse Lane
Red Hook, NY 12571

Some format issues inherent in the e-media version may also appear in this print version.

Printed by Curran Associates, Inc. (2018)

For permission requests, please contact the Institute of Physics at the address below.

Institute of Physics
Dirac House, Temple Back
Bristol BS1 6BE UK

Phone: 44 1 17 929 7481
Fax: 44 1 17 920 0979

techtracking@iop.org

Additional copies of this publication are available from:

Curran Associates, Inc.
57 Morehouse Lane
Red Hook, NY 12571 USA
Phone: 845-758-0400
Fax: 845-758-2633
Email: curran@proceedings.com
Web: www.proceedings.com

Table of contents

Volume 987

Applied Nanotechnology and Nanoscience International Conference 2017 (ANNIC 2017)

18–20 October 2017, Rome, Italy

Accepted papers received: 14 March 2018
Published online: 5 April 2018

Preface

Applied Nanotechnology and Nanoscience International Conference 2017 (ANNIC 2017)

Peer review statement

Papers

Investigation on nonlinear optical properties of MoS_2 nanoflake, grown on silicon and quartz substrates

S Bayesteh, S Z Mortazavi and A Reyhani.....1

Metal nanolayer deposited highly stable Ag thin films and their optical properties

M Kawamura, T Kiba, Y Abe, K H Kim and H Murotani.....7

Pd surface functionalization of 3D electroformed Ni and Ni-Mo alloy metallic nanofoams for hydrogen production

A Petica, O Brincoveanu, F Golgovici, A C Manea, M Enachescu and L Anicai.....12

The method of synthesizing of superhydrophobic surfaces by PECVD

Sagi Orazbayev, Maratbek Gabdullin, Tlekkabul Ramazanov, Merlan Dosbolayev, Askar Zhunisbekov, Dulat Omirbekov and Zhuldyz Otarbay.....20

Detection of organic vapors on sputtered and annealed thin Au films

O Kvitek, V Kopacek, A Reznickova and V Svorcik.....26

Niobium oxide nanocolumns formed via anodic alumina with modulated pore diameters

A Pligovka, A Zakhlebayeva and A Lazavenka.....32

Formation of various types of nanostructures on germanium surface by nanosecond laser pulses

S I Mikolutskiy, R R Khasaya, Yu V Khomich and V A Yamshchikov.....36

Effects of (Ce, Cu) Co-doping on the Structural and Optical Properties of ZnO Aerogels Synthesized in Supercritical Ethanol

D Djouadi, O Slimi, L Hammiche, A Chelouche and T Touam.....41

Thermally induced fading of Mn-doped YAP nanoceramics

Ya Zhydachevskyy, N Martynyuk, A I Popov, D Sugak, P Bilski, S Ubizskii, M Berkowski and A Suchocki.....50

Design and analysis of non-linear properties of photonic crystal fiber with Various Doping Concentration

M. Benhaddad, F. Kerrour and O. Benabbes.....53

Effect of Processing Parameters on the Morphology of PVDF Electrospun Nanofiber

M A Zulfikar, I Afrianingsih, M Nasir and A Alni.....58

Thermal and non-thermal explosion in metals ablation by femtosecond laser pulse: classical approach of the Two Temperature Model

Ahmed Abdelmalek, Zeyneb Bedrane and El-Hachemi Amara.....63

Characterization of Polycaprolactone and Rice Husk Silica Composite (PCL-SiO_2) by E-Spinning to Apply Supporter for Drug Release

Sinae Song, Askwar Hilonga and Hee Taik Kim.....70

Indium-saving effect and physical properties of transparent conductive multilayers

M Kawamura, T Kiba, Y Abe and K H Kim.....77

Rapid thermal process by RF heating of nano-graphene layer/silicon substrate structure: Heat explosion theory approach

M Sinder, J Pelleg, V Meerovich and V Sokolovsky.....82

Electrical properties of thin films deposited from TMS/O_2 in Microwave Multipolar Plasma reactor

M Kihel, S Sahli, Y Fermi, P Raynaud and M Benhaddad.....87

A built-in sensor with carbon nanotubes coated by Ag clusters for deformation monitoring of glass fibre/epoxy composites

P. Slobodian, P. Riha, J. Matyas, R. Olejnik, S. Lloret Pertegás, R. Schledjewski and M. Kovar.....92

Fabrication and characterization of Aerogel-Polydimethyl siloxane (PDMS) Insulation Film

Yeoung ah Noh, Sinae Song and Hee Taik Kim.....98

Shift of semimetal-semiconductor bond direction on "0 1 1" to "1 1 1" Bismuth quazi-two-dimension system

Ahmad Yazdani and Sajad Hamreh.....105

Kinetics of the melting front movement in process of centrifugal induction surfacing of powder material with nanoscale modificaters

I Sasnouski and A Kurylionak.....110

Stretchable and flexible thermoelectric polymer composites

P. Slobodian, P. Riha, J. Matyas and R. Olejnik.....114

Using graphene/styrene-isoprene-styrene copolymer composite thin film as a flexible microstrip antenna for the detection of heptane vapors

Robert Olejnik, Jiri Matyas, Petr Slobodian and Pavel Riha.....119

Study of spectroscopic properties of nanosized particles of core-shell morphology

T N Bzhalava and P J Kervalishvili.....124

Suppression and enhancement of deep level emission of ZnO on Si^{4+} & V^{5+} substitution

T Srivastava, G Bajpai and S Sen.....130

Hydrogenization of fullerene as a method of storage of hydrogen

Zh.E. Otarbay, M.T. Gabdullin, Kh.A. Abdullin, D.V. Shur, D.V. Ismailov, Ye. Yerlanuly and D.S. Kerimbekov.....135

Comparison between layering $NbSe_2$ and rod characteristic of MgB_2 by investigation of elastic constants

Asiye Shokri, Ahmad Yazdani and Behrad Barakati.....141

Dynamic Consolidation and Investigation of Nanostructural W-Cu / W-Y Cylindrical Billets

B. Godibadze, A. Dgebuadze, E. Chagelishvili, G. Mamniashvili and A. Peikrishvili.....147

Decoration of wide bandgap semiconducting materials for enhancing photoelectrochemical efficiency of PEC systems.

N Bakranov, A Zhabaikhanov, S Kudaibergenov and N Ibraev.....153

Structural and optical properties of $SiC-SiO_2$ nanocomposite thin films

I. Bozetine, A. Keffous, S. Kaci, H. Menari and A. Manseri.....161

Scattering Resonances in bilayer graphene

Michele Pisarra, Antonio Sindona and Pierfrancesco Riccardi.....167

Characterization of local atomic structure in Co/Zn based ZIFs by XAFS

Yulia Podkovyrina, Vera Butova, Elena Bulanova, Andriy Budnyk, Maria Kremennaya, Alexander Soldatov and Carlo Lamberti.....172

Wireless digital pressure gauge based on nanomaterials

Dilyara Abay, Zhuldyz Otarbay, Madengul Token, Nazim Guseinov, Mukhit Muratov, Maratbek Gabdullin and Daniyar Ismailov.....176

All optical programmable logic array (PLA)

Dawit Hiluf.....181

Graphene like ZnO 2D structure from semiconductor to conductor behaviour by doping F

Ahmad Yazdani, Behrad Barakati and Asiye Shokri.....194

Investigation of the phase formation from Ni-modified nanostructured silicon

O V Volovlikova, Y I Shilyaeva, A Y Berezkina, D I Smirnov and S A Gavrilov.....198

Magnetic Thermometer: Thermal effect on the Agglomeration of Magnetic Nanoparticles by Magnetic field

Daeseong Jin and Hackjin Kim.....204

Determination of permeability of ultra-fine cupric oxide aerosol through military filters and protective filters

E Kellnerová, Z Večeřa, J Kellner, T Zeman and J Navrátil.....210

Magnetic, thermal and luminescence properties in room-temperature nanosecond electron-irradiated various metal oxide nanopowders

S Yu Sokovnin, M E Balezin and V G Il'ves.....218

Investigation of the Morphological Evolution and Etching Kinetics of black Silicon During Ni-Assisted Chemical Etching

OV Volovlikova, G O Silakov, S A Gavrilov, A A Dudin, G O Diudbin and Y I Shilyaeva.....224

Synthesis of nanocrystalline α-Fe2O3 by using thermal oxidation of Fe Films

G Fortas, I Saidoun, H Abboud, N Gabouze, N Haine, A Manseri, M Zergoug, H Menari, S Sam, H Cheraga and I Bozetine.....232

Simple morphological control over functional diversity of SERS materials

A A Semenova and E A Goodilin.....239

Characterization and Comparison of Photocatalytic Activity Silver Ion doped on TiO_2(TiO_2/Ag+) and Silver Ion doped on Black TiO_2(Black TiO_2/Ag^+)

Jin Yi Kim, Ho Hyung Sim, Sinae Song, Yeoung Ah Noh, Hong Woon Lee and Hee Taik Kim.....245

On low-dimensional models at NMR line shape analysis in nanomaterial systems

M M Kucherov and O V Falaleev.....251

Adherence of amino acids functionalized iron oxide nanoparticles on bacterial models *E. Coli* and *B. subtilis*

W. Trujillo, J. Zarria, J. Pino, L. Menacho, M. Coca and A. Bustamante.....257

IOP Conf. Series: Journal of Physics: Conf. Series **987** (2018) 011001 doi:10.1088/1742-6596/987/1/011001

Preface

Introduction:

The Applied Nanotechnology and Nanoscience International Conference is an international meeting organized in a different European country each year. The Chairs Committee and Plenary Speakers change each year in order to enrich the content and provide the participants with different views.

After the success of Paris (2015) and Barcelona (2016), the Applied Nanotechnology and Nanoscience International Conference 2017 was held in Rome, on October 18th -20th.
The main achievements have been to:

- Attract high-quality papers in different fields.
- Offer the opportunity to be updated on the latest research outputs on several Nanotechnology and Nanoscience topics.
- Deliver specific workshops around the most attractive and current issues.
- Gather worldwide experts as conference speakers.

Group Picture:

Main Sponsor: NFFA (http://www.nffa.eu/)

IOP Conf. Series: Journal of Physics: Conf. Series **987** (2018) 011001 doi:10.1088/1742-6596/987/1/011001

Peer review statement

All papers published in this volume of *Journal of Physics: Conference Series* have been peer reviewed through processes administered by the proceedings Editors. Reviews were conducted by expert referees to the professional and scientific standards expected of a proceedings journal published by IOP Publishing.

IOP Conf. Series: Journal of Physics: Conf. Series **987** (2018) 012001 doi:10.1088/1742-6596/987/1/012001

Investigation on nonlinear optical properties of MoS_2 nanoflake, grown on silicon and quartz substrates

S Bayesteh[1], S Z Mortazavi[1*] and A Reyhani[1]

[1] Physics Department, Faculty of Science, Imam Khomeini International University, P.O.Box 34149-16818, Qazvin, Iran
*Email: z.mortazavi@sci.ikiu.ac.ir

Abstract. In this study, MoS_2 was directly synthesized by one-step thermal chemical vapour deposition (TCVD), on different substrates including Si/SiO_2 and quartz, using MoO_3 and sulfide powders as precursor. The XRD patterns demonstrate the high crystallinity of MoS_2 on Si/SiO_2 and quartz substrates. SEM confirmed the formation of MoS_2 grown on both substrates. According to line width and frequency difference between the E^1_{2g} and A_{1g} in Raman spectroscopy, it is inferred that the MoS_2 grown on Si/SiO_2 substrate is monolayer and the MoS_2 grown on quartz substrate is multilayer. Moreover, by assessment of MoS_2 nanoflake band gap via UV-visible analysis, it verified the formation of few layer structures. In addition, the open-aperture and close-aperture Z-scan techniques were employed to study the nonlinear optical properties including nonlinear absorption and nonlinear refraction of the synthesized MoS_2. All experiments were performed using a diode laser with a wavelength of 532 nm as light source. The monolayer MoS_2 synthesized on Si/SiO_2, display considerable two-photon absorption. However, the multilayer MoS_2 synthesized on quartz displayed saturable absorption (SA). It is noticeable that both samples demonstrate obvious self-defocusing behaviour.

1. Introduction

Two-dimensional (2D) transition metal dichalcogenides (TMD) semiconductors such as MoX_2 (X = S, Se, Te), have recently received increasing attention research and industrial interest because of their prominent electronic and optical properties which are a natural result of the quantum confinement effect related with their thin 2D structure [1] and also their potential applications for the next-generation of flexible field-effect transistors (FET) [2] and solar cells [3]. MoS_2 semiconductor has been illustrated to be viable for many device applications such as ultrafast lasers [4]. The layer-dependent properties of MoS_2 has recently drawn wide attention, e.g., bulk or multilayer MoS_2 with an indirect band gap of 1.2 eV and monolayer MoS_2 with a direct band gap of 1.8 eV [5]. The indirect-to-direct transition results in the exhibitive enhancement of photoluminescence [6]. Also the optical properties of MoS_2 show interesting layer affiliation. This optical property makes MoS_2 a potential optical material with various nonlinear optical responses like optical limiting, saturable absorption, two-photon absorption and second and third harmonic generations [7, 8]. This optical materials are used for photonic devices such as optical limiters and Q-switchers [9]. The layer structure of MoS_2 consists of covalently bonded S-Mo-S layers, held together through interlayer Van der Waals interactions. For growth of 2D-MoS_2, different methods have been reported, including mechanical exfoliation, chemical vapour deposition (CVD) and physical vapour deposition [10, 11]. Among them, CVD has potential for large-scale, low cost manufacturing of 2D-MoS_2. There are several reports on the effect of substrate for growing of 2D-MoS_2 by CVD. These suggest a significant effect of the substrate on growth of 2D-MoS_2 [12, 13]. Also, Very recently, a wide range of conclusions has been reported in this regard conveying saturable absorption, nonlinear absorption, nonlinear scattering and nonlinear refraction of few layers MoS_2 [14, 15]. In this work, a simple method for growing MoS_2 layered structures through thermal evaporation of MoO_3 nano-powder along with sulfur on different substrates (Si/SiO_2 and quartz) was investigated. The synthesis of the thin films was carried out by using one-step thermal chemical vapour deposition (TCVD). The resulting thin films were then characterized using scanning electron microscopy (SEM), micro Raman spectroscopy, X-ray diffraction (XRD), UV-Visible and photoluminescence (PL). Then, the nonlinear optical properties of the obtained nanoflakes were studied by Z-scan technique. The results are demonstrating saturable absorption (SA) performance in multilayer MoS_2 nanoflakes growth on quartz substrate and two-photon absorption (TPA) performance in monolayer MoS_2 nanoflakes growth on

Si/SiO_2 substrate. It is noticeable that both samples demonstrate obvious self-defocusing behaviour. The real and the imaginary parts of the third-order nonlinearity were specified by means of the closed and open aperture Z-scan.

2. Experimental setup

The growth of the nanoflakes were performed in a horizontal furnace with a diameter of 2 inch quartz tube at atmospheric pressure. The schematic of our experiments is shown in figure 1. At the beginning, for removing native oxides, the silicon substrate were soaked in hydrofluoric acid (HF) (10%). then, 200 nm SiO_2 was sputtered on the silicon substrates by RF magnetron sputtering (DST3 model). Then, before each experiment, the both substrates (Silicon and quartz) were sonicated in acetone, ethanol and DI water, respectively, for duration of 10 minutes and they were then dried for 3 minutes at 80 °C. A ceramic boat containing sulfur powder (TITRACHEM 99, 0%, 500 mg) was placed outside the hot zone at the lower-temperature entrance of the tube (150-200 °C). Another ceramic boat containing MoO_3 powder (MERCK 99.5%, 50 mg) was placed at the highest-temperature of the tube (900 °C) where the reaction occurred (the distance between the ceramic boats was about 25 cm). Afterward, the substrate was placed face down on top of the boat containing MoO_3 at the hot zone. The temperature of the furnace was gradually increased to 900 °C within 20 minutes (45 °C/min) in the Ar environment. The growth time was set at 60 min for all of the samples to react the precursors via Ar flow rate of 200 sccm as carrier gas and finally the temperature was ramped down to room temperature. The growth films were analyzed by Micro Raman spectroscopy (confocal PerkinElmer spectrum version 10.03.06) with exciting laser wavelength at 785 nm, XRD (X'Pert PRO MPD), SEM (XL30), UV- Visible (stellarNet EEP- 2000) (in the range of 200-800 nm) and PL (Perkin Elmer LS55) excited with xenon lamp at wavelength of 525 nm.

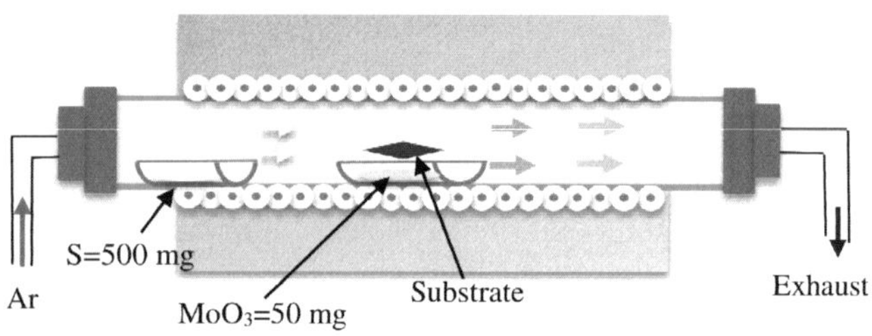

Figure 1. Schematic of the one-step TCVD experimental setup. Ceramic boats containing sulfur (500 mg) powder and MoO_3 powder (50 mg) was placed respectively at the higher-temperature center of the tube and at the lower-temperature. The substrate (Si/SiO_2, quartz) was placed face down on top of the boat containing MoO_3 powder at the hot zone.

3. Result and Discussion

3.1. structure

The structural properties of the nanoflakes was studied using XRD analysis. The XRD patterns of the nanoflakes that were grown on different substrates are shown in figure 2a. It can be observed that both of the XRD patterns present three peaks at 2θ=14.4°, 29.2° and 44.4° (Ref Code: 01-077-0341) corresponding to the (003), (006) and (009) plane of 3R-MoS_2, respectively. Beside, three peaks were observed at 2θ=18.4°, 26° and 37.3° (Ref Code: 01-086-0135) corresponding to the (100), (011) and (200) plane of MoO_2, respectively. Also, the intensity of the peaks corresponding to MoS_2 (003), MoS_2 (006) and MoS_2 (009) increases on quartz substrate. It seems that the growth of the MoS_2 nanoflakes on quartz substrate were more than Si/SiO_2 substrate. The morphology of the nanoflakes grown on different substrates, clearly found out by the SEM images as shown in figure 2b and 2c. According to SEM images, the nanoflakes MoS_2 grown on Si/SiO_2 substrate can be obtained the same size with relative uniform distributions. However, the nanoflakes MoS_2 were obviously grown on quartz substrate with vertically oriented, larger size and more thickness. For investigating the optical properties of the thin films grown, UV-visible spectroscopy was carried out. figure 2d shows the absorption spectra of the obtained thin films on different substrates. Two separate excitonic absorption peaks can be seen at 658nm (1.88 eV) and 611 nm (2.02 eV) for Si/SiO_2 substrate, at 679 nm (1.82 eV) and 630 nm (1.96 eV) for quartz substrate. These peaks are signs of the A and B exciton transitions and they are also reported as a direct transition at the K point of the Brillouin zone [16].

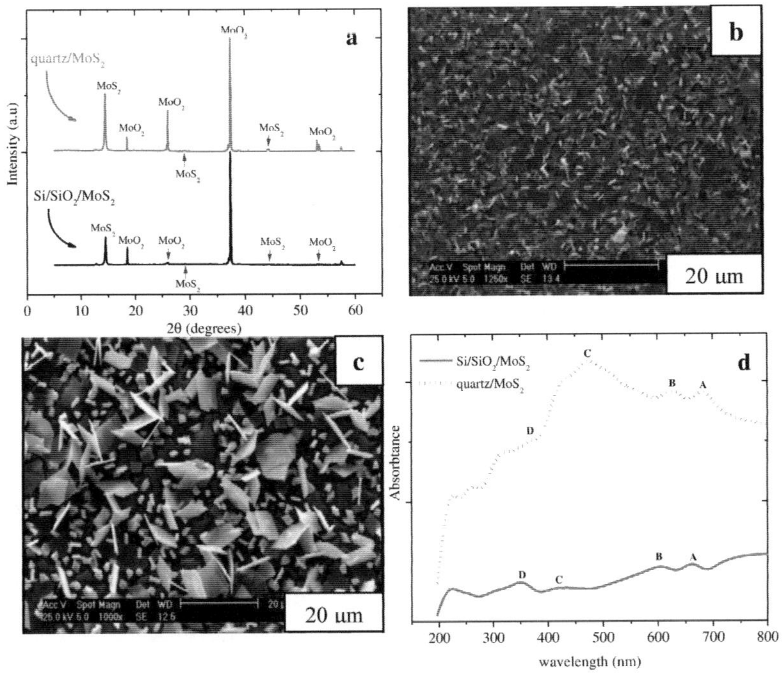

Figure 2. a) XRD patterns of the thin films grown on different substrates, b) SEM images of the thin films grown on Si/SiO$_2$ substrate, c) SEM images of the thin films grown on quartz substrate, d) UV-visible spectra of the thin films grown on different substrates.

In addition, figure 2d shows C and D absorption peaks associated with van Hove uniqueness [17]. The existence of van Hove uniqueness makes the monolayer MoS$_2$ suitable for photovoltaic cells. The difference between the two exciton transitions, due to the spin-orbital splitting of the valence band, was found to be 0.14 eV for both substrates, consistent with the theoretical values for 2D-MoS$_2$ (monolayer MoS$_2$) [18]. The results demonstrate that the MoS$_2$ nanoflakes grown on different substrates have a direct band gap of about 1.8 eV.

We characterized the synthesized of the films using Raman spectroscopy and also to affirm the thickness of the thin films. In figure 3a and b, the two characteristic modes can be obtained from the spectra: the A_{1g} mode corresponds to the out-of-plane vibration of sulfur atoms and the E_{2g}^1 mode is linked to the in-plane vibration of Mo and sulfur atoms [19]. The frequency difference between these modes (ΔK) of the MoS$_2$ thin film is linked to the thickness of the MoS$_2$ film and the number of layers. The E_{2g}^1 and A_{1g} modes for the nanoflakes grown on Si/SiO$_2$ substrate were observed at 386 cm^{-1} and 406 cm^{-1} (ΔK=20 cm^{-1}) (monolayer MoS$_2$), respectively, while that for the nanoflakes grown on quartz substrate, the E_{2g}^1 and A_{1g} modes were observed at 385 cm^{-1} and 410 cm^{-1} (ΔK=25 cm^{-1}) (multilayer MoS$_2$), respectively.

The photoluminescence spectrum of MoS$_2$ thin film grown on Si/SiO$_2$ substrate is presented in figure 3c confirm the uniformity and thickness of the monolayer MoS$_2$ film. Two PL peaks A and B, can be found around 676 nm (1.83 eV) and 619 nm (2 eV), respectively. These two peaks commonly are fingerprints of the band edge emission of MoS$_2$ from valence band splitting due to spin-orbit split [20].

3.2. Nonlinear optics

The single beam Z-scan technique was used to investigate on the nonlinear optical properties of the nanoflakes containing nonlinear absorption and nonlinear refraction. The experimental setup is shown in figure 4. A diode laser with wavelength at 532 nm was used as the light source. As the sample is translated through the focal region of the beam, the photodiode measures the fraction of intensity passing through the aperture, which is transmitted through the sample. For determining the nonlinear absorption, Open-aperture (OA) was employed, whereas closed-aperture (CA) was utilized to determine the nonlinear refraction of the nanoflakes. The intense laser beam was sent through a "long" focal length lens with a focal length of 10 cm. While the sample was consecutively moved along the z-axis via the beam waist, the OA Z-scan measured total transmittance via the

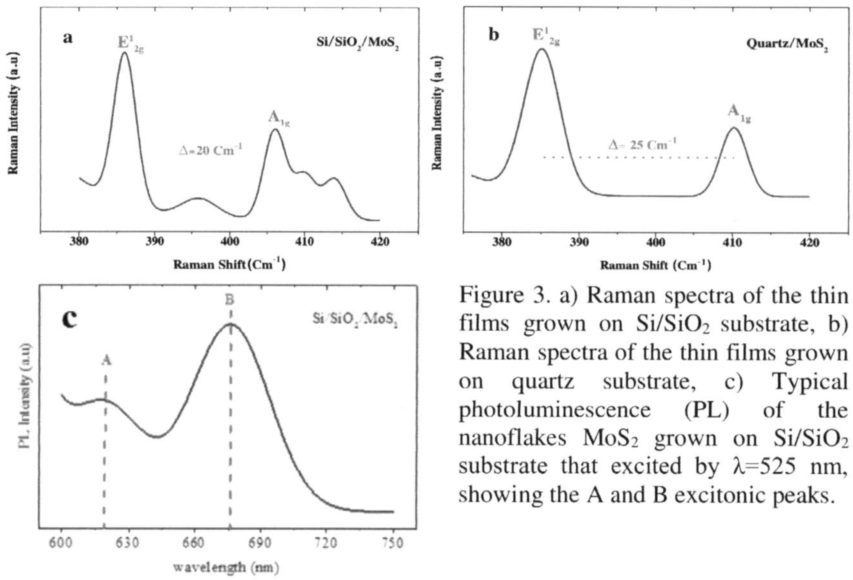

Figure 3. a) Raman spectra of the thin films grown on Si/SiO$_2$ substrate, b) Raman spectra of the thin films grown on quartz substrate, c) Typical photoluminescence (PL) of the nanoflakes MoS$_2$ grown on Si/SiO$_2$ substrate that excited by λ=525 nm, showing the A and B excitonic peaks.

-sample as a function of incident laser intensity and the photodiode measured the changes in the absorbance of light. For CA Z-scan a small aperture for limiting the transmitted beam incident on the detector had used. Finally, the technique gives the real and imaginary parts of the third-order susceptibility, χ^3.

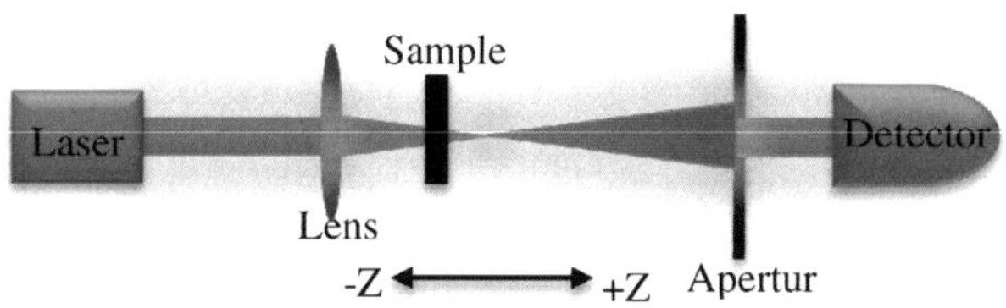

Figure 4. Schematic diagram of the Z-scan experimental setup.

Figure 5 shows the normalized transmittance of closed and open aperture Z-scan of the nanoflakes grown on different substrates (Si/SiO$_2$ and quartz). As demonstrated in figure 5a and b, the nanoflakes were displayed a pre-focal peak followed by a post-focal valley. Therefore, the negative value of the nonlinear refractive index, observed obvious self-defocusing behaviour in both samples. The change in normalized transmittance between peak and valley, $\Delta T_{P\text{-}V} = T_P - T_V$, can be applied for calculating the nonlinear refractive index (n_2) using Equation.1:

$$\Delta T_{P-V} = 0.406\,(1-S)^{0.27}|\Delta\varphi_0| \quad \text{for } |\Delta\varphi_0| \le \pi$$
$$\text{and} \quad n_2 = \left(\frac{\lambda}{2\pi}\right)\frac{\Delta\varphi_0}{I_0 L_{eff}} \qquad (1)$$

Where S is the fraction of beam transmitted via the aperture that is defined by S=1-exp($^{-2r_a^2}/_{w_a^2}$) that r_a and w_a show the radius of aperture and the beam radius at the aperture, respectively, $\Delta\varphi_0$ is the magnitude of phase distortion, I_0 is the intensity of the laser beam at focal point that is equal to I_0=7269.093 W/cm^2 and $L_{eff} = [(1 - exp(-\alpha_0 l)) / \alpha_0]$ is effective thickness with linear absorption coefficient $\alpha_0 = -\frac{1}{l}ln(\frac{I}{I_0})$. Where, I_0 and I are the intensity of the incident and the transmitted radiation, respectively and l is the thickness of sample. The calculated values of the nonlinear refractive indexs for the nanoflakes are tabulated in Table 1.

The open-aperture Z-scan curves of the monolayer MoS$_2$ grown on Si/SiO$_2$ substrate, in figure 5c, is illustrated a minimum transmission when the samples achieved the focal point (Z=0). Normalized transmittance peaks in figure 5c, exhibit obvious two-photon absorption (TPA) response, corresponding to positive sign nonlinear absorption. However, the curve in figure 5d, the multilayer MoS$_2$ grown on quartz substrate, is illustrated a maximum transmission when the sample achieved the focal point and a normalized transmittance peak, display the presence of saturable absorption (SA) response, corresponding to negative sign nonlinear absorption.

Interestingly, in comparison with the monolayer MoS_2 grown on Si/SiO_2 substrate, the multilayer MoS_2 grown on quartz substrate demonstrates a SA response under the same excitation conditions. The normalized change in transmitted intensity *($\Delta T(Z) = T(Z)-1$)* can be estimated by using the Equation 2:

$$\Delta T(z) \approx \frac{q_0}{2\sqrt{2}}\left[\frac{1}{1+Z^2/Z_0^2}\right] \qquad (2)$$

Where $Z_0 = kw_0^2/2$ is the diffraction length of the focused beam. Nonlinear absorption coefficient, β, can be determined using the Equation 3:

$$q_0=\beta I_0 L_{eff} \qquad (3)$$

The obtained values of nonlinear absorption coefficient β are shown in Table 1. The results showed that the nonlinear absorption coefficient of multilayer MoS_2 grown on quartz substrate was impressively increased. The real and imaginary parts of the third-order nonlinear optical susceptibility χ^3 are computed by using the following equations:

$$Re\,\chi^3(esu) = (10^{-4}\varepsilon_0 c^2 n_0^2/\pi)n_2\,(Cm^2/W) \qquad (4)$$

and

$$Im\,\chi^3(esu) = (10^{-2}\varepsilon_0 c^2 n_0^2\lambda/4\pi^2)\beta\,(Cm/W) \qquad (5)$$

Where n_2 is linear refractive index, β is nonlinear absorption coefficient, ε_0 is the vacuum permittivity, c is the speed of light in vacuum and n_0 is the linear refractive index of the MoS_2 that is equal to 4.77. Based on this theory, results of experiments are used to calculate susceptibility (Re χ^3 and Im χ^3), which are listed in Table 1.

Table 1: Nonlinear optical parameters of the MoS_2 thin films.

Sample	n_2 (Cm^2/W)	β (Cm/W)	$\lvert Re\,\chi^3\rvert$ (esu)	$\lvert Im\,\chi^3\rvert$ (esu)
$Si/SiO_2/MoS_2$	-3.11×10^{-4}	31.18	0.1798×10^{-4}	7.6383×10^{-4}
$Quartz/MoS_2$	-5.16×10^{-4}	-145.72	0.2×10^{-4}	35.69×10^{-4}

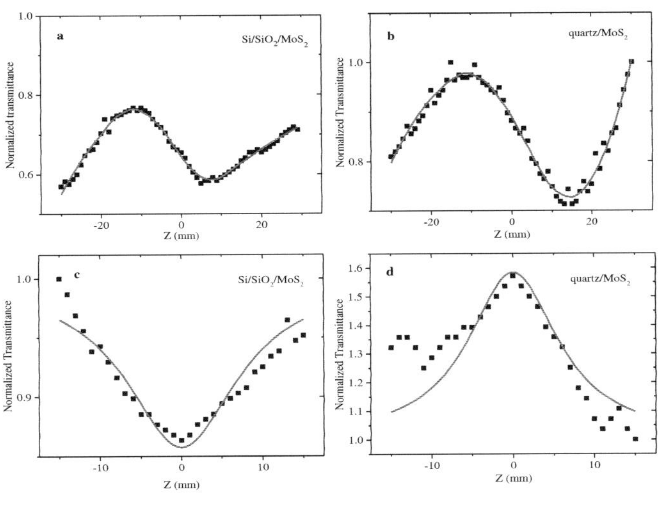

Figure 5. Z-scan results of the nanoflakes grown on different substrates. a) Closed aperture (CA) Z-scan result of the nanoflakes grown on Si/SiO_2 substrate, b) CA Z-scan results of the nanoflakes grown on quarts substrate, c) open-aperture (OA) of the nanoflakes grown on Si/SiO_2 substrate and d) open-aperture (OA) of the nanoflakes grown on quartz substrate.

4. Conclusion

Thin films containing 2D structures of MoS_2 were grown by one-step thermal chemical vapour deposition on Si/SiO_2 and quartz substrates through thermal evaporation of MoO_3 powder along with sulfur. In the XRD patterns of the obtained thin films that were grown on Si/SiO_2 and quartz substrates, the crystalline structures of 3R-MoS_2 were observed. It seems that the intensity of the MoS_2 peaks increases on quartz substrate. By assessment of line width and frequency difference between the E^1_{2g} and A_{1g} in Raman spectroscopy and the nanoflakes' band gap via UV-visible and photoluminescence analysis, it was confirmed that the nanoflakes grown on Si/SiO_2 substrate were monolayer structures of MoS_2 while the nanoflakes grown on quartz substrate were multilayer. In addition, the nonlinear optical properties of the obtained nanoflakes grown on Si/SiO_2 and quartz substrates, were studied by Z-scan technique. Close-aperture of the nanoflakes were displayed a pre-focal peak followed by a post-focal valley. Therefore, the negative value of the nonlinear refractive index, observed obvious self-defocusing behaviour in both samples. The open-aperture Z-scan curves of the MoS_2 grown on Si/SiO_2 substrate exhibit obvious two-photon absorption response. However, the multilayer MoS_2 grown on quartz substrate, display the presence of saturable absorption response under the same excitation conditions.

5. References

[1]Q H Wang, K Kalantar-zadeh, A Kis, J N Coleman and M S Strano 2012 *Nature Nanotechnol* **7,** 699–712.
[2] H Qiu, L Pan, Z Yao, J Li, Y Shi and X Wang 2012 *Appl. Phys. Lett.* **100**, 123104.
[3] J M Yun, Y J Noh, J S Yeo, Y J Go, S I Na, H G Jeong, J Kim, S Lee, S S Kim, H Y Koo, T W Kim and D Y Kim 2013 *J. Mater. Chem. C.* **1** 3777–3783.
[4] S Wang, H Yu, H Zhang, A Wang, M Zhao, Y Chen, L Mei and J Wang 2014 *Adv. Mater.* **26** 3538–3544.
[5] Y Zhan, Z Liu, S Najmaei, P M Ajayan and J Lou 2012 *Small.* **8** 966-971.
[6] J N Coleman, M Lotya, A O'Neill, S D Bergin, P J King, U Khan, K Young, A Gaucher, S De, R J Smith, I V Shvets, S K Arora, G Stanton, H Y Kim, K Lee, G T Kim, G S Duesberg, T Hallam, J J.Boland, J J Wang, J F Donegan, J C Grunlan, G Moriarty, A Shmeliov, R J Nicholls, J M Perkins, E M Grieveson, K Theuwissen, D W McComb, P D Nellist, and V Nicolosi 2011 *Science* **331** 568–571.
[7] R Wang, H C Chien, J Kumar, N Kumar, H Y Chiu and H Zhao 2014 *ACS Appl. Mater. Interfaces* **6** 314–318.
[8] N Kumar, S Najmaei, Q Cui, F Ceballos, P M Ajayan, J Lou and H Zhao 2013 *Phys. Rev. B: Condens. Matter Mater. Phys.* **87** 161403.
[9] Q L Bao, H Zhang, Y Wang, Z Ni, Y Yan, Z X Shen, K P Loh, and D Y Tang 2009 *Adv. Funct. Mater.* **19** 3077–3083.
[10] S Najmaei, Z Liu, W Zhou, X Zou, G Shi, S Lei, B I Yakobson, J C Idrobo, P M Ajayan and J Lou 2013 *Nat. Mater.* **12** 754–759.
[11] Y Huang, L Liu, W Zhao and Y Chen 2017 *Thin Solid Films.* **624** 101–105.
[12] J Shi, X Zhang, D Ma, J Zhu, Y Zhang, Z Guo, Y Yao, Q Ji, X Song, Y Zhang, C Li, Z Liu, W Zhu and Y. Zhang 2015 *ACS Nano* **9** 4017–4025.
[13] C Li,Y Zhang, Q Ji, J Shi1, Z Chen, X Zhou, Q Fang1 and Y Zhang 2016 *2D Mater.* **3** 035001.
[14] K Wang, Y Feng, C Chang, J Zhan, C Wang, QZhao, J N Coleman, L Zhang, W J Blau and J Wang 2014 *Nanoscale* **6** 10530.
[15] Y Li, N Dong, S Zhang, X Zhang, Y Feng, K Wang, L Zhang and J Wang, 2015 *Laser Photonics Rev.* **9** 427–434.
[16] J Seo, Y Jun, S Park, H Nah, T Moon, B Park, J Kim, J Youn and J Cheon 2007 *Angew. Chem., Int. Ed.* **46** 8828–8831.
[17] R Coehoorn, C Haas, J Dijkstra and C J F Flipse 1987 *Phys Rev B* **35** 6195.
[18] X Y Zhu, Y C Cheng and U Schwingenschlogl 2011 *Phys. Rev. B.* **84** 153402.
[19] H Li, Q Zhang, C C Ray Yap, B K Tay, T H Tong Edwin, A Olivier and D Baillargeat 2012 *Adv. Func. Mater.s.* **22** 1385–1390.
[20] Y Xie, Z Wang, Y Zhan, P Zhang, R Wu, T Jiang, S Wu, H Wang, Y Zhao, T Nan and X Ma 2016 *IOP science* **28** 084001.

Acknowledgments

The authors gratefully acknowledge the financial support of the Iran Science Elites Federation under grant 11/66332 dated 2015/05/20, and the Research Council of Imam Khomeini International University.

IOP Conf. Series: Journal of Physics: Conf. Series **987** (2018) 012002 doi:10.1088/1742-6596/987/1/012002

Metal nanolayer deposited highly stable Ag thin films and their optical properties

M Kawamura[1], T Kiba[1], Y Abe[1], K H Kim[1], H Murotani[2]

[1]Kitami Institute of Technology, Kitami, Hokkaido 090-8507 Japan
[2]Tokai University, Faculty of Engineering, Hiratsuka, Kanagawa, 259-1292, Japan

E-mail: kawamumd@mail.kitami-it.ac.jp

Abstract. We have reported that deposition of several-nm-thick Al surface layer on Ag thin films improved thermal stability remarkably. In the present study, we investigated the influence of the deposition of Al nanolayer on the reflectance of the Al/Ag multilayer. 100-nm-thick Ag film and 1–5-nm-thick Al surface layer were deposited successively by vacuum evaporation, and their optical properties were measured. The reflectance of the samples on which a 1- or 3-nm-thick Al layer was deposited on the Ag film was almost the same as that of the Ag single-layer film. However, the samples on which a 5-nm-thick Al layer was deposited on the Ag film showed decreased reflectance. It was considered that the deposited Al nanolayer was fully oxidized in the former samples, but metallic Al and oxidized Al coexisted in the latter sample, which coincided with the simulated reflectance data. As a result, we revealed that thermally stable Ag films formed by 1- or 3-nm-thick Al nanolayer deposition have also high optical reflectance.

1. Introduction

Silver (Ag) is known for its excellent electrical and optical properties, such as the lowest electrical resistivity, highest optical reflectance, and lowest emissivity. However, a drawback is that Ag thin films agglomerate readily under heat treatment because of low adhesion to glass (oxide) substrates and Ag atoms migrate easily as shown in Fig. 1a. This leads to the deterioration of film properties, namely, an increase in electrical resistivity and decrease in reflectance. Alloying Ag films with other metals is well known as a useful way to suppress agglomeration [1, 2], but it also influences the physical properties of Ag films and causes the increase of electrical resistivity.

We reported thermally stable Ag thin films (Fig. 1b) sandwiched between surface and interface nanolayers of metals such as Al, Ti and so on [3-6]. We also found the features required for the nanolayer materials, namely, high cohesive energy or high free energy of oxide formation and low solid solubility in Ag [5, 6]. It is well known that Al is readily oxidized by exposure to air. Approximately 2.5-nm-thick porous oxide layer on Al nanoparticles was confirmed by high-resolution transmission electron microscopy [7]. We confirmed that a 1-nm-thick Al layer (nominal thickness) was oxidized to form an approximately 3-nm-thick Al oxide layer by ellipsometery [3]. Therefore, the current method is simple and facile method to form very thin metal oxide layer on Ag thin film, making use of feature of Al. In a sense, this is favourable way to form oxide nanolayers because a method such as reactive sputtering often damages the underlying metal layer.

In the present study, we investigate optical properties of the Al nanolayer deposited Ag films. It is known that Al film have lower reflectance than Ag film, except in the ultraviolet region. Even if the Al layer is several nanometers thick, it might influence the reflectance. By contrast, if the surface layer is

Al oxide, it may be optically transparent owing to its wide energy band gap about 8.4 eV [8]. Therefore, it is worthwhile to investigate the optical properties of highly stable Ag thin films fabricated by the simple deposition of Al and Ag thin films. Influence of thick (>100 nm) protective coatings has been reported [9, 10]. But our current purpose is to clarify the influence of deposited several-nanometer-thick Al surface layer on the multilayer.

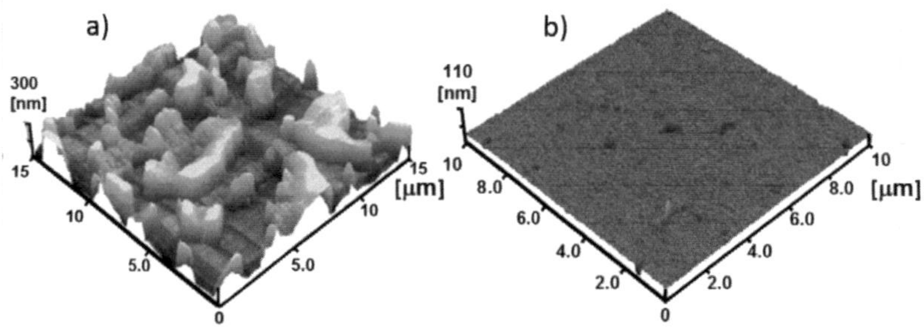

Figure 1. 3D AFM images of samples after annealing at 600°C for 1h in vacuum.
a) 100-nm-thick Ag single film (scanning area: 15×15μm^2), b) 100-nm-thick Ag film sandwiched between 1-nm-thick Al nanolayers (scanning area: 10×10μm^2)

2. Experimental procedure

We prepared multilayers consisting of an Al nanolayer (1-, 3-, 5-nm-thick) and 100-nm-thick Ag thin films on glass substrates (Corning EAGLE XG). The multilayers were deposited by means of vacuum evaporation of Ag wire (99.99%) and Al wire (99.99%) successively without breaking vacuum after evacuated below 4.0 × 10^{-4} Pa. The deposition rates of Ag and Al were about 0.4 nm/s and 0.05 nm/s, respectively. Here, only the surface nanolayer was deposited for the current purpose. The electrical resistivity of the prepared films was measured using the four-point-probe method. The surface morphology was observed using an atomic force microscope (AFM). The optical reflectance spectra were measured by a spectrophotometer equipped either with an absolute reflectance measurement unit or an integrated sphere at wavelengths ranging from 200 to 2000 nm at a scanning rate of 20 nm/min. We compared the experimental results with simulation results obtained using Fresnel equations for multilayers [11].

3. Results and discussion

The resistivity of the obtained Ag single films was 2.5 μΩcm, which is almost the same as the value obtained in our previous study [3].

AFM measurements revealed that the surfaces of all films were very smooth. 2D images of the samples are shown in Fig. 2. The root mean square (RMS) roughness of the Ag single layer was 2.2 nm and that of the Al/Ag films was 1.2 nm. These changes suggest the film surface became smoother upon Al deposition. However, even though the RMS value of 2.2 nm is extremely small compared to the wavelength of light, it is believed to be adequately smooth to achieve high reflectance.

The absolute specular reflectance spectra of the Ag single layer and the Al/Ag multilayers at an incidence angle of 5° are shown in Fig. 3. Inset is enlarged view of the 300–700 nm region of these spectra. The reflectance spectrum of the Ag single layer is close to other high-quality films formed by vacuum evaporation and sputtering [10, 12]. In terms of the influence of the deposition of Al nanolayer on Ag films, we found that the spectra of Al/Ag multilayers, where Al thickness is 1 or 3 nm are similar to that of the Ag single layer. The deposition of the 1- or 3-nm-thick Al layer did not affect the reflectance because the layer was oxidized immediately after it was removed from the deposition chamber. The reflectance of the films obtained herein is higher than that the values reported in the literature for thicker (100-nm-thick) Al_2O_3 layer [10] or SiN layer [13] deposited on an Ag film. However, the multilayer sample in which 5-nm-thick Al was deposited shows remarkably lower reflectance. It is considered that the deposited 5-nm-thick Al layer remained partly in the metallic state, which has a larger extinction coefficient than that of Ag. Therefore, it led to a noticeable decrease in reflectance. The total reflectance of the films measured using an integrated sphere was slightly (about 2 %) higher than the specular reflectance spectra. This suggests that the films show negligible level of diffuse reflectance.

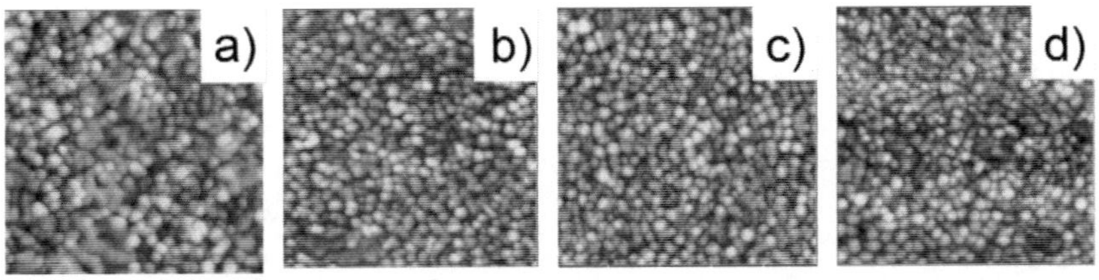

Figure 2. 2D AFM images (1×1 μm^2) of the samples: a) Ag single film, b) Al(1nm)/Ag film, c) Al(3nm)/Ag film, d) Al(5nm)/Ag film.

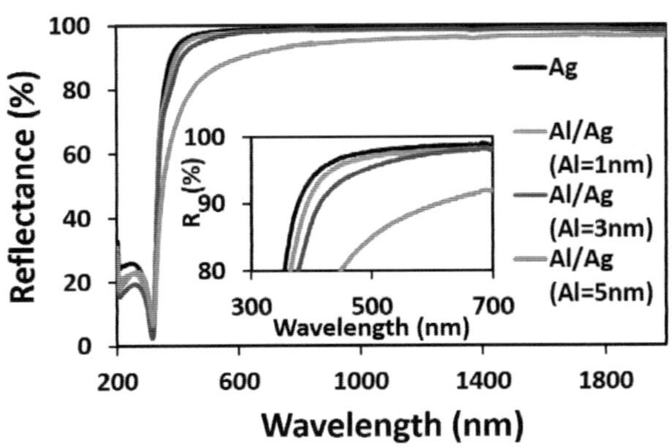

Figure 3. Absolute specular reflectance spectra of Ag film and 1–5 nm thick Al surface layer deposited Ag films. Inset: enlarged view of the 300–700 nm region.

Then, we compared the measured specular reflectance spectra and the simulated spectra, which were based on the complex-matrix form of the Fresnel equations [11], where optical constants were referred to from separate articles [14, 15]. As models for the simulation, Al metal (M), Al oxide (O), and Al oxide/Al metal (O+M) layers were assumed on Ag film as shown in Fig. 4 (a). For Al(1 nm)/Ag multilayers, a model in which the surface is 1-nm-thick Al metal does not fit well, but better coincidence is obtained by assuming a surface comprising 3-nm-thick Al oxide because of oxidation of the Al layer (Fig. 4(b)). Similarly, for Al (3 nm)/Ag multilayers, simulation using the model in which the surface layer is not 3-nm-thick Al metal but 4-nm-thick Al oxide leads to better coincidence, as shown in Fig. 4(c). Here, the thickness of the Al oxide layer was varied from 3 nm to 5 nm, and the best match was obtained for the 4-nm-thick layer. By contrast, neither the metal layer nor the oxide layer fits the measured result for the Al (5nm)/Ag multilayer, as shown in Fig. 4(d). However, the best match was obtained for a model with a surface layer consisting of 4-nm-thick oxide and 1-nm-thick Al layer, though slight difference remained. These results also support the assertion that the decrease in reflectance is caused by the metallic Al in the surface layer.

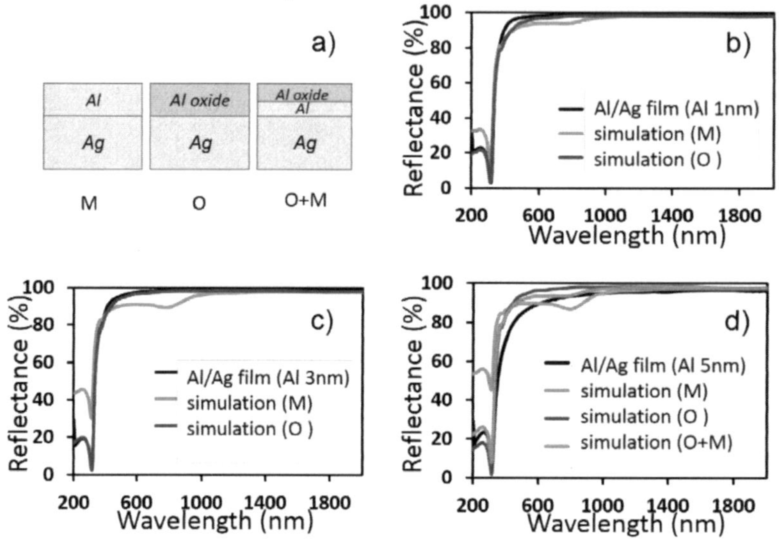

Figure 4. Comparison of measured reflectance spectra and simulated spectra of Al surface layer deposited Ag films: a) simulation models for the surface layer on Ag films, b) the spectrum of 1-nm-thick Al deposited Ag film, simulation spectra of 1-nm-thick Al layer on Ag film (M) and 3-nm-thick Al oxide layer on Ag film (O), the spectrum of 3-nm-thick Al deposited Ag film, simulation spectra of 3-nm-thick Al layer on Ag film (M) and 4-nm-thick Al oxide layer on Ag film (O), d) the spectrum of 5-nm-thick Al deposited Ag film, simulation spectra of 5-nm-thick Al layer on Ag film (M) and 5-nm- thick Al oxide layer on Ag film (O), and 4-nm-thick oxide and 1-nm-thick Al layer on Ag film (O+M).

4. Conclusions

In conclusion, we investigated the influence of Al surface nanolayer deposition on 100-nm-thick Ag thin films. The deposition of an Al layer measuring up to 3 nm in thickness leads to the preservation of high reflectance, in addition to imparting high thermal stability and low electrical resistivity, which has been already confirmed. The multilayer is prepared by simple metal layer deposition by means of the vacuum evaporation method, accompanied with natural oxidation of the surface Al nanolayer, which also helps avoid any possible damage while depositing the metal oxide layer on Ag films.

Acknowledgment

The authors deeply thank Mr. Kohei Yamaguchi for his assistance on sample preparation.
This work was supported by JSPS KAKENHI grant number 16H04503.

References

[1] Kim H C and Alford T L 2003 *J. Appl. Phys.* **94** 5393
[2] Sugawara K, Minamide Y, Kawamura M, Abe Y and Sasaki K 2008 *Vacuum* **83** 610
[3] Kawamura M, Inami Y, Abe Y and Sasaki K 2008 *Jpn. J. Appl. Phys.* **47** 8917
[4] Kawamura M, Fukuda D, Inami Y, Abe Y and Sasaki K 2009 *J. Vac. Sci. and Technol.* A **27** 975
[5] Zhang Z, Kawamura, Abe Y and Kim K H 2013 *Jpn. J. Appl. Phys.* **52** 975
[6] Zhang Z, Kawamura, Abe Y and Kim K H 2014 *Jpn. J. Appl. Phys.* **53** 48003

[7] Ramaswamy A L and Kaste P 2005 *J. Energ. Mater.* **23** 1
[8] Tahir D, Kwon H L, Shin H C h, Oh S K, Kang H J, Heo S, Chung J G, Lee J C h and Tougaard S 2010 *J. Phys. D: Appl. Phys.* **43** 255301
[9] Hass G, Heaney J B, Herzig H, Osantowski J F and Triolo J J 1975 *Appl. Opt.* **14** 2639
[10] Jobst P J, Stensel O, Scurmann M, Modsching N, Yulin S, Wilbrandt S, Gaebler D, Kaiser N and Tunnermann A 2014 *Adv. Opt. Techn.* **3** 91
[11] http://www.filmetrics.com/reflectance-calculator
[12] Vergöhl M, Malkomes N, Szyszka B, Neumann F, Matthée T and Bräuer G 2000 *J. Vac. Sci. Technol.* A **18** 1632
[13] Boccas M, Vucina T, Araya C, Vera E and Ahhee C 2006 *Thin Solid Films* **502** 275
[14] Winsemius P, van Kampen F F, Lengkeek H P and van Went C G 1976 *J. Phys.* **F 6** 1583
[15] Rakić A D 1995 *Appl. Opt.* **34** 4755

Pd surface functionalization of 3D electroformed Ni and Ni-Mo alloy metallic nanofoams for hydrogen production

A Petica[1], O Brincoveanu[1], F Golgovici[1,2], A C Manea[1,3], M Enachescu[1] and L Anicai[1]

[1] Center of Surface Science and Nanotechnology, University Politehnica of Bucharest, Splaiul Independentei 313, Bucharest, 060042, Romania
[2] Department of General Chemistry, Faculty of Applied Chemistry and Materials Science, POLITEHNICA University of Bucharest, Bucharest, 060042, Romania
[3] Department of Inorganic Chemistry, Physical Chemistry and Electrochemistry, Faculty of Applied Chemistry and Materials Science, POLITEHNICA University of Bucharest, Bucharest, 060042, Romania

E-mail: lanicai@itcnet.ro

Abstract. The paper presents some experimental results regarding the functionalization of 3D electroformed Ni and Ni-Mo alloy nanofoams with Pd nanoclusters, as potential cathodic materials suitable for HER during seawater electrolysis. The electrodeposition from aqueous electrolytes containing $NiCl_2$ and NH_4Cl has been applied to prepare the 3D Ni nanofoams. Ni-Mo alloys have been electrodeposited involving aqueous ammonium citrate type electrolytes. Pd surface functionalization has been performed using both electroless and electrochemical procedures. Pd content varied in the range of 0.5 – 8 wt.%, depending on the applied procedure and the operation conditions. The use of a porous structure associated with alloying element (i.e. Mo) and Pd surface functionalization facilitated enhanced performances from HER view point in seawater electrolyte (lower Tafel slopes). The determined Tafel slope values ranged from 123 to 105 $mV.dec^{-1}$, suggesting the Volmer step as rate determining step. The improvement of the HER catalytic activity may be ascribed to a synergistic effect between the high real active area of the 3D electroformed metallic substrate, Ni alloying with a left transition metal and surface modification using Pd noble metal.

1. Introduction

The hydrogen production through electrochemical water splitting under the assistance of electrocatalysts attracted an increased interest as one of the most promising routes among the existing processes. Moreover water represents an abundant resource, about 75% of the globe surface being covered by water, however mostly found as seawater. To use the abundant seawater instead of fresh water for hydrogen production through electrolysis may be a significant advantage, especially if the required electrical energy should be obtained involving sea waves generated power [1,2,3].

Hydrogen evolution reaction (HER) and oxygen evolution reaction (OER) represent the two significant electrochemical processes occurring during water electrolysis and Pt species are the most preferred electrocatalysts. Although Pt electrode exhibits the highest efficiency and long-term stability, its high cost and limited occurrence restricts the commercial applications. Cheaper but high electrochemically active electrodes are needed in order to optimize the efficiency of hydrogen evolution reaction (HER) and the overall efficiency of electrolysis [4 and included references].

Nickel and nickel-based alloys obtained as coating layers involving electrochemical deposition process have been attracted an increased interest due to the low cost and promising electrocatalytic activity for HER. In addition, the use of electrodes possessing high surface area, including metallic nanofoams, may contribute to an improvement of the catalytic activity [5,6].

Among the electrodeposited Ni alloys, Ni-Mo electrocatalysts have been intensively investigated due to their excellent corrosion resistance and long-term stability, besides the high HER activity [4,7,8]. Wang et al. [8] reported the electrodeposition of 3D porous Ni-Mo alloy layers under super gravity field which presented enhanced electrocatalytic activity toward HER in 10% NaOH solution and good long-term stability. The extremely high real active area of the deposits facilitated the high catalytic activity. Gonzalez-Buch et al. [9] reported the electrodeposition of 3D macroporous NiMo cathodes for alkaline water electrolysis on a stainless steel AISI 304 substrate involving a double-template electrochemical process. 3D electroformed Cu foam served as a template for the Ni-Mo co-deposition. According to the obtained Tafel results the authors concluded that the NiMo electrodes exhibit higher intrinsic catalytic activity for HER than the pure Ni electrode as a consequence of alloying hypo–hyper-d-electronic transition metals.

The electrocatalytic HER characteristics of nickel and nickel based alloys may be improved through surface functionalization using noble metals (i.e. Pd, Ru, Rh or Pt) involving electrodeposition, spontaneous deposition or chemical reduction processes [10,11,12]. In addition, the research interest is focused on developing more active and stable Pd-based electrocatalyts in order to partially or completely replaces costly Pt [13].

Only few studies can be found related to the development of cathode materials aimed to catalyse HER process during seawater electrolysis. Jiang et al. [2] investigated the HER activity of the electrodeposited Ni–Fe–C films onto A3 steel substrate during simulated seawater (3.5 wt% NaCl) electrolysis. They found a non-linear correlation between the average coating grain sizes in the range of 4.3 – 6.4 nm and the hydrogen evolution overpotential. Moreover, the optimum hydrogen overpotential of Ni-Fe-C cathode was about 65 mV in 3.5% NaCl at 90°C and pH 12.

Zheng [14] studied the synthesis of $PtNi_x$ electrocatalysts exhibiting high efficiency hydrogen evolution during seawater splitting. $PtNi_5$ was found to provide the best catalytic behavior, with the Tafel slope of 119 mV dec^{-1}. The synthesized alloy electrodes showed a remarkable stability at -1.2V over 12 h. Binary platinum alloy electrodes, PtM (M = Fe, Co, Ni, Pd) electrodeposited on Ti foil involving a cycle voltammetry strategy have been also very recently reported as promising cathodic materials for HER during seawater electrolysis [15]. The author showed that by tuning M species, the Ti supported PtPd alloy electrode displayed a maximal efficiency, yielding an onset potential of −52 mV in HER. The long-term stability has also been increased by alloying Pt with M.

Therefore, the aim of this work was the preparation and characterization of Pd-modified 3D electroformed Ni and Ni-Mo alloy nanofoams using both electrochemical and electroless procedures, as potential cathodic materials suitable for HER during natural seawater electrolysis.

2. Experimental

2.1. Chemicals and materials

All used chemical reagents were of analytical grade and supplied by Sigma Aldrich and Fluka. They were used as received without any further purification. The seawater was taken from Black Sea – Constanta location (Romania) and filtrated to remove any suspended species.

2.2. Electrodeposition of 3D Ni and Ni-Mo alloy coatings

The Ni and Ni-Mo alloy coatings have been electrodeposited onto Cu metallic substrates under galvanostatic conditions using a DC power supply and a two-electrode cell configuration. The used Cu metallic substrates (70x35x0.2 mm) have been initially subjected to a mechanical polishing using 1000 and 2000 abrasive paper, followed by chemical pickling in 1:1 HNO_3: H_2O solution at 25°C for 20–30 s, then rinsed with running water and distilled water and air dried.

The Ni foams have been electrodeposited on Cu using the dynamic hydrogen bubble template [16, 17] from an aqueous electrolyte based on $NiCl_2.7H_2O$ and NH_4Cl at 25°C against a Ni plate as anode. The Ni-Mo alloy coatings have been electrochemically deposited onto electroformed Ni foams as metallic substrates, against platinized titanium as anode involving aqueous citrate-ammonium type electrolytes. Table 1 summarizes the bath compositions and the experimental conditions used in the synthesis of the above mentioned coatings.

Table 1. Electrolytes composition and operating conditions for 3D Ni and Ni-Mo alloy coatings electrodeposition

Coating type	Electrolyte composition	Operating conditions
Ni foam (Ni_f)	0.1M $NiCl_2.7H_2O$ 2M NH_4Cl	pH = 4-4.5 T = 25°C i = 0.2-2 A/cm^2 time= 1-5 min.
Ni-Mo alloy (NiMo)	0.4M $NiSO_4.6H_2O$ 0.025-0.064M $Na_2MoO_4.2H_2O$ 0.2M $Na_3C_6H_5O_7$ 0.5M H_3BO_3 1-3 g/L saccharine 0.2 g/L tensioactive agent	pH =6-6.5 T = 50-60°C i = 1-5 A/dm^2 time = 10-20 min.

2.3. Pd surface functionalization

To modify the 3D Ni and Ni-Mo alloy coatings using Pd, several procedures have been applied, involving either electrochemical or chemical sequences.

Therefore Ni-Pd alloy foams have been electrochemically synthesized through addition of 5-10 mM $PdCl_2$ to the $NiCl_2$-NH_4Cl electrolyte used to obtain Ni foams. Ni-Pd alloy coatings have been also applied involving a classical Ni based electrolyte with 5-10 mM $PdCl_2$ additions.

The chemical Pd deposition has been performed by dipping the pre-deposited Ni and Ni-Mo alloy foam electrodes in tetraammine palladium (II) chloride based solution for 1-10 min. at 50°C. The composition and the operation conditions applied during Pd surface functionalization are presented in Table 2.

The surface morphology and composition of the prepared coatings has been examined using scanning electron microscopy (SEM) associated with energy dispersive X-ray spectroscopy (EDX) (SU8230, HITACHI High-Technologies Corporation, Japan). The phase composition and structure were determined using X-ray diffractometry (XRD) involving a Rigaku Ultima IV equipment with graphite diffracted beam monochromator for CuK_α radiation, at a speed of 2 s/step (1step = 0.05°).

Table 2. Solution composition and operating conditions for Pd surface functionalization

Coating type	Solution composition	Operating conditions

Ni-Pd alloy foam ($NiPd_f$)	0.1M $NiCl_2.7H_2O$ 2M NH_4Cl 5-10 mM $PdCl_2$	T = 25°C i = 0.2-2 A/cm^2 time= 1-5 min.
Pd chemical deposition (Pd_{chem})	5.4 g/L $[Pd(NH_3)_4]Cl_2$ $2H_2O$ 67.2 g/L Na_2EDTA 650 mL/L NH_3 (28%) 0.5 mL/L N_2H_4. H_2O	T = 50°C immersion time= 1-10 min.

2.4. Electrochemical measurements

All the investigations related to the assessment of the catalytic activity for HER on the prepared electrodes acting as working electrode (with a constant geometrical surface of 0.63 cm^2) have been performed in a three-electrode configuration cell using natural seawater as electrolyte at room temperature. A platinum mesh was auxiliary electrode and the reference electrode was the Ag/AgCl electrode. The cell was connected to PARSTAT 4000 potentiostat controlled with VersaStudio software. HER polarization curves were recorded at a scan rate of 1 mV s^{-1}, from -1.7 V *vs* Ag/AgCl up to the equilibrium potential. Tafel slopes were determined from HER polarization curves.
Before the tests, the working electrode was held at -1.70 V *vs* Ag/AgCl for 1000 s in the same solution, in order to reduce the oxide film present on the porous surface electrode layer.
EIS measurements were also recorded after obtaining the polarization curves in the same cell configuration with 10 mV *ac* voltage within 100 kHz–50 mHz frequency range. The fitting of the impedance data has been performed using ZView 2.4 software from Scribner Association Inc., Derek Johnson.

3. Results and Discussion

The Ni foams electrodeposited onto Cu substrate involving the experimental conditions as shown in Table 1 exhibit a porous structure characterized by the presence of pores having different diameters, depending on the applied current density and deposition time. Regardless the applied operating conditions, uniform, black layers have been formed. The hydrogen bubbling occurring simultaneously to the electrochemical reduction of nickel ions promotes the development of this type of morphology. The final coating possesses a high surface area, which may enhance the hydrogen evolution during water electrolysis [10,11]. In addition it is suitable for further deposition of other metals and alloys which will take the 3D shape of the porous substrate with a very high accuracy [17].

Figure 1 presents an example of the obtained morphology for Ni-Pd alloy foams ($NiPd_f$) electrodeposited on Cu using a Pd-modified Ni electrolyte (see Table 1) and the spatial distribution profiles for the main elemental components.

Figure 1. SEM micrographs of Ni-Pd alloy foam (1 wt. % Pd) electrodeposited at 25°C for 3 min. (c.d. = 1 A/cm^2) and EDX maps showing the distribution profiles of elements within the deposited alloy

As shown in figure 1, the addition of $PdCl_2$ in the electroforming electrolyte facilitates the co-deposition of Pd, while keeping the porous morphology of genuine Ni foam. Pd element is quite uniformly distributed within the alloy deposit. Usually the Pd content was in the range of 0.3-3 wt.%

with higher values on the increase of Pd concentration in the electrolyte or of the applied current density.

Quite similar morphology with a uniform Pd distribution has been also noticed in the case of Pd chemical deposition on previously electrodeposited Ni foam, as exemplified in figure 2.

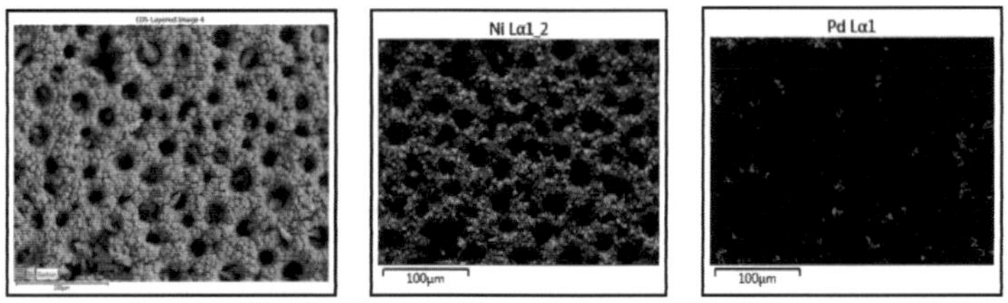

Figure 2. EDX maps showing the distribution profiles of elements for Pd_{chem} onto Ni foam (6 wt.% Pd) using chemical deposition for 5 min. at $50^{o}C$

As expected, an increased Pd content has been noticed as the immersion time was longer, in the range of 3-7 wt.% for 1-10 min. deposition periods. The Ni based porous structures usually consisted in macropores of 20-40 μm and micropores of 3-5 μm, with a nodular growth, leading to a cauliflower-like morphology. The XRD analysis (not shown here) evidenced the characteristic peaks of $Ni_{0.82}Pd_{0.12}$ phase, besides those of Ni. Based on XRD and using Scherrer equation, average sizes of crystallites between 22-62 nm have been estimated.

Electrodeposition of Ni-Mo alloys represents a classical example of induced co-deposition [18], when the reduction reaction of nickel enhances the co-deposition of molybdenum. The use of NiMo electrolyte (see Table 1) allowed the obtaining of different values of Mo content in the alloy (in the range of 10 – 30 wt.% Mo) by varying the applied operation conditions, mainly the current density and temperature. In addition the Ni-Mo alloy coatings possess promising protective characteristics in chloride aggressive environment as reported in [19,20]. Generally NiMo electrolyte facilitates the electrodeposition of uniform, bright-grey deposits, with an excellent adhesion to the Cu or Ni substrates. It is well-known that the electrochemical deposition is an atomic/molecular level process. Therefore a Ni-Mo deposited layer entirely takes the 3D shape of the Ni foam substrate with a very high accuracy and adhesion.

Figure 3 shows an example of the morphology and EDX maps for Ni-Mo alloy electrodeposited onto Ni-Pd foam.

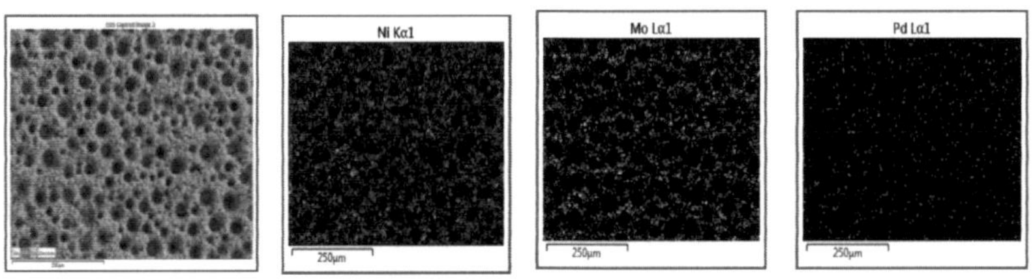

Figure 3. EDX maps showing the distribution profiles of elements for Ni-Mo alloy electrodeposited onto $NiPd_f$ (0.5 % Pd) at $60^{o}C$ for 20 min. (c.d = 4 A/dm^2)

The porous matrix of the $NiPd_f$ is clearly observed in figure 3. The metallic elements are quite uniformly distributed within the deposit as EDX maps show.

A similar morphology and homogeneity has been also evidenced in the case of Pd chemical deposition onto Ni-Mo alloy/Ni_f substrate, as exemplified in figure 4.

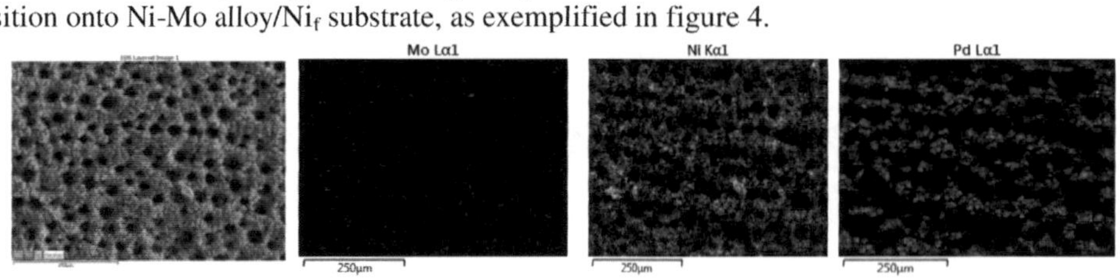

Figure 4. EDX maps showing the distribution profiles of elements for Pd_{chem} onto Ni-Mo alloy electrodeposited onto Ni_f (6 wt. % Pd) at 60°C for 20 min. (c.d = 4 A/dm^2)

The electrochemical activity of the electrodeposited electrodes towards HER has been investigated in natural seawater at room temperature. Figure 5 presents the recorded polarization curves of the different Pd-modified electroformed Ni and Ni-Mo alloy nanofoams. The cathodic Tafel slopes (b), exchange current densities (i_o), overpotentials at 1 and 10 mA.cm^{-2} were determined from the corresponding cathodic current-potential curves and are listed in Table 3. The apparent exchange current densities were estimated by extrapolating the Tafel lines (not shown here) to the corresponding zero current potentials.

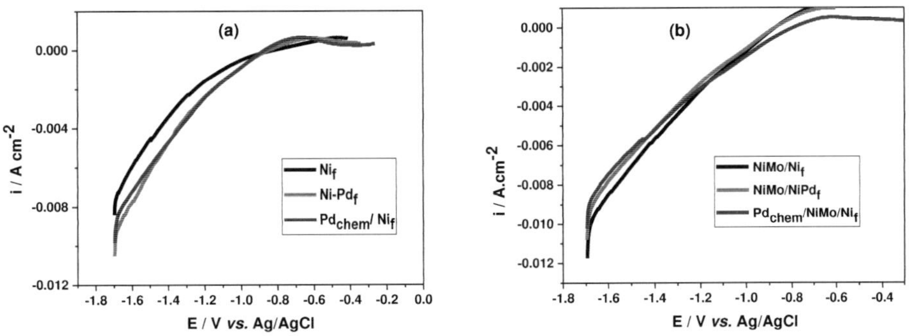

Figure 5. Comparative linear polarization curves of HER in seawater for different systems of Pd functionalized Ni foams (a) and NiMo/ Ni foams (b)

Table 3. Cathodic Tafel slopes, exchange current densities, overpotentials at 1 and 10 mA.cm^{-2} for the prepared electrodes

Working electrode type	b / mV dec^{-1}	i_o / A cm^{-2}	η/mV (i = 1 mA cm^{-2})	η/mV (i = 10 mA cm^{-2})
Ni_f	185	6.64.10^{-5}	289	948
$NiPd_f$	123	8.53.10^{-5}	172	851
Pd_{chem}/Ni_f	120	1.31.10^{-4}	146	827
NiMo/Ni_f	118	1.57.10^{-4}	108	813
NiMo/$NiPd_f$	115	1.35.10^{-4}	113	820
Pd_{chem}/NiMo/Ni_f	105	9.91.10^{-5}	152	903

It is clear from Table 3 and figure 5 that the HER activity of Ni_f may be enhanced through Pd surface modification. The corresponding Tafel slopes for $NiPd_f$ and Pd_{chem}/Ni_f are of 123 and 120 mV.dec^{-1}, respectively, suggesting that the HER process obeys Volmer reaction mechanism [7,14 and included references]. The higher value of Ni_f may be related to the presence of oxides films on the surface [7]. The Pd functionalization of Ni-Mo alloy based systems determines only a slight improvement of the catalytic activity. The significant effect is mainly due to the presence of Mo as alloying element. The determined Tafel slope values ranged from 118 to 105 mV.dec^{-1}, indicating the Volmer step as rate determining step.

Figure 6 shows examples of EIS spectra recorded on the developed Pd modified Ni and Ni-Mo alloy foams. As seen from figure 6a, a single semicircle was observed in the Nyquist plots of Pd modified Ni foams. This suggests that the HER is mainly controlled by a charge transfer process [7,21]. Therefore, one-time constant electric equivalent circuit model was used to fit the EIS experimental data as shown in the inset of figure 6a. During fitting, double layer capacitances were replaced by constant phase element (CPE). T (F s^{n-1} cm^{-2}) was used as capacitive parameter.
The recorded of EIS spectra on Pd modified Ni-Mo alloys foam present in the Bode plots (not shown here) a two-time constant response. A slightly modified equivalent circuit initially proposed by Armstrong and Henderson [22] has been used to fit the response of the HER, as shown in the inset of figure 6b.
The real surface area of the Pd modified Ni and Ni-Mo alloy foams was calculated through EIS results by means of the surface roughness factor (R_f) which was determined for each coating type by dividing C_{dl} by the double-layer capacitance of a smooth electrode, 20 μF.cm^{-2} [7,8 and included references].

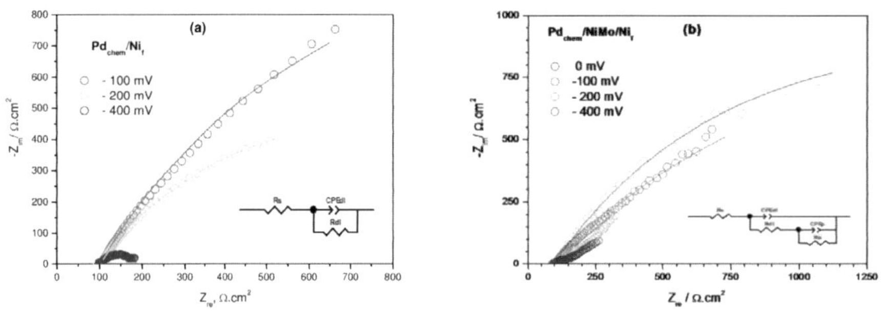

Figure 6. EIS data recorded in seawater electrolyte at 25°C for Pd-modified Ni and Ni-Mo alloy foams as Nyquist plots (inset: the proposed equivalent circuit to fit the measured points)

Table 4 presents the double layer capacitance, real surface area and corresponding surface roughness values for the investigated Pd modified foams calculated from the EIS data.

Table 4. Double layer capacitance, real surface area and corresponding surface roughness values for the investigated Pd modified foams calculated from the EIS data

Working electrode type	CPE_{dl} / mF	A_{real}/cm^2	σ
Ni_f	2.47±0.2	123.6	196
$NiPd_f$	3.09±0.2	151.5	240
Pd_{chem}/Ni_f	2.3±0.6	115.9	184
$NiMo/Ni_f$	4.9±0.9	248.5	395
$Pd_{chem}/NiMo/Ni_f$	4.5±0.9	224.5	356

Obviously, the improvement of catalytic activity may be ascribed to the increase of real surface area. Current densities with respect to real surface area at a potential of -1.5V vs.Ag/AgCl were also calculated to compare the intrinsic catalytic activity of $NiMo/Ni_f$ deposits. The values of $i_{-1.5V}/\sigma$ were 14.4 μA cm^{-2} and 17.1 μA cm^{-2} for $NiMo/Ni_f$ and $Pd_{chem}/NiMo/Ni_f$ respectively, suggesting a quite constant intrinsic catalytic activity. The improvement of catalytic activity may be however ascribed to the Pd surface functionalization. A more pronounced influence of Pd functionalization was observed in the case of Ni foam deposits. The determined values of the current densities with respect to real surface area at the same potential were 23.4, 26.6 and 32 μA cm^{-2} for Ni_f, $NiPd_f$ and Pd_{chem}/Ni_f electrodes, thus suggesting an enhanced HER catalytic activity.

4. Conclusions

As a result of the performed investigations easy and effective methods to prepare Pd-modified 3D electroformed Ni and Ni-Mo alloy nanofoams have been developed involving both electrochemical and electroless procedures. Pd content varied in the range of 0.5 – 8 wt.%, depending on the applied procedure and the operation conditions.

The use of a porous structure associated with alloying element (i.e. Mo) and Pd surface functionalization facilitated enhanced performances from HER view point in seawater electrolyte (lower Tafel slopes). The determined Tafel slope values ranged from 123 to 105 $mV.dec^{-1}$, indicating the Volmer step as rate determining step.

The improvement of the HER catalytic activity may be ascribed to a synergistic effect between the high real active area of the 3D electroformed metallic substrate, Ni alloying with a left transition metal and surface modification using Pd noble metal.

These preliminary results showed that the Pd-modified 3D electroformed Ni and Ni-Mo alloy nanofoams may represent a promising technological approach to build cathodic materials suitable for seawater electrolysis. Moreover the investigations are considered to be a step forward toward real application of seawater for hydrogen production.

Acknowledgements: Part of this work was supported by the NANOFOAM project, Contract No. 37/2016, M-Era.NET Program.

References

[1] Abdel-Aal HK, Zohdy KM and Abdel Kareem M 2013 *The Open Fuel Cells Journal* **3** 1-7
[2] Jiang N, Meng H M, Song L J and Yu H Y 2010 *Int. J. Hydrogen Energy* **35** 8056–62
[3] Badea G E, Caraban A, Cret P and Corbu I 2007 *University of Oradea, Fascicle of Management and Technological Engineering* VI(XVI)
[4] Sapountzi F M, Gracia J M, Weststrate C J, Fredriksson H O A and Niemantsverdriet J W 2017 *Prog. Energ. Combust.* **58** 1–35
[5] Gong M, Wang D Y, Chen C C, Hwang B J and Dai H 2016 *Nano Res.* **9** 28–46
[6] Safizadeh F, Ghali E and Houlachi G 2015 *Int. J. Hydrogen Energ.* **40** 256-74
[7] Navarro-Flores E, Chong Z and Omanovic S 2005 *J. Mol. Catal. A Chem.* **226** 179–97
[8] Wang M, Wang Z, Yu X and Guo Z 2015 *Int. J. Hydrogen Energy* **40** 2173-81
[9] González-Buch C, Herraiz-Cardona I, Ortega E, García-Antón J and Pérez-Herranz V 2016 *J.Appl.Electrochem.* **46** 791-803
[10] Pierozynski B, Mikolajczyk T and Kowalski I M 2014 *J. Power Sources* **271** 231-38
[11] Pierozynski B, Mikolajczyk T, Turemko M, Czerwosz E and Kozlowski M 2015 *Int. J. Hydrogen Energy* **40** 1795-99
[12] Rasouli H, Tabaian S H and Rezaei M 2016 *RSC Adv.* **6** 22500-51
[13] Zhang L, Chang Q, Chen H and Shao M 2016 *Nano Energy* **29** 198-219
[14] Zheng J 2017 *Appl. Surf. Sci.* **413** 360-5
[15] Zheng J 2017 *Appl.Surf.Sci.* **413** 72-82
[16] Shin H C, Dong J and Liu M 2003 *Adv. Mater.* **15** 1610–14
[17] Eugénio S, Silva T M, Carmezim M J, Duarte R G and Montemor M F 2014 *J. Appl. Electrochem.* **44** 455-65
[18] Brenner A 1963 *Electrodeposition of Alloys* vols. 1 and 2 (New York: Academic Press)
[19] Prioteasa P, Anicai L and Visan T 2010 *UPB Sci. Bull. Series B* **72** 11-24
[20] Prioteasa P, Golgovici F, Sbircea G, Anicai L and Visan T 2010 *Rev. Chim. (Bucharest)* **61** 1046-53.
[21] Solmaz R and Kardas G 2009 *Electrochim.Acta* **54** 3726–34
[22] Armstrong R D and Henderson M H 1972 *J.Electroanal.Chem.* **39** 81-90

IOP Conf. Series: Journal of Physics: Conf. Series **987** (2018) 012004 doi:10.1088/1742-6596/987/1/012004

The method of synthesizing of superhydrophobic surfaces by PECVD

Sagi Orazbayev[2] , Maratbek Gabdullin[2], Tlekkabul Ramazanov[1] , Merlan Dosbolayev[2], Askar Zhunisbekov[1], Dulat Omirbekov[2, *], Zhuldyz Otarbay[2]

[1]IETP, Al-Farabi KazNU, Almaty 050040, Kazakhstan
[2]NNLOT, Al-Farabi KazNU, Almaty 050040, Kazakhstan

*Corresponding author e-mail: omirbekov@physics.kz

Abstract. The aim of this work was to obtain superhydrophobic surfaces in a plasma medium. The experiment was carried out using the PECVD method in two different modes: constant and pulsing. The surface roughness was obtained by applying nanoparticles synthesized in plasma in a mixture of argon and methane. The resulting particles were deposited on the surface of silicon and glass materials. The contact angle increased linearly depending on the number of cycles, until it reached 160° at 150-160th cycles, after that the increase in cycles does not affect the contact angle, since the saturation process is in progress. Also the effect of the working gas composition on the hydrophobicity of the surface was studied. At low concentrations of methane (1%) only particles are synthesized in the working gas, and hydrophobicity is unstable, with an increase in methane concentration (7%) nanofilms are synthesized from nanoclusters, and surface hydrophobicity is relatively stable. In addition, a pulsing plasma mode was used to obtain superhydrophobic surfaces. The hydrophobicity of the sample showed that the strength of the nanofilm was stable in comparison with the sample obtained in the first mode, but the contact angle was lower. The obtained samples were examined using SEM, SPM, optical analysis, and their contact angles were determined.

1. Introduction

As a rule, if a solid surface has $\alpha > 90^0$ contact angle (CA) with water, it is considered hydrophobic, and if it has $\alpha < 90^0$, then it is hydrophilic. There is also a superhydrophobic surface, that has a contact angle greater than 150^0 and a superhydrophilic with contact angle less than 10^0. The wettability of the surface as a whole depends on two factors: the chemical composition and roughness of the surface. If both parameters are used properly, it is possible to create superhydrophobic surface with contact angle close to 180° (lotus effect) [1]. Hydrophobic or hydrophilic properties of the solid surface are key aspects for creating anti-icing, anti-corrosion, anti-fogging, medicine, water harvesting, oil/water separation, self-cleaning, etc.

At present, many methods have been developed, including physical, chemical and a combination of physical and chemical methods, to produce superhydrophobic surfaces in a simple and environmentally friendly way [2]. Surface modification methods such as plasma treatment, vapor deposition method, electrohydrodynamics/ electrospinning, sol–gel methods, phase separation method, spraying method, one-step method, spin-coating method, ion-assisted deposition method and electrochemical methods are used to create a hydrophobic topography of the surface. This can be a complex multi-step processes. The authors of [3] made tape-like superhydrophobic films by precipitating fluorocarbons on silicon in the RF glow discharge. Since the rough surface obtained by plasma treatment is easily tuned, this method is unique for obtaining superhydrophobic coatings with good optical properties. For example: in [4] transparent hydrophobic surfaces with a roughness of 20 nm were synthesized. The Khan group in [5] adjusting the LDPE crystallization parameters (crystallization time and growth rate) prepared superhydrophobic surfaces with different structures. The crystallization time and the nucleation rate were increased by evaporation of the solvent at a low temperature, so the contact angle also increased. Then, adding cyclohexane and drying at room temperature, a superhydrophobic surface with a contact angle of 173° was obtained. The polymeric material has a number of advantages, such as relatively inexpensive, transparent, easily manufactured, can be treated with a solvent, etc. Park and others, using a new copolymer, obtained a super-hygrophobic surface by spin-coating method [6]. This method can be applied to different substrates with different complexities. After surface damage, the hydrophobicity of such coatings is easily restored by spraying of dispersion. In [7] the authors obtained structured films, consisting of nanoclusters and nanoparticles on a glass substrate, by the sol-gel method. Surface structure can be regulated by controlling the hydrolysis and condensation. The method includes two steps: a sol-gel process for obtaining a surface roughness, after surface modification with fluorides to achieve transparency. Cho and others prepared superhydrophobic surfaces with a contact angle of 160^0 by the electrochemical reaction of a copper alloy [8]. The non-electric chemical coating of Cu on steel sheets led to microstructures on these surfaces, while nanotubes of copper sulfide on microstructures were obtained by electrochemical reaction of Cu in a sulfur-containing medium at a temperature of 150° C. The LBL method can be combined with others to control surface structures. In [9] the authors obtained a stable superhydrophobic film with a modification by the combination of LBL and electrochemical deposition methods. Miller et al. in [10] obtained superhydrophobic surfaces with a contact angle of ~ 150° using a single-phase PVD. Photolithography is also used to produce a rough surface. In work [11] scientists obtained superhydrophobic surfaces by photolithography. As a result, the obtained surface appears superhydrophobic with low adhesion, and the size of the square bars X-Y is 2 and 32 µm. But when the dimensions of the columns reached 64 and 128 µm, water droplets were fixed to the surface of the substrate.

This paper presents some experimental results on the production of superhydrophobic and superhydrophilic surfaces by the PECVD method in the RFCD plasma.

2. Experimental study

Carbon nanoparticles were synthesized in RFC discharge plasma in gas mixtures of argon and methane. The main part of the experimental installation is an electrode system, which consists of plane-parallel discs with a diameter of 10 cm. Interelectrode distance is 3 cm. The upper electrode is connected to the RF generator and the lower electrode is grounded. The experimental setup is presented and described in [12,13]. The gas flow was observed and controlled via MFC Bronkhorst. The experiment was performed at different plasma parameters (discharge power, gas pressure, the percentage of the mixture gas flow and maintain the plasma, etc.).

3. Results and discussion

At ignition of plasma at certain parameters nanoparticles are synthesized due to the carbon-containing gas. When plasma is excluded, depending on time a certain number of these nanoparticles is deposited on the surface of the substrate that is located on the lower electrode. This process is carried out cyclically. Thus, the amount of particles on the substrate can be regulated by the number of cycles used to produce a surface with different roughness. With increasing numbers of cycles increases the density of particles on the substrate, which leads to an increase in contact angle. This effect for different cycles of deposition of nanoparticles on the surface of a glass substrate is shown in Figure 1.

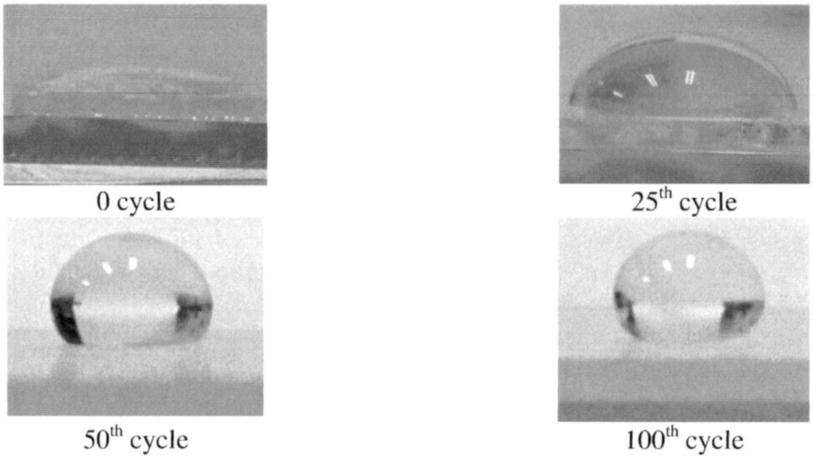

Fig. 1 – Photographs of obtained hydrophobic surfaces at different cycles of deposition of nanoparticles

Using the obtained experimental data the dependence of the contact angle on the deposition of nanoparticles was found, it is shown in Figure 2. As it can be seen from the graph the contact angle increases linearly depending on the number of cycles, until it reaches a value of about 160° at 150 cycles. After this an increase in cycles does not affect the contact angle, since saturation is reached. The graph also shows the curves for different cycle times, which directly affects the contact angle of the surface. This effect can be explained by the fact that at the same plasma parameters depending on time nanoparticles with specific sizes are synthesized, for example, in 10 seconds nanoparticles with diameter of ~70-90 nm are synthesized, in 15 seconds nanoparticles with diameter of ~ 120-170 nm are synthesized. Depending on the size of the precipitated nanoparticles the surface roughness changes affecting the contact angle.

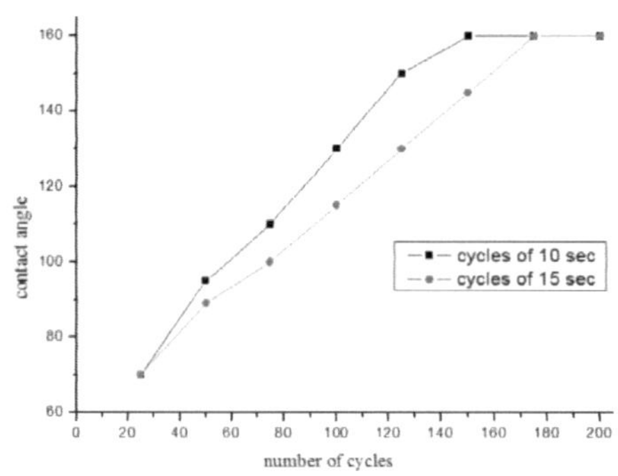

Fig. 2 – Dependence of the contact angle on the deposition cycle of nanoparticles

In addition, the dependence of the hydrophobicity of the surface on the composition of the working gas was investigated. At low concentrations of methane (1%) only nanoparticles are synthesized in the working gas, and with increasing methane concentration (7%) simultaneously with the nanoparticles nanofilm is synthesized. It is shown by the SEM, optical images and photos of the obtained samples shown in Figure 3 and Figure 4. When only nanoparticles are deposited on the surface, they interact with water, which leads to the destruction of hydrophobicity. When surface of substrate with nanoparticles and nanofilm is obtained, the hydrophobicity during contact with water is not destroyed due to the fact that the nanofilm is the connecting element between nanoparticles, due to which surface roughness appears.

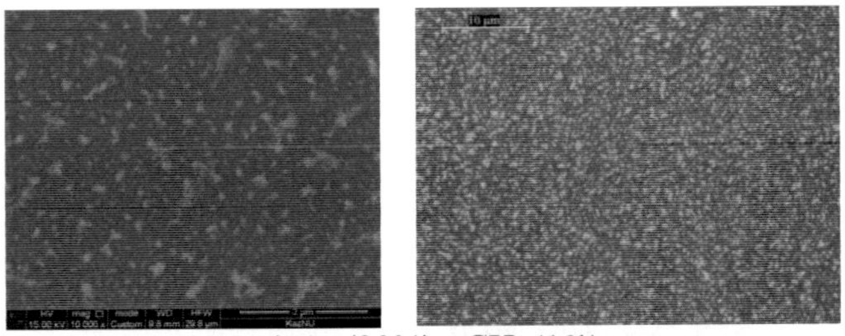

a) Ar (99%)+ CH_4 (1%)

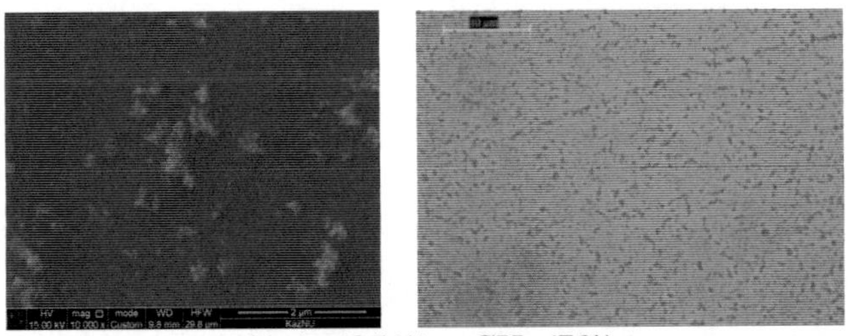

b) Ar (95%) + CH_4 (7%)

Fig.3 – SEM and optical images of the samples obtained at different gas compositions

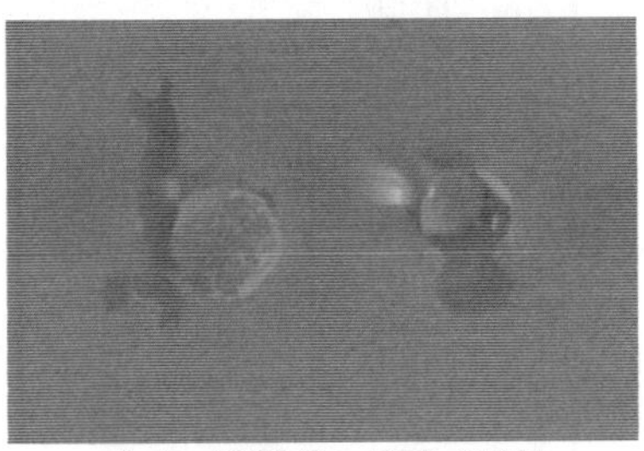

a) Ar (99%)+ CH_4 (1%)

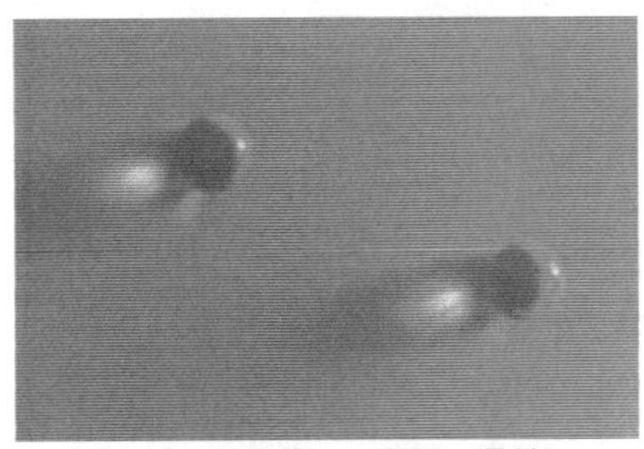

b) Ar (95%) + CH_4 (7%)

Fig.4 – Photos of the samples obtained at different gas compositions

To obtain stable hydrophobic surface, consisting of nanoparticles and nanofilms, plasma pulse mode can also be used. In this mode hydrophobic surfaces with small contact angles in comparison to the previous method results were obtained. In this method of obtaining hydrophobicity plasma confinement time is significantly low, therefore nanoparticles with diameter of about 4-10 nm and nanofilms that affect surface roughness are synthesized. The obtained sample was investigated with SZM, which indicates that the surface is rough and has an ordered structure. Figure 5 shows the results of the SZM analysis.

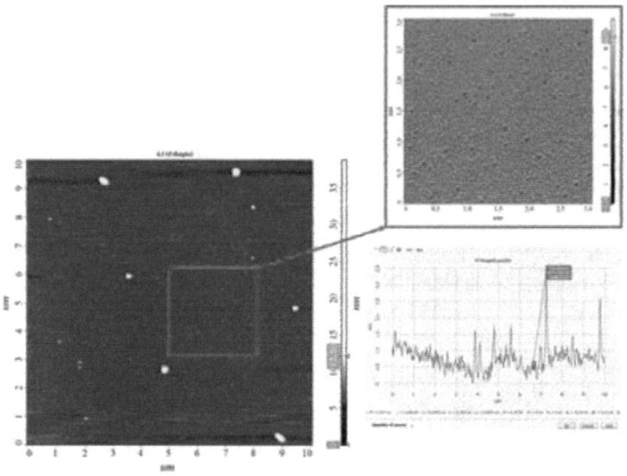

Fig.5 – SZM analysis of the obtaining samples in pulse mode

When superhydrophobic surface is treated with a buffer plasma (argon or nitrogen), the hydrophobicity is destroyed and the surface acquires superhydrophilic properties. Figure 6 shows the transformation from superhydrophobicity to superhydrophilicity. It should also be noted that the contact angle decreases with increasing plasma processing time.

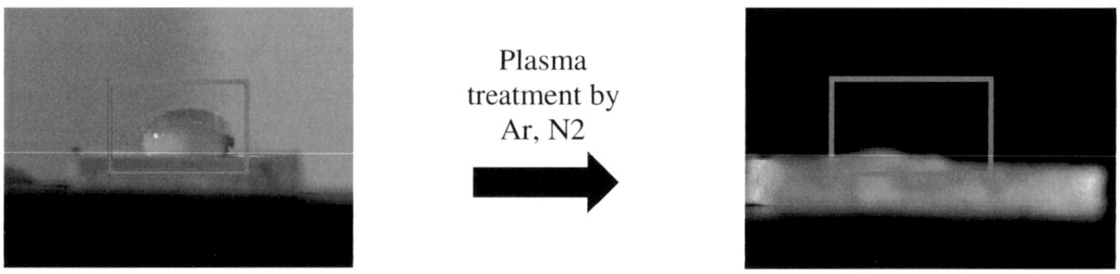

Fig.6 – Transformation from superhydrophobicity to superhydrophilicity

In this work, firstly, we got superhydrophobic surfaces using a simple one-step process, based on polymerizing carbonaceous nanoparticles in plasma and deposition on silicon wafer. Nanoparticles arising in plasma polymerization process are typical example of plasma polymers, i.e. materials, which in difference to conventional structured polymers, do not consist of repeating units. In addition, we researched how wetting characteristics were changing when processed in different plasma environments. We focused on the process of growth of nanoparticles and their subsequent deposition. Specific feature of this process is the negative charge of the nanoparticles. Once the particles reach a size of several nanometers, they quickly collect a negative charge (due to the high mobility of the electrons in the plasma). As a result, the particles are held in positive plasma potential, i.e. they are levitating in the discharge, where they continue to grow due to the accumulation of neutral radicals and positive ions. After the plasma turns off, particles lose the negative charge and fall down to the lower electrode. Every time plasma is turned off a certain amount of nanoparticles will fall onto the substrate, which is located on the lower electrode.

The SEM image shows that the synthesized materials have a rough surface with contact angles of from 100^0 to 165^0, depending on the plasma parameters. Experimentally it has been found, that the number of particles on a silicon substrate depends on the number of cycles, i.e. the film's hydrophobic feature (contact angle) increases with the number of cycles.

4. Conclusion

In this paper, as a result of the conducted experiments, superhydrophobic and superhydrophilic surfaces were obtained. The obtained samples were examined by SEM, photos and optical microscopy, the results of which indicate surface roughness. It was found that the hydrophobicity of the surface varies depending on the nanoparticles application cycle, the cycle time and the resulting surface structure.

5. Acknowledgments

This work was supported by the Ministry of Education and Science of the Republic of Kazakhstan (project 3222/GF4).

References

[1]. D. Ebert, B. Bhushan, *J. Colloid Interface Sci. 368,* 584−591 (2012)

[2]. Sanjay Subhash Latthe et. al, *J. of Surface Engineered Mat. and Advanced Tech. 2*, 76-94 (2012)

[3]. Favia, P., Cicala, G., Milella, A., Palumbo, F., Rossini, P., *Surf. Coat. Technol., 169−170* (0), 609−612 (2003)

[4]. Fresnais, J., Chapel, J. P., *Surf. Coat. Technol., 200* (18,19), 5296−5305 (2006)

[5]. Lu, X.; Zhang, C.; Han, Y. *Macromol. Rapid Commun., 25* (18), 1606−1610 (2004)

[6]. Zhang, H.; Zeng, X.; Gao, Y.; Shi, F.; Zhang, P.; Chen, J.-F. *Ind. Eng. Chem. Res., 50,* 3089−3094 (2011)

[7]. Shang, H. M.; Wang, Y.; Limmer, S. J.; Chou, T. P.; Takahashi, K.; Cao, G. Z. *Thin Solid Films, 472* (1−2), 37−43 (2005)

[8]. Han, J. T.; Jang, Y.; Lee, D. Y.; Park, J. H.; Song, S.-H.; Ban, D.-Y.; Cho, K. *J. Mater. Chem., 15,* 3089−3092 (2005)

[9]. Zhang, X., Shi, F., Yu, X., Liu, H., Fu, Y., Wang, Z., Jiang, L., Li, X. *J. Am. Chem. Soc., 126,* 3064−3065 (2004)

[10]. Miller, J. D.; Veeramasuneni, S.; Drelich, J.; Yalamanchili, M. R.; Yamauchi, G. *Polym. Eng. Sci., 36* (14), 1849−1855 (1996)

[11]. Öner, D.; McCarthy, T. J. *Langmuir, 16,* 7777−7782 (2000)

[12]. Omirbekov D. B., Zhunisbekov A.T., Ramazanov T.S., Orazbayev S.A., Dosbolayev M.K., Gabdullin M.T., Zhumadylov R.E., *The 5th International conference on Nanomaterials and Advanced Energy Storage Systems. (INESS2017)* – Astana, Kazakhstan, 9-11 August, – P.98. (2017)

[13]. Zhunisbekov A.T., Ramazanov T.S., Orazbayev S.A., Dosbolayev M.K., Gabdullin M.T., Omirbekov D.B. *International Conference Strongly Coupled Coulomb Systems (SCCS).* – Kiel, Germany, 30 July-4 August, – P.146. (2017)

Detection of organic vapors on sputtered and annealed thin Au films

O Kvitek, V Kopacek, A Reznickova and V Svorcik

Department of Solid State Engineering, University of Chemistry and Technology Prague, Technicka 5, 166 28 Prague, Czech Republic

E-mail: kviteko@vscht.cz

Abstract. Unique optical properties of metal nanostructures enable construction of new types of chemical sensors. Nanostructures composed of Au on glass substrate were prepared by annealing of 2-20 nm thick sputtered Au films at 300 °C for 1 h. The annealing leads to transformation of the as sputtered continuous Au layers to a nanoisland structure. The forming nanostructure shows a strong, well defined surface plasmon resonance absorption band in UV-Vis spectrum, which is useful for construction of a chemical sensor. The samples were used to detect vapors of acetone and water in an experimental testing apparatus. The achieved signal-to-noise ratio was 583 and 386 for acetone and water vapors, respectively on the nanostructure prepared from 4 nm thick Au layer. The nanostructured sensitive layers, however, showed poor signal stability; therefore a polymer overlayer was introduced to protect it. The employed polystyrene film prepared by spin-coating improved sensitivity and selectivity of the sensor, while the dynamic properties of the sensing influenced only slightly.

1. Introduction

Metal nanostructures are recently a highly sought after material thanks to their unique properties. Decreasing the dimensions of structure of a material to nanoscale level leads to changes of its properties. A nanomaterial then possesses different properties than a bulk material of the same chemical composition. This phenomenon is caused by effects of two categories – the surface effects and the quantum-size effects. Surface effects are related to increasing ratio of surface atoms in the material with decreasing particle size. The surface atoms have generally higher energy than those in a bulk of a material, are bonded more weakly to the material and are usually more reactive [1]. In sensor technology, the increasing surface to volume ratio with decreasing particle size is also desirable because of bigger area of interaction of the material with the analyte. The quantum-size effects originate in changes of the electron energy levels is the nanomaterial. As the particle size decreases the density of electron states decreases, which leads to disruption of the energy band structure of the material [2]. A manifestation of this in the case of metal nanostructures is evolution of surface plasmon resonance (SPR) absorption band in their optical absorption spectrum [3].

Thanks to their electromagnetic nature, the plasmon states are dependent on the optical parameters of the environment, which has been used in construction of sensor devices. SPR sensors are employed to detect molecular interactions on a surface of a continuous metal layer. The most significant application of these sensors is in biosensing and analysis of interactions of biomolecules bound to the metal surface [4]. A certain disadvantage of the method is its dependence on a specific transducing medium and an optical aperture to introduce the analytical beam into the metal layer. For a more general use as a chemical sensor, localized SPR (LSPR) sensors seem to be more practical. The LSPR sensors exploit metal nanostructures to make use of non-propagating – localized – surface plasmons [5]. Thanks to the nature of the nanostructure, there is no need for the optical aperture and transducing medium [6].

The usual methods to achieve a metal nanostructure on solid substrate include either some sort of lithographical process on a continuous layer or anchoring of pre-prepared metal nanoparticles [7,8]. A simple method to transform a continuous metal layer to a nanostructure is solid-state dewetting, which utilizes a configuration, where the interfacial energy between the metal and the substrate is high. With usual physical vapor deposition methods the film growth speed is quite high and the metal atoms are not able to migrate over the surface to occupy the lowest energy positions. As such, the as deposited metal film is usually continuous, but metastable and its annealing leads to its transformation to a configuration with lower total surface energy [9]. In case of noble metals on glass substrate the interfacial energy is high and the lowest energy configuration according to the Young equation is where the metal and the glass are in as low contact as possible. Therefore, with annealing of such structure at sufficient temperatures, where surface diffusion is enabled, the morphology of the metal layer transforms from continuous coverage to a structure of isolated islands [10,11].

In this work, we employed the method of solid-state dewetting to prepare nanostructures of Au on glass substrate from thin Au films produced by cathode sputtering. The successful formation of nanostructures was monitored by atomic force microscopy (AFM). UV-Vis spectroscopy was used to observe the appearance of SPR absorption bands necessary for the samples to be used as sensitive layers for chemical detection. The samples of the Au nanostructures were then tested for their sensitivity to presence of vapors of acetone and water in an experimental sensor testing apparatus.

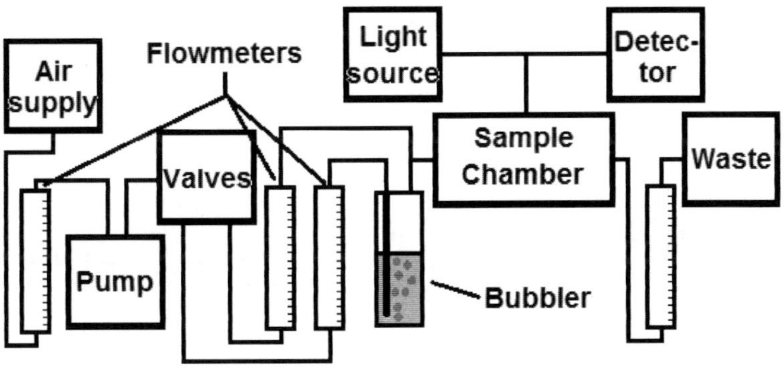

Figure 1. A schematic representation of the experimental sensing apparatus.

2. Materials and Methods

2.1. Sample preparation methods

Au films of thickness 2-20 nm were prepared by cathode sputtering from Au target of 99.95 % purity (Safina) on SCD 050 sputter coater (BalTec) at 20 mA for 10-100 s on 18×18 mm^2 borosilicate glass substrates of 0.13-0.16 mm thickness (Menzel-Gläser) rinsed in methanol and dried in a stream of nitrogen gas before the deposition process. The pre-deposition vacuum was 2 Pa, the deposition was carried out at Ar work gas pressure of 5 Pa (99.99 % purity, SIAD Czech). The Au films were subsequently annealed at 300 °C for 1 h in FED 23 oven (Binder) and cooled in air to room temperature. Polystyrene (PS) overlayer of 200 nm nominal thickness was prepared by spin-coating method from a 5 % solution in toluene (PS powder supplied by Goodfellow) at 2000 rpm. A reflective ca 100 nm thick Ag mirror was prepared on the backside of the samples for the gas sensing studies by cathode sputtering at 40 mA for 200 s on the SCD 050 device.

2.2. Methods of analysis

Optical properties of the nanostructured Au samples were studied by UV-Vis spectroscopy on Lambda 25 spectrophotometer (Perkin-Elmer) in 10×10 mm^2 area. The optical absorption was measured in 300-800 nm spectral range at 240 nm·min^{-1} scanning rate and 1 nm data collection interval with halogen (350-800 nm) and deuterium (300-350 nm) lamps. The absorption of the glass substrate was subtracted from the spectra of the samples.

Surface structure was observed by AFM on Digital Instruments CPII microscope (Bruker). Silicon P-doped probes RTESPA-CP with spring constant of 20-80 N·m^{-1} were used in tapping mode. Area of 2.4×2.4 μm^2 was scanned at 512×512 px resolution. The obtained morphological scans were used for arithmetic average surface roughness (R_a) calculation.

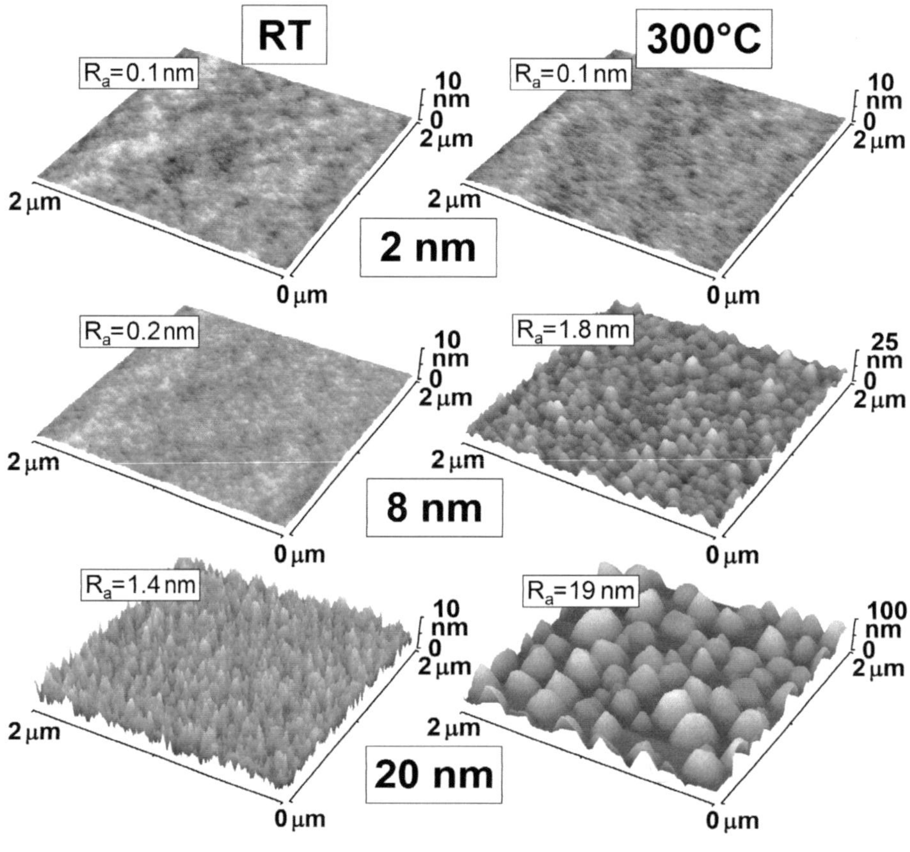

Figure 2. AFM images of as deposited (RT) and annealed (300 °C) Au layers 2-20 nm thick, values of average surface roughness (R_a) in nm are provided.

The prepared Au nanostructures were tested for their optical response to introduction of vapors of acetone, water, ethanol and toluene in an experimental apparatus (Fig. 1). The gas circuit of the apparatus was designed to push dry air (80:20 mixture of N_2 and O_2) from the air supply with aquarium pump at the rate of 35 ml·min^{-1} alternately directly through the sample chamber and through a bubbler with the analyte and the sample chamber using computer controlled valves. The flow of the gas was controlled in each part of the apparatus with adjustable flowmeters FLDA3428ST (Omega Engineering). The optical system was designed to guide light beam with optic fibers from the halogen lamp of DH-2000 source (Ocean Optics) to the sensitive layer placed in the sample chamber, where it reflected from the backside Ag mirror and to the RedTide USB650 (Ocean Optics) detector, which analyzed the modified emission spectrum of the light source in real time. The regulatory valves were controlled by PC with control software self-designed in LabView environment and the system was set to change the atmosphere from air to the analyte and vice versa every 10 min. The system was designed for the spectra evaluation as well. Signal was obtained as the area under the detected light intensity spectrum in the wavelength range of 475-575 nm, where the SPR peak of the Au nanostructures is located and where the greatest variation of the signal with introduced analyte was identified. The analytes employed in this study were acetone and water.

3. Results and Discussion

Morphology of the thin Au films as observed by AFM is in Fig. 2. Surface roughness of the as deposited films increases with increasing film thickness, but the surface still remains relatively smooth and the Au layers are continuous. Annealing leads to a dramatic change of the surface structure. Due to the high interfacial energy between glass and the Au and the metastable nature of the as deposited Au films, the continuous Au coverage transforms into an island-like structure (this process is called "solid state dewetting"). This structural transformation is more apparent in the case of the thicker layers since a more homogeneous Au layer is formed during the deposition. The surface roughness increases significantly and as the isolated Au islands are formed the underlying glass substrate gets exposed in certain areas.

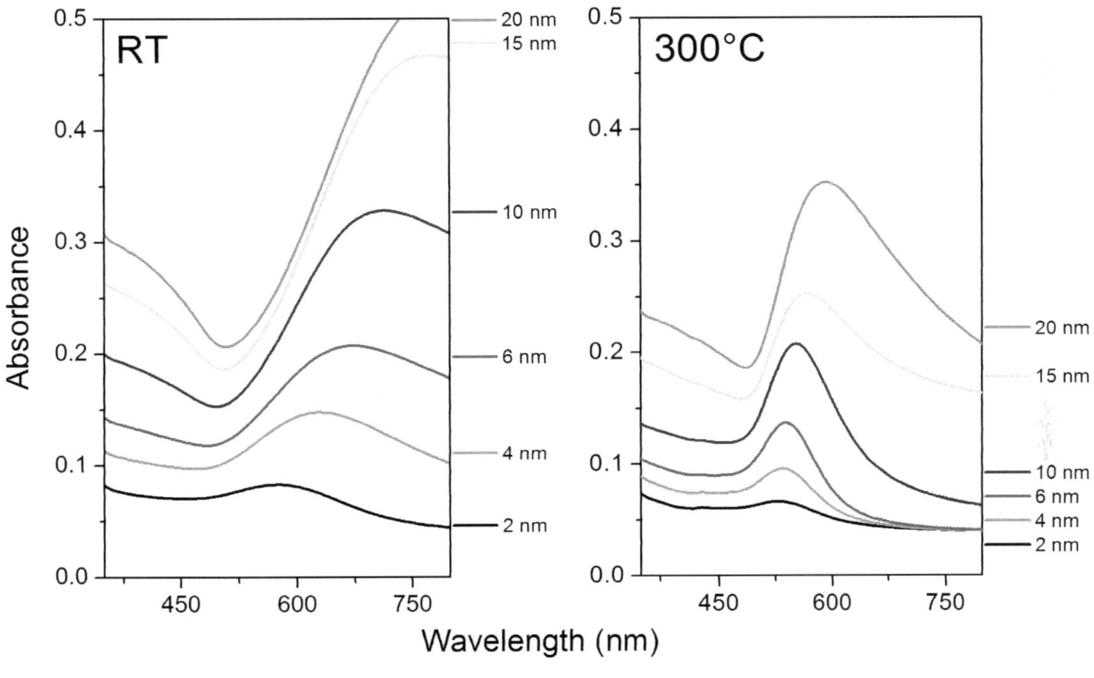

Figure 3. UV-Vis spectra of as deposited (RT) and annealed (300 °C) Au layers 2-20 nm thick.

The structural transformation of the Au layer has a great impact on the optical properties of the samples as well (Fig. 3). The as deposited layers show UV-Vis spectra characteristic for bulk Au with absorption minimum close to 525 nm, where the gap between plasmonic and interband transition absorption lies. Only in the case of the thinnest studied layers a somewhat defined SPR peak forms. With increasing thickness of the film, the values of absorption generally increase as the analytical beam passes through more material. With annealing the values of absorption generally decrease, because of the decreased effective coverage of the substrate in case of the isolated Au islands. The character of the spectra changes dramatically as well. A well-defined SPR peak forms, which confirms the formation of nanostructures. Maximum of the SPR peak lies at ca 530 nm in the case of the thinner films and shifts towards higher wavelengths with increasing film thickness as a result of the increasing size of the Au nanostructures. The well-defined nanostructure with strong SPR absorption bands is a vital requirement for utilization of the Au nanostructures as sensitive layers for chemical sensors since the absorption of the SPR band is highly sensitive on variation of refractive index of the medium surrounding the nanostructure.

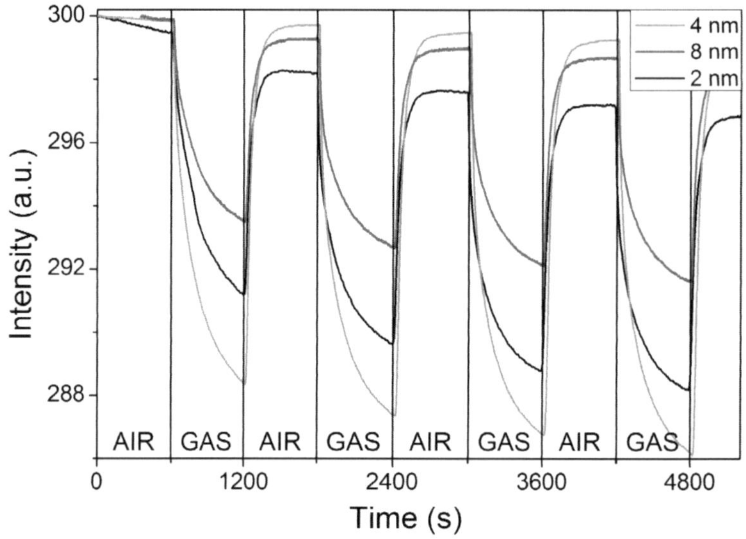

Figure 4. Sensorgrams of response of 2-8 nm thick Au nanostructures to acetone vapors.

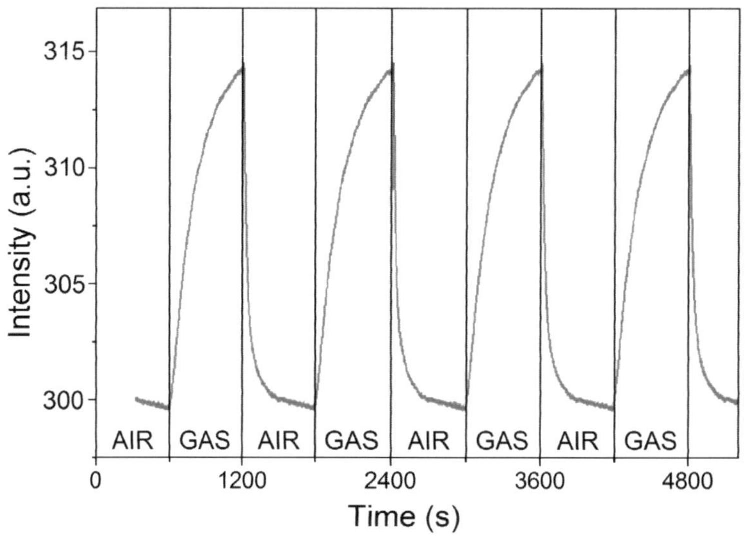

Figure 5. Sensorgram of response of 4 nm thick Au nanostructure with PS overlayer to acetone vapors.

The results of the sensitivity tests of the Au nanostructures against acetone vapors are in Fig. 4. Nanostructures prepared by annealing from films of various thicknesses were tested. The comparison of the sensitivities of the different thicknesses showed the highest sensitivity for the samples with the 4 nm thick Au layer. As the refractive index of the medium around the sensitive layer increases with introduction of the acetone vapors, the signal on the optical detector decreases, because the absorption of the SPR band of the sensitive layer is higher. The sensitivity towards the acetone vapors expressed as signal-to-noise ratio (SNR) reached the value of SNR(acetone) = 583. In case of the water vapors the sensitivity was somewhat lower at SNR(water) = 386. The response of the sensor is quite slow, after 10 min of analyte exposure the signal values still seem to be quite far from the minimal value. On the other hand, the regeneration of the layer after re-introduction of air is quicker and the signal reaches plateau within the 10 min set in the valve program. The dynamic properties of the sensor could be improved by increasing the temperature. The measurements were carried out at standard laboratory conditions, in case of an elevated temperature the sorption/desorption process of the analyte would be accelerated. The value of the signal in the sensorgram seems to be decreasing with ongoing measurement cycles. This is a problem since the layer is probably degraded during the detection process.

To improve the durability and time stability the thin Au films were covered by a PS overlayer. The overlayer changes the optical properties of the structure, naturally. The sensorgram for acetone vapor detection on this sample is in Fig. 5. Due to the PS having higher refractive index than air and the analyte absorption effectively decreasing it, the introduction of the analyte in this case leads to increase of the value of the signal on the optical detector. The sensorgram clearly shows the stability of the sensitive layer during the sensing process increases considerably with the dynamic characteristics not being influenced too much. The PS layer even leads to increased sensitivity at SNR(PS, acetone) = 706. The polymer overlayer could be as well used to improve selectivity of the sensor as it modifies affinity of the sensor surface to various analytes.

4. Conclusions

Nanostructures of Au on glass substrate were prepared by solid-state dewetting of sputtered thin films. The post-deposition annealing leads to formation of well-defined structure of Au nanoislands, which shows strong SPR absorption bands in UV-Vis spectrum. That is a good presumption to use the structures as sensitive layers in chemical sensors. In an experimental apparatus the samples showed sensitivity towards presence of acetone and water vapors. From testing different sizes of the Au nanostructures the samples prepared from 4 nm thick Au films appeared to be the most sensitive to presence of the vapors. The achieved SNR was 583 and 386 for acetone and water vapors, respectively. Furthermore, we used a PS overlayer, which seemed to improve sensitivity and stability of the Au layers, with mostly indistinguishable impact on dynamic parameters. The polymer overlayer could be also used to improve selectivity of the sensor. In future, we would like to build on the achieved results and test the samples against more analytes and use various overlayer polymers.

Acknowledgment

Authors would like to thank the Czech science foundation (GACR) for the financial support under the project 17-00939S.

References

[1] Wang N, Rokhlin S I and Farson D F 2008 *Nanotechnology* **19** 415701
[2] Roduner E 2006 *Chem. Soc. Rev.* **35** 583
[3] Barnes W L, Dereux A and Ebbesen T W 2003 *Nature* **424** 824
[4] Homola J, Yee S S and Gauglitz G 1999 *Sensor Actuat. B* **54** 3
[5] Liz-Marzan L M 2004 *Mater. Today* **7** 26
[6] Mayer K M and Hafner J H 2011 *Chem. Rev.* **111** 3828
[7] Cheng C-S, Chen Y-Q and Lu C-J 2007 *Talanta* **73** 358.
[8] Liu M, Ji Z and Shang L 2010 Top-down Fabrication of Nanostructures *Nanostructured Surfaces (Nanotechnology* vol 8*)* ed L Chi (Weinheim: Wiley-VCH) chapter 1 pp 3-47
[9] Thompson C V 2012 *Annu. Rev. Mat. Res.* **42** 399
[10] Svorcik V, Kvitek O, Lyutakov O, Siegel J and Kolska Z 2011 *Appl. Phys. A* **102** 747
[11] Tesler A B, Chuntonov L, Karakouz T, Bendikov T A, Haran G, Vaskevich A and Rubinstein I 2011 *J. Phys. Chem. C* **115** 24642

IOP Conf. Series: Journal of Physics: Conf. Series **987** (2018) 012006 doi:10.1088/1742-6596/987/1/012006

Niobium oxide nanocolumns formed via anodic alumina with modulated pore diameters

A Pligovka, A Zakhlebayeva, A Lazavenka

R&D Lab 4.10 "Nanotechnologies", Belarusian State University of Informatics and Radioelectronics, Brovka Str. 6, Minsk 220013, Belarus

E-mail: pligovka@bsuir.by

Abstract. Niobium oxide nanocolumns with modulated diameters were formed for the first time. An Al/Nb bilayer specimen was prepared by successive sputter-deposition of 300 nm niobium layer and 1200 nm aluminum layer onto silicon wafer. Regular anodic alumina matrix with modulated pore diameters was formed by sequential anodization of initial specimen in tartaric acid at 180 V, and in oxalic acid at 37 V. Further potentiodynamic reanodization of the specimen up to 400 V causes the simultaneous growth of 440 nm continuous niobium oxide layer beneath the alumina film and two types of an array of oxide nanocolumns (thick – with 100 nm width and 630 nm high and thin – with 25 nm width and 170 nm high), which are the filling of the alumina pores. The morphology of the formed anodic niobium oxide nanocolumns with modulated diameters was determined by field emission scanning electron microscopy. The formed nanostructures can be used for perspective devices of nano- and optoelectronics such as photonic crystals.

1. Introduction

The porous anodic alumina (PAA) is characterized by uniform cylindrical pores, which allow it to be used as a matrix for template synthesis of arrays of different nanostructures. Manipulation of the size, position and regularity of the pores allow forming conceptually new structures on its base with a wide range of applications [1]. The most promising are PAA with modulated pore diameters and structures based on it. PAA templates with unusual and complex shape of channels can be synthesized by anodic oxidation of aluminum substrates [2]. Works of the some authors [3, 4] revealed that anodic processing of a niobium layer covered with a relatively thick layer of aluminum in certain electrolytes for PAA formation under appropriate conditions results to the formation of metal oxide "nanocolumns" which represent nanostructured niobium oxide. Such nanostructures have unique properties and can be used for functional applications in devices with improved characteristics [4].

This paper presents the first result of formation of niobium oxide nanocolumns with variable diameters by high-voltage reanodization via PAA matrix with modulated pore diameters. The morphology properties of the synthesized niobium oxide nanostructures are determined by the means of field emission scanning electron microscopy (SEM).

2. Experimental

The initial substrate was double-side polished *n*-type Si wafer, 4" in diameter, 300 μm thick, with (100) crystal orientation and 4-40 Ω/□ resistivity, covered by 200 nm layer of thermally grown SiO_2.

Layers of 300 nm niobium and 1200 nm aluminum were successively deposited on the Si wafer by radio-frequency magnetron sputtering (Figure 1, a). The wafer was cut into pieces of ca. 1×2 cm which were individually anodized in specially designed two electrode cell. The programmable power supply Keysight N5751A was used as the anodization unit. Regular PAA matrix with modulated pore diameters was formed by sequential anodization of the aluminum layer in aqueous solutions of 0.2 mol·dm^{-3} tartaric acid ($C_4H_6O_6$) at 180 V (Figure 1, b) and in 0.4 mol·dm^{-3} oxalic acid ($H_2C_2O_4$) at 37 V (Figure 1, c). Array of niobium oxide nanocolumns was formed by high-voltage reanodization of the niobium layer via PAA matrix in the mixed solution of 0.5 mol·dm^{-3} boracic acid (H_3BO_3) and 0.05 mol·dm^{-3} sodium tetraborate ($Na_2B_4O_7$) in potentiodynamic mode at increase of potential until 400 V (Figure 1, d, e). The temperature of electrolytes was maintaining within ±1°C of the set value during anodization process. After formation of niobium oxide nanocolumns PAA matrix was selective dissolved in aqueous solution of 50% phosphoric acid (H_3PO_4) at 50°C (Figure 1, f).

The main stages of the growth of Nb_2O_5 nanocolumns via PAA matrix, with the relation between the film layers, are depicted in Figure 1.

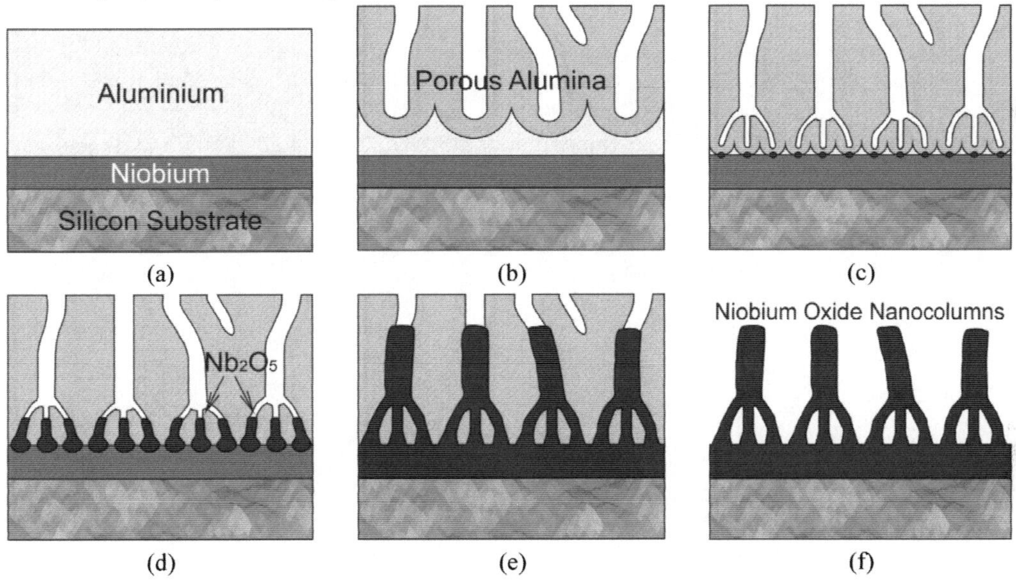

Figure 1. The main phases of the formation process of Nb_2O_5 nanocolumns via anodic alumina matrixes with modulated pore diameters: sputter-deposition of Al/Nb bilayer on SiO_2/Si substrate (a), growth of porous alumina film (b), anodization of the niobium underlayer through the alumina nanopores (c), growth of thin niobium oxide nanocolumns in the alumina pores (d), growth of thick niobium oxide nanocolumns in the alumina pores (e), dissolution of the alumina ("alumina-free" sample) (f).

3. Results and discussion

Figure 2, a–c shows SEM images of the surface and cross-section view of the anodic niobium oxide nanocolumns with modulated diameters which were formed from the initial Al/Nb bilayer sputtered on SiO_2/Si substrate. Three layers can be distinguished on the SEM image. An upper layer comprises the thick nanocolumns of niobium oxide. A middle layer comprises the thin niobium oxide nanocolumns. The lower layer is a strip of continuous niobium oxide, approximately 440 nm in thickness, is lying between the SiO_2/Si substrate and the columnar layers. The material of nanocolumns is about 800 nm in height is pulled out of the band outwards at intervals and generally similar in contrast to the band.

The deposited niobium was completely consumed as a result of the anodization. A schematic view of the array of niobium oxide nanocolumns, which elucidate the layered film structure, is drawn in Figure 2, d. The average diameters (diameters of the equivalent circle) and the height of the niobium oxide nanocolumns were estimated using the electron micrographs in Figure 2 and other numerous images of re-examined samples (not shown). The average diameters of the thick niobium oxide nanocolumns are about 100 nm. The height of the upper layer which comprises these thick nanocolumns is about 630 nm. The average diameter of the thin niobium oxide nanocolumns which is located in the middle layer is about 25 nm and their height is about 170 nm.

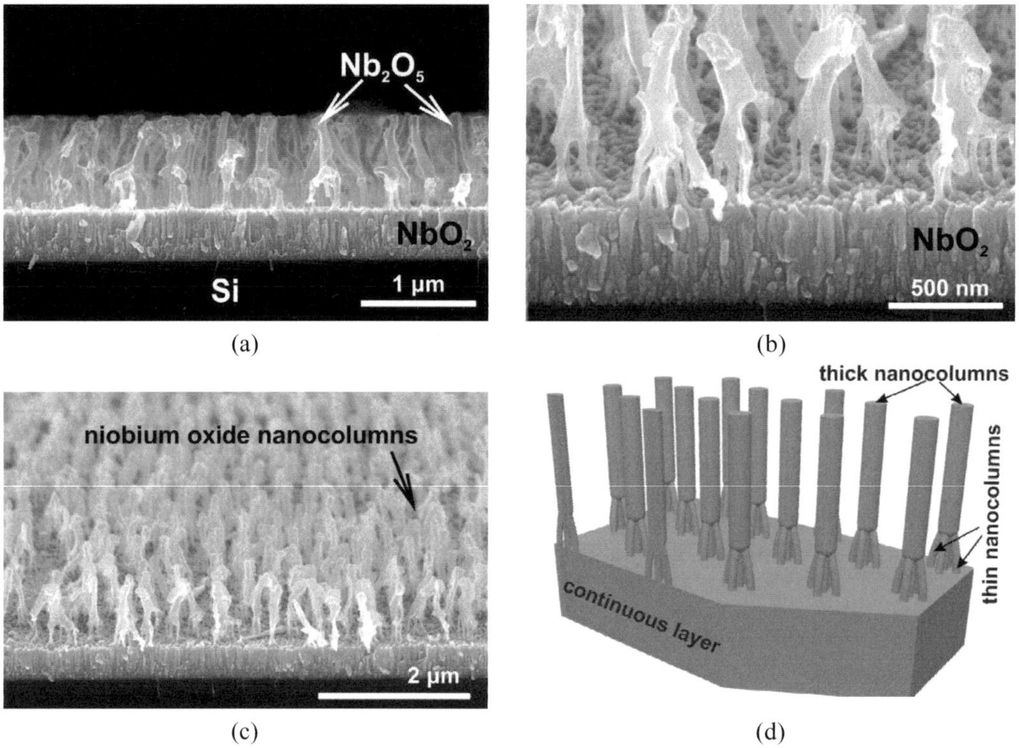

Figure 2. Scanning electron microscope images (a–c) and schematic view (d) of arrays of niobium oxide nanocolumns on Si-substrate formed by sequential anodization at 180 V in 0.2 mol·dm^{-3} $C_4H_6O_6$ and at 37 V in 0.4 mol·dm^{-3} $H_2C_2O_4$ with subsequent reanodization in 0.05 mol·dm^{-3} H_3BO_3+$Na_2B_4O_7$ at 400 V; all images of shown nanostructures were obtained after the alumina layer has been dissolved away ("alumina-free" samples).

Most of the nanocolumns appear to be broken or have an irregular structure and an uneven diameter. The three-step anodizing method, which is generally used in accordance with the procedure set forth in Ref. [5], will improve the regularity of niobium oxide nanocolumns.

4. Conclusions

In our work to date we had successfully formed anodic niobium oxide nanocolumns with modulated diameters by sequential anodization and high-voltage reanodization of the sputter-deposited Al/Nb metal layers via the anodic alumina matrix. The formed nanocolumns have the unordered structure because anodization of aluminum was carried out in one step without preliminary structuring of the surface. The additional ordering of the aluminum layer by the three step anodization will make it possible to form regular matrixes of the anodic aluminum oxide and produce high-ordered nanocolumns on their basis. Such approach to the formation of nanostructures have great practical importance, because used methods are technically simple, well reproducible, cost-effective and environmentally responsible, ecologically-friendly. Shown methods also can be used for the growing of metal-oxide nanostructures on transition metals and their alloys for various potential applications. The formed structures can be used as photonic crystals, autoemission elements and functional applications of perspective devices of nano- and optoelectronics.

References

[1] Jani A M M, Losic D and Voelcker N H 2013 *Prog. Mater Sci.* Nanoporous anodic aluminium oxide: Advances in surface engineering and emerging applications **58** (5) 636–704

[2] Zaraska L, Jaskuła M and Sulka G D 2016 *Mater. Lett.* Porous anodic alumina layers with modulated pore diameters formed by sequential anodizing in different electrolytes **171** 315-318

[3] Mozalev A, Sakairi M, Saeki I and Takahashi H 2003 *Electrochim. Acta* Nucleation and growth of the nanostructured anodic oxides on tantalum and niobium under the porous alumina film **48** 3155–70

[4] Mozalev A, Vazquez R M, Bendova M, Pytlicek Z, Llobet E and Hubalek J 2015 *Procedia Eng.* Porous-alumina-assisted formation of 3-D nanostructured niobium oxide films for advanced sensing applications **120** 435–438

[5] Mozalev A, Smith A J, Borodin S, Plihauka A, Hassel A W, Sakairi M and Takahashi H 2009 *Electrochim. Acta* Growth of multioxide planar film with the nanoscale inner structure via anodizing Al/Ta layers on Si **54** 935–945

IOP Conf. Series: Journal of Physics: Conf. Series **987** (2018) 012007 doi:10.1088/1742-6596/987/1/012007

Formation of various types of nanostructures on germanium surface by nanosecond laser pulses

S I Mikolutskiy, R R Khasaya, Yu V Khomich and V A Yamshchikov

Institute for Electrophysics and Electric Power RAS, Dvorzovaya Naberezhnaya 18, 191186, Saint- Petersburg, Russia

E-mail: mikolserg@mail.ru

Abstract. The paper describes the formation of micro- and nanostructures in different parts of irradiation zone on germanium surface by multiple action of nanosecond pulses of ArF-laser. It proposes a simple method using only one laser beam without any optional devices and masks for surface treatment. Hexa- and pentagonal cells with submicron dimensions along the surface were observed in peripheral zone of irradiation spot by atomic-force microscopy. Nanostructures in the form of bulbs with rounded peaks with lateral sizes of 40-120 nm were obtained in peripheral low-intensity region of the laser spot. Considering experimental data on material processing by nanosecond laser pulses, a classification of five main types of surface reliefs formed by nanosecond laser pulses with energy density near or slightly above ablation threshold was proposed.

1. Introduction

Nanostructures in solids are important object of studies because of their manifold possibilities for applications in science and technology. Surface nanostructuring gives rise to an improvement of thermal, tribological and electrical material properties [1-3]. It increases biocompatibility with living tissues of implants and prostheses used in orthopedics and stomatology [4]. Nanostructures on semiconductor surfaces applied in microelectronics, super-high density storaging of information, nanophotonics for development of light-emitting devices and for spectroscopy [5, 6]. Therefore it is of a great interest to develop new effective methods for obtaining surface structures with characteristic sizes less than 100 nm.

Previously various techniques have been used for nanostructuring with laser radiation. These are the masks projected with a demagnification on the treated surface [7], surface screening from incident radiation by micro-, and nanoparticles [8], and interference of two or more laser beams on the surface or in a volume of the material [9]. The combination of laser beam with a tip of an atomic force or scanning tunneling microscope has also been used, and that has allowed induce the change of material surface profile in a number of consequent superficial nanoscale regions [10].

In the given paper for creation of surface nanoreliefs we use only one laser focused beam without any masks and without any auxiliary atomic force microscope tip. It is a so-called "direct" surface nanostructuring. We report the results of experiments, where the formation of micro- and nanostructures was observed on germanium surface under the multiple action of nanosecond pulses of ArF-laser with radiation wavelength of 193 nm.

2. Materials and methods

A scheme of experimental setup is shown on figure 1. The main source of coherent radiation is an excimer ArF-laser marked by 1. Its radiation having a wavelength of 193 nm is strongly absorbed by a large number of materials. The laser radiation passes through the reflector 2, which directs a small part of the energy to NOVA II marked by 3. The other part of the laser energy is directed to the optical forming system consisting of a condenser 4, a forming diaphragm 5 and projection lens 6. The system creates a laser spot with a diameter of several hundred micrometers on the sample surface.

The sample 8 is placed in a gas chamber 9 with fluorite glasses 7, 7'. The gas chamber is connected to a vacuum pump 13 and a gas balloon 15 through the system of gas valves and pressure regulators. If it is necessary the gas chamber allows sample processing at pressures from 50 to 2000 mbar. Also for laser beam scanning it is possible to use a three-dimensional moving platform 10 [11]. In this work the samples were irradiated at atmospheric air in a stationery laser beam.

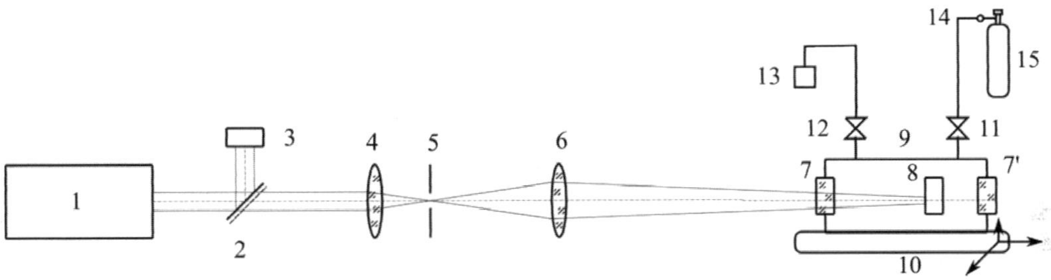

Figure 1. Experimental setup for surface nanostructuring. 1 - ArF-laser; 2 – reflector; 3 – NOVA II; 4 – condenser; 5 – forming diaphragm; 6 – projection lens; 7,7' – fluorite glasses; 8 – sample; 9 – gas chamber; 10 – three-dimensional moving platform; 11,12 – gas valves; 13 – vacuum air-pump; 14 – filter; 15 – gas balloon.

In the experiment on direct laser nanostructuring we used germanium samples with transverse dimensions of 8x8 mm and a thickness of 2 mm. After necessary preparatory procedures the samples were placed in the chamber of experimental setup and irradiated by 20 nanosecond laser pulses at rate frequency of 2 Hz. The duration of laser pulses was about 20 ns. The energy density in the center of laser spot was varied from 0.5 to 5 J/cm^2 with a step of 0.5 J/cm^2 for each laser spot. All five types of surface reliefs were observed at energy density of about 3 J/cm^2 on each germanium sample used in experiment. An analysis of irradiated germanium surfaces is carried out by an atomic-force microscope (AFM). Typical AFM views of main reliefs obtained at energy density of 3 J/cm^2 are represented in this work.

3. Results and discussions

The intensity of the laser radiation takes the maximum value in the center of the spot and falls to its periphery (figure 2a). As a consequence, the crater appears in the center of the spot at the multipulse irradiation of the same surface plot. There is a removal of the material from this crater (figure 2b). As we move from the center of the irradiation spot to its periphery, the AFM analysis showed the presence of various types of micro- and nanostructures on germanium surface.

Figure 3 shows a three-dimensional AFM image of the relief of the initial non-irradiated germanium surface. One can see a roughness in the form of protuberances with characteristic dimensions along the surface of 40-90 nm and a height of about 40 nm. Also traces of mechanical surface treatment are observed in the form of long linear scratches with a width of 40-80 nm.

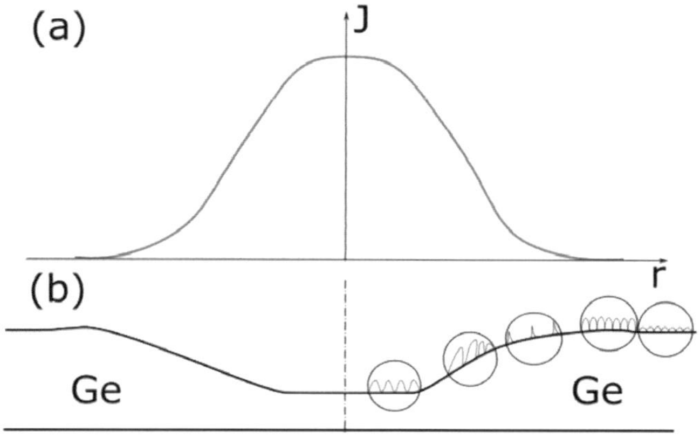

Figure 2. Schematic views of (a) laser pulse energy spatial distribution and (b) irradiation spot with five main types of surface reliefs on germanium surface.

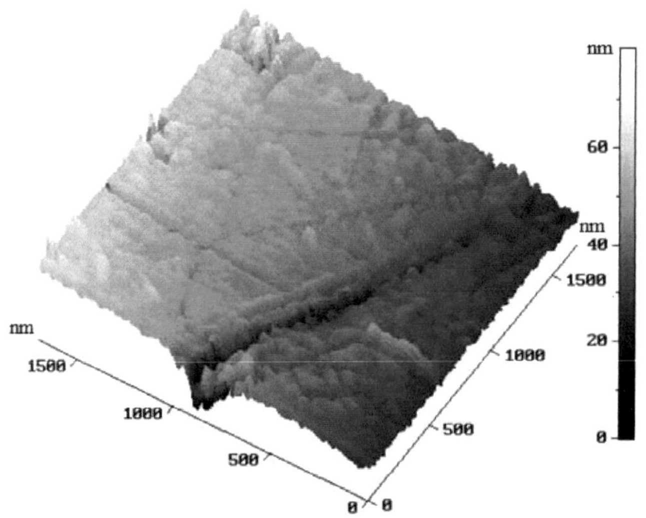

Figure 3. AFM image of non-irradiated germanium surface.

Figure 4 represents main types of reliefs obtained by irradiation of germanium surface with nanosecond pulses of ArF-laser. The pulse energy density in the centre of laser spot was about 3 J/cm^2. Considering that pulse energy had quasi-Gaussian distribution and germanium absorption capacity at wavelength of 193 nm was about 0.35, we can estimate deposited energy density in different places of laser spot.

In the central part (deposited energy density was about 1.1 J/cm^2) of the irradiation spot (figure 4a), which is ablation zone, a wavelike relief with periods of 200-400 nm and amplitude of about 150 nm was observed. Figure 4b shows an example of a wavy relief with the periods of about 1500 nm and the amplitude with a value of about 700 nm in the zone of a "deep" melt without significant ablative removal of the material. Estimation for deposited energy was from 0.9 to 1.1 J/cm^2. This wavy micron relief corresponds to mechanism of capillary waves, that has been studied before [12].

Figure 4c represents relief in the form of hexa- and pentagonal cells with characteristic dimensions along the surface of 300-500 nm and the height of the edges between the cells of 20-25 nm in the region of "shallow" melt (deposited energy density was about 0.4 J/cm^2). The formation mechanism of this type of relief on germanium surface under the action of laser pulses is unknown and it is an interesting scientific problem to solve.

Figure 4d shows nanostructures in the form of bulbs with rounded peaks with characteristic dimensions along the surface of 40-120 nm and height of 40-70 nm. They are observed in peripheral low-intensity region of the laser spot and its size makes it possible to use such structures as quantum dots. Estimation for deposited energy was about 0.1 J/cm^2.The studies for the structures in the form of microcones are presented in [13], where the mechanism of p-n junction formation by laser radiation and thermogradient effect in the elementary semiconductor is proposed.

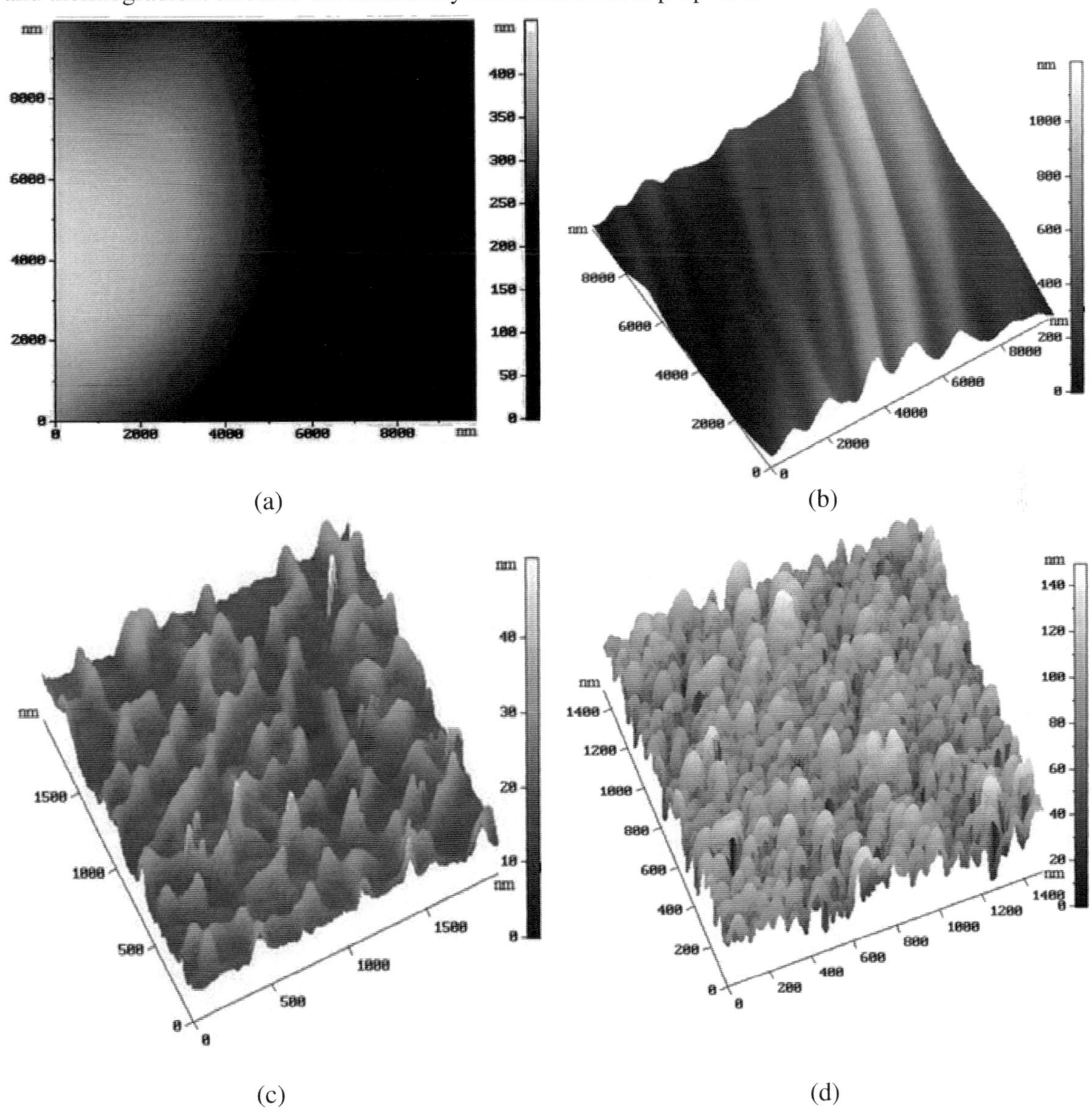

Figure 4. AFM images of irradiated germanium surface in different parts of irradiation spot: (a) central part (ablation zone); (b) zone of a "deep" melt without significant ablative removal of the material; (c) low-intensity peripheral zone of the “shallow" melt; (d) peripheral low-intensity spot zone outside the visible melt area.

Considering our experimental data for germanium and data from other works [14-17] concerning material processing by nanosecond laser pulses, we propose five main types of surface reliefs formed by nanosecond laser pulses with energy density near or slightly above ablation threshold:

(I) undulating relief with micron/submicron periods in the ablation zone in the central part of spot;

(II) wave-like relief with periods of one or several microns in the "deep" melt zone (melt thickness is about several hundred nanometers) without significant ablative removal of the material, what corresponds well known effect of excitation of capillary waves [18, 19];

(III) submicron relief with periods of several hundred nanometers in the low-intensity peripheral zone of the "shallow" melt (melt thickness is about several tens nanometers), when energy density is near or slightly above the melting threshold;

(IV) a small-scale relief with periods of several tens of nanometers in the peripheral low-intensity spot zone outside the visible melt area;

(V) micron/submicron relief of the original non-irradiated surface.

These types of relief considered may appear in the form of corresponding superpositions ((I) + (II) or (III) + (IV)) in different transition regions between the zones mentioned above. As it was shown above all five types of reliefs were observed on the germanium surface irradiated by nanosecond pulses of ArF-laser.

4. Conclusions

Formation of micro- and nanostructures on germanium surface by the multiple action of nanosecond pulses of ArF-laser has been demonstrated. A simple method of direct laser nanostructuring without any optional devices and masks has been proposed for surface treatment. The analysis of surfaces by an atomic-force microscope has shown various types of surface structures in different parts of irradiation zone. Hexa- and pentagonal cells with characteristic dimensions along the surface of 300-500 nm and the height of the edges between the cells of 20-25 nm were observed in peripheral zone of shallow melt and its formation mechanism requires following studies. Nanostructures in the form of bulbs with rounded peaks with characteristic dimensions along the surface of 40-120 nm and height of 40-70 nm were obtained in peripheral low-intensity region of the laser spot and could be used for quantum surface effects. Considering experimental data on material processing by nanosecond laser pulses, a classification of five main types of surface reliefs formed by nanosecond laser pulses with energy density near or slightly above ablation threshold was proposed.

References

[1] Quintana H A, Song E, Wang G T and Martinez J A 2013 *Chem. Eng. Process Tech.* **1** 1008.
[2] Fominskii V Y, Grigor'ev S N, Romanov R I and Nevolin V N 2012 *Tech. Phys.* **57** 516–23.
[3] Guo L, Chen G, Zhu Y, Duan X and Ye H 2015 *J. Cryst. Growth* **426** 49–53.
[4] Remeeva E A, Rozanova I B, Elinson V M and Sevastianov V I 2005 *International J. Artif. Organs* **28** 89.
[5] Claeys C 2007 *Germanium-Based Technologies: from Materials to Devices* (London: Elsevier)
[6] Medvid' A, Onufrijevs P, Dauksta E and Kyslyi V 2011 *Adv. Mater. Res.* **222** 44–47
[7] Takao H, Okoshi M and Inoue N 2005 *Jap. J. Appl. Phys.* **44** 241
[8] Vestentoft K, Olesen J A, Christensen B H and Balling P 2005 *Appl. Phys. A* **80** 493
[9] Kawamura K, Sarukura N, Hirano M, Ito N and Hosono H 2001 *Appl. Phys. Lett.* **79** 1228
[10] Gorbunov A A and Pompe W 1994 *Phys. Status Solidi A* **145** 333
[11] Khasaya R R, Khomich Yu V, Malinskiy T V, Mikolutskiy S I, Tokarev V N, Yamshchikov V A and Zheleznov Yu A 2014 *Appl. Phys.* **3** 83–87
[12] Kolasinski K W 2007 *Curr. Opin. Solid State Mater. Sci.* **11** 76-85
[13] Medvid' A, Dmytruk I, Onufrijevs P and Pundyk I 2007 *Phys. Status Solidi C* **4** 3066–69
[14] Mikolutskiy S I, Khomich V Yu, Shmakov V A and Yamshchikov V A 2011 *Nanotechnologies in Russia* **6** 733–38
[15] Tokarev V N, Khomich V Yu, Shmakov V A and Yamshchikov V A 2008 *Dokl. Phys.* **53** 206
[16] Ganin D V, Mikolutskiy S I, Tokarev V N, Khomich V Yu, Shmakov V A and Yamshchikov V A 2014 *Quant. Electr.* **44** 317–21
[17] Tokarev V N, Shmakov V A, Khasaya R R, Mikolutskiy S I, Nebogatkin S V, Khomich V Yu and Yamshchikov V A 2010 *Proc. Int. Cong. on Applications of Lasers and Electrooptics* (Anaheim, USA) pp 1257–65
[18] Van Driel H M, Sipe J E and Young J F 1985 *J. Lumin.* **30** 446–71
[19] Sakai K, Tachibana K, Mitani S and Takagi K 2003 *J. Colloid Interface Sci.* **264** 446–51

IOP Conf. Series: Journal of Physics: Conf. Series **987** (2018) 012008 doi:10.1088/1742-6596/987/1/012008

Effects of (Ce, Cu) Co-doping on the Structural and Optical Properties of ZnO Aerogels Synthesized in Supercritical Ethanol

D Djouadi[1], **O Slimi**[1], **L Hammiche**[1], **A Chelouche**[1] **and T Touam**[2]

[1]Laboratoire de Génie de l'Environnement, Faculty of Technology, University of Bejaia, Algeria
[2]Laboratoire des Semi-Conducteurs, Université Badji Mokhtar-Annaba, BP 12, Algeria
e-mail: djameldjouadi@yahoo.fr

Abstract. Undoped, Ce-doped, Cu-doped and (Ce,Cu) co-doped ZnO aerogels were synthesized by sol-gel process in supercritical conditions of ethanol. [Cu]/ [Zn] and [Ce]/[Zn] atomic ratios were fixed at 0.02 (2%). The aerogels were investigated without any additional treatments by using X-ray diffraction (XRD), UV–visible spectrophotometry, scanning electron microscopy (SEM), Energy-dispersive X-ray spectroscopy (EDS), Fourier transforms infrared spectroscopy (FTIR) and photoluminescence spectroscopy (PL). XRD results revealed that all the samples are well crystallized in hexagonal wurtzite structure. EDS measurements showed that highly pure aerogels are prepared. SEM analysis indicated that the morphology of the samples is dependent on Cu and Ce dopants. From UV-visible spectroscopy analyses, it was shown that the absorption and the band gap of the aerogels are strongly affected by Ce and Cu dopants. FTIR spectra demonstrated that co-doping induces a shift of Zn-O bond vibration band toward low wavenumbers. The room temperature photoluminescence spectra put into evidence that the visible emission intensity is influenced by Ce and Cu doping. In particular, the co-doping leads to the appearance of a blue emission band at 443 nm.

1. Introduction

Zinc oxide (ZnO) has been extremely studied as a multifunctional material for spintronics, optoelectronics, catalytic and photovoltaics applications [1-6]. Doping is a well known method to improve the physical properties of large gap materials. The introduction of extrinsic elements frequently affects the structural and optical properties of the material, but it is difficult to prove if these changes are due to the dopant elements or to the variation of native defects concentration induced by doping. Because of its high optical transmittance in the visible region and its important electrical conductivity, zinc oxide is highly recommended for photovoltaic applications [7].

ZnO particles are generally doped with transition metals or rare earths atoms in order to exalt its physical properties. Cu-doped ZnO nanopowder has been extensively investigated for use in particular applications [8-10]. Furthermore, Ce-doped ZnO was also studied in order to alter its optical characteristics [11-13]. Assadi et al. [14] have reported the enhancement of ZnO optical and magnetic properties by co-doping. Amiri et al. [15] have also demonstrated that (Ce, Cu) co-doped ZnO creates donor impurity bands producing a super-exchange interaction between the two atoms. Moreover, it has been reported that the interaction between Ce and Cu ions results in synergetic effects involving an easy creation of surface oxygen vacancies, leading to an improvement in catalytic activity [16, 17]. Zhang et al. [18] also reported the improvement of the optical activity of ZnO by Ce doping.

(Ce, Cu) co-doped ZnO nanoparticles have been synthesized with different processes such as solution combustion [19], microwave combustion [16], microemulsion reaction in mild conditions [20]. Up to now and to the best of our knowledge, (Ce, Cu) co-doped ZnO aerogels synthesized in superficial fluids has not been yet reported. In this context, the present work aims to complete our previously reported studies [21-23]. The choice of the sol-gel process followed by drying in supercritical conditions of organic fluids as synthesis route is justified by its simplicity, non-toxic chemicals, and low cost and large-scale nanostructures productions.

In this paper, undoped, Ce-doped, Cu-doped and (Ce, Cu) co-doped ZnO aerogels have been successfully synthesized via sol-gel method in supercritical drying conditions of ethanol. Structural, morphological and optical properties of the prepared samples have been investigated.

2. Experimental procedures

Undoped zinc oxide aerogel was synthesized by dissolving 16 g of zinc acetate dihydrate [[Zn $(COOCH_3)_2$, $2H_2O$]: 95%, Aldrich] in 112 mL of methanol under constant magnetic stirring for 15 min.Copper acetate monohydrate [(CH_3COO) $_2Cu$, H_2O), Proanalysis, 99%], cerium nitrate ($CeNO_3$,$6H_2O$) were added to the mixture in order to get Cu-doped, Ce-doped and (Ce, Cu) co-doped ZnO aerogels. The atomic dopants concentrations were fixed at 2 % in all the sols ([Cu]/[Zn] = 0.02, [Ce]/[Zn] = 0.02 and [Ce]/[Zn] = [Ce]/[Zn] = 0.02). After complete dissociation of these sources, each sol was poured into an autoclave (Parr 4848 Reactor) containing ethanol. Then, the mixture underwent a drying at supercritical conditions of ethanol (Tc = 243 ° C , Pc = 6,3 10^6 Pa). Once the supercritical conditions were reached, the solvent was removed and the autoclave was allowed to cool spontaneously to room temperature. The obtained aerogel was characterized without any chemical or heating treatments.

The crystalline structure of the as-prepared aerogels was studied using a PanAlytical diffractometer where the X-rays are produced from a radiation source CuKα (λ= 0.154 nm), with an acceleration voltage of 40 kV and a current of 30 mA. Infrared spectra (FT-IR) were recorded with a Shimadzu's IRAffinity-1 spectrometer by KBr pellet technique. SEM micrographs were recorded by a Quanta 2000 microscope with energy dispersive X-ray (EDS) spectroscope. EDS was used to identify the composition of the obtained products. UV–visible absorption spectra were recorded at room temperature using a Safas $UVmc^2$ spectrophotometer. The photoluminescence properties were measured at room temperature using Xenon lamp operating at 350 nm. The incident beam was focused perpendicularly to the samples and the fluorescence was collected in a guiding mode using a TRIAX 550 Horiba-Jobin Yvon monochromator equipped with a nitrogen cooled CCD camera.

3. Results and discussion

XRD patterns of undoped, Cu-doped, Ce-doped and (Ce, Cu) co-doped ZnO aerogel powders are shown in figure 1. All the peaks can be indexed to ZnO hexagonal wurtzite structure (JCPDS 36-1451). As can be seen, no diffraction peaks of Ce, Cu or their related oxides are detected indicating the introduction of the dopants in the ZnO crystal structure. XRD diffraction peaks have been found to be quite sharp and intense which implies high crystallinity of the prepared aerogels. It can also be observed that doping ZnO with Cu atoms leads to the increase in XRD peaks intensities indicating the enhancement of the aerogel crystalline quality. This result implies, in the one hand, the occupation of Zn^{2+}(0.074 nm) vacancies by Cu^{2+} ions (0.072 nm) [24]. In the other hand, the strong covalency of Cu-O makes the interaction of Cu with O stronger than that of Zn with O leading to the decrease of the oxygen defects in the lattice, which promotes the crystalliniy of the aerogels nanoparticles [25].

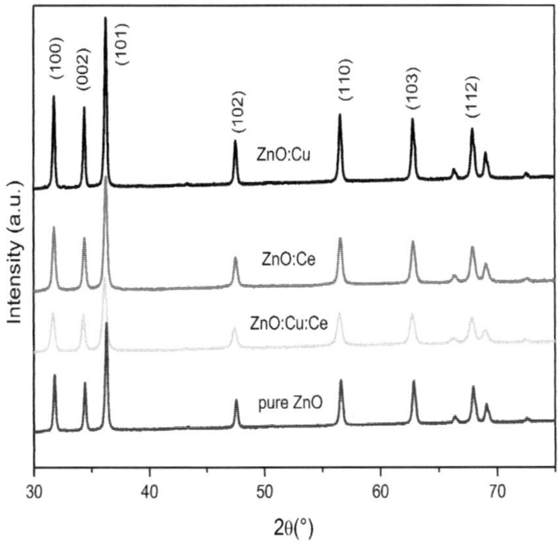

Figure 1. XRD spectra of pure, doped and codoped ZnO aerogels

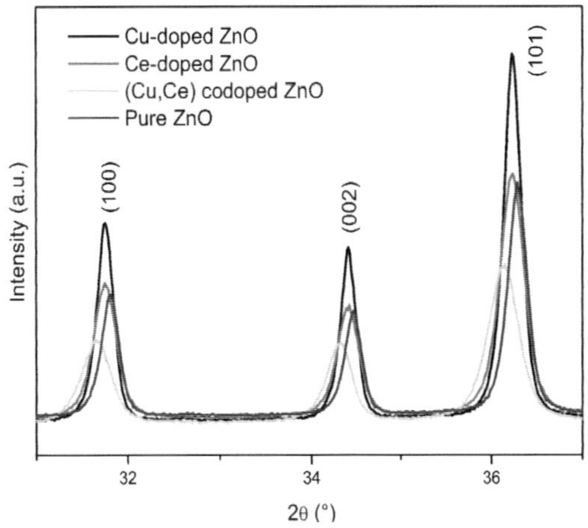

Figure 2. Most pronounced XRD peaks of the aerogels

The XRD spectra of the most pronounced peaks of figure 1, from which the principal structural properties of ZnO aerogels were determined, are shown in figure 2. Moreover, as can be seen from figure 2, the Ce doping induces the broadness and the shift to the lower diffraction angles without affecting the intensities of the XRD peaks of ZnO aerogel. This result indicates the increase of both the ZnO lattice parameters and the decrease in the crystallite size with Ce-doping. Furthermore, the introduction of Ce^{3+} and Cu^{2+} ions simultaneously in the lattice of ZnO deteriorates the crystalline quality of the aerogel. It was already reported that the intensity of the diffraction peaks decreases after the introduction of Ce and Cu atoms as compared to that of undoped ZnO due to the random occupation of the intrinsic defects in ZnO lattice by dopant ions and the difference in the ionic radii of zinc, cerium and copper [16,26].
ZnO lattice parameters a $(a= \lambda/(3^{1/2} \sin\theta))$ and c $(c= \lambda/\sin\theta)$ were calculated from the positions of the (100) and (002) peaks respectively. The degree of crystallinity Xc is calculated using the following empirical relation [27]:

$$X_c = (0.24/\beta_{002})^3 \quad (1)$$

where β_{002} is the full width at half maximum (in degrees) of (002) Miller's plane. The crystallites size and lattice micro-stress have been calculated using Williamson –Hall (W-H) method. In the case when both size and lattice micro-stress effects on the XRD line broadening are operative, Williamson and Hall have suggested a method combining the two effects to calculate these parameters. The W–H equation is expressed in the form [28]:

$$\beta cos\theta = \frac{0.89\lambda}{D} + 4\varepsilon \sin\theta \quad (2)$$

Where ε is the stress associated with the nanoparticle, D is the crystallite size, λ is the X-ray wavelength (1.5418Å), β is the full width at the half-maximum of the XRD line (in radians) and θ is the diffraction angle. The stress and the particle size can be obtained by plotting ($\beta\ cos\theta$) versus (4 $sin\theta$) for the 6 most important DRX peaks as shown in figure 3. The slope of the line (4ε)gives the stress and the intercept ($0.89\lambda/D$)of this line on y-axis gives the particle size(D). The cell parameters *a* and *c*, average crystallites size *D*, Zn-O bound length (*L*), microstrain (ε) and the degree of crystallinity are calculated for all the aerogels and are shown in Table 1.

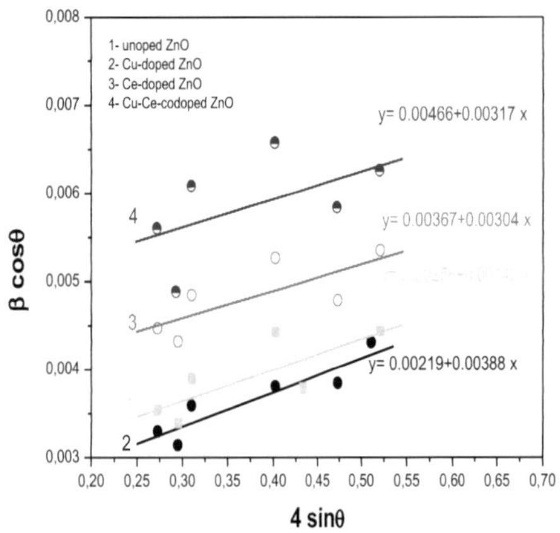

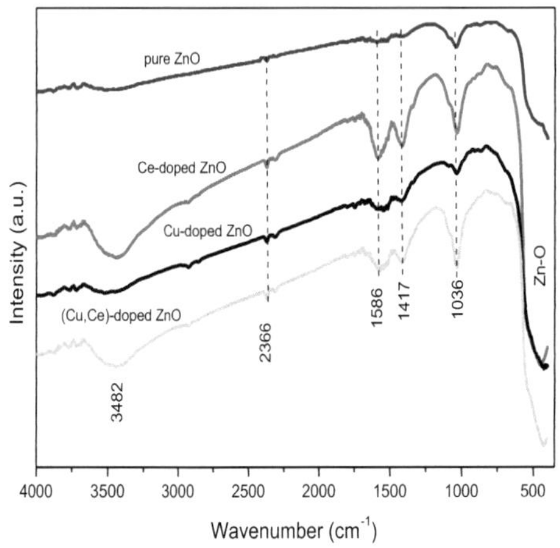

Figure 3. W-H analysis of the pure, doped and codoped aerogels.

Figure 4. FTIR curves of pure and codoped aerogels.

Table 1. Cell parameters *a* and *c*, average crystallites size *D*, Zn-O bound length (*L*), micro-strain (ε) and the degree of crystallinity X_C.

Sample	*a (Å)*	*c (Å)*	*L (Å)*	*D (nm)*	*Micro-strain (ε)*	X_c
Undoped ZnO	3.2451	5.1996	1.9749	53	$3.6\ 10^{-4}$	1.094
Cu-doped ZnO	3.2506	5.2072	1.9781	62	$9.7\ 10^{-4}$	1.375
Ce-doped ZnO	3.2509	5.2076	1.9783	37	$7.6\ 10^{-4}$	0.555
(Ce,Cu)codoped ZnO	3.2606	5.2221	1.9841	29	$7.9\ 10^{-4}$	0.281

The average crystallites size of the particles has been found to be 53, 62 , 37 and 29 nm in undoped, Cu-doped, Ce-doped and (Cu,Ce)-codoped ZnO aerogels , respectively. This results indicates that the simultaneously presence of Cu and Ce atoms in the aerogel leads to the diminution of the particles size due probably to the decrease of molecular concentration at the crystallites surface of the aerogel [29]. Also, It has been reported that smaller size of the codoped samples is caused by the formation of Zn–Ce–Cu–O in the crystal lattice, which plays an important role in hindering the crystal growth [30]. The best crystallinity has been found in the Cu-doped ZnO aerogel and the worst one has been found in codoped aerogel. This result confirms the XRD measurements. For Cu-doped ZnO aerogel, the dilation of the cell parameters can be due to the tensile stress exerted by the solvant during the supercritical drying process on the Cu-O bond. Kuluk et al. [31] reported that for low Cu

concentration, the substitution of Cu^{2+} ions instead of Zn^{2+} ions at their lattice sites increases the lattice constants and the interplanar distance, which would lead to the decrease of the diffraction angle compared with undoped ZnO. The increase of the cell parameters in Ce-doped aerogel can be explained by the difference in ionic radii of Zn^{2+}(0.074 nm) and Ce^{3+}(0.114 nm) when Ce ions occupy Zn^{2+} sites in ZnO lattice. The results show that in the codoped ZnO aerogels Ce and Cu ions can simultaneously substitute Zn vacancies in the same cell and the tensile strain exerted by the solvent during the drying process can participate to the increase of Zn-O length bond. This suggestion can be explained by the value of microstrain obtained in ZnO aerogel in which Ce and Cu atoms are simultaneously introduced (Table1). In the other hand, the incorporation of cerium ions increases the lattice parameters in (Ce, Cu) codoped ZnO compared to Ce-doped ZnO aerogel. This indicates that Ce^{4+} (0.101 nm) and Ce^{3+} (0.114 nm) ions occupy Zn^{2+} (0.074 nm) sites in Ce-doped ZnO and (Ce, Cu) codoped ZnO, respectively. The presence of Cu^{2+} ions in the codoped aerogel favors the replacement of Zn^{2+}sites by Ce^{3+} thus dilating the cell parameters of the ZnO lattice. Again, ZnO resultant crystalline structure was the hexagonal wurtzite structure for all the synthesized ZnO aerogels.
The recorded FT-IR spectra of pure, Ce-doped, Cu-doped and (Cu, Ce) codoped ZnO aerogels are shown in figure 4. The vibrational bands with different intensities are observed in all spectra approximately at the same positions (3482, 2366, 1586, 1417 and 1036 cm^{-1}). No bonds related to the dopants or dopant-based compounds are found in the FT-IR curves. The small intensities of the lines testify the high chemical purity of the elaborated aerogels. Compared to those of the undoped aerogel, the bands located at 1586, 1417 and 1036 cm^{-1} are amplified in Ce-doped and (Cu,Ce) codoped ZnO aerogels. In Cu-doped aerogel the intensity of these bands is diminished. The absorption band at 3482 cm^{-1} is due to the stretching mode of O-H group which reveals the existence of small amount of water absorbed by the ZnO aerogels. The peak at 1036 cm^{-1}can be attributed to aromatic C=C stretching mode [31]. The absorption band at 1586 cm^{-1} results from the C=O asymmetric stretching vibration of zinc monoacetate [33]. The band located at 1417 cm^{-1} is assigned to C-C stretching bond peak [35]. A very strong band between 400 and 460 cm^{-1} due to the vibration of the Zn-O bond characterizes the spectra. This band is observed at 453, 436 and 428 cm^{-1} in Ce-doped ZnO, Cu-doped ZnO and (Cu, Ce) codoped ZnO aerogels, respectively. The presence of Ce atoms in the aerogels leads to the intensification of Zn-O bond. The simultaneously presence of Ce and Cu ions in ZnO lattice induces the shift of Zn-O bond vibration band toward low wavenumbers.

Figure 5 shows SEM images and EDS result of Ce-doped, Cu-doped and (Cu,Ce)-codoped ZnO aerogels. The Ce-doped ZnO particles agglomerate to form torus-like shape grains with a high density. The image of Cu-doped ZnO aerogel shows homogeneous quasi-spherical-like nanostructures. The particles are less agglomerated and appear as "snow"or a "frozen fume". In (Ce, Cu)co-doped aerogel SEM micrograph, we can observe the presence simultaneously of the two previous morphologies (torus-like and spherical-like). The changes in the grains morphologies could be related to the nucleation mechanism in the growth process of the particles. This is because a restrained nucleation rate and a subsequent lower growth rate of the aerogels may appear when the ionic radii of Ce^{3+} and Cu^{2+} ions are different from that of Zn^{2+}[16].
The EDS data of these aerogels show the presence only O, Zn, Cu and C atoms. The C atoms come from the grid used in the experimentation. Without taking into a count C atomic percentage, [Cu] /[Zn] atomic ratio has been found to be 2.7 % and 3.1 % in Cu-doped and (Ce, Cu)-codoped ZnO aerogels, respectively. However, [Ce] / [Zn] has been found to be 3.4 % and 2.3 % in Ce-doped and (Ce,Cu)-codoped ZnO aerogels, respectively. This result indicates that the percentage of Ce, Cu and Zn values obtained did not strongly deviate from their initial stoichiometry.

IOP Conf. Series: Journal of Physics: Conf. Series **987** (2018) 012008 doi:10.1088/1742-6596/987/1/012008

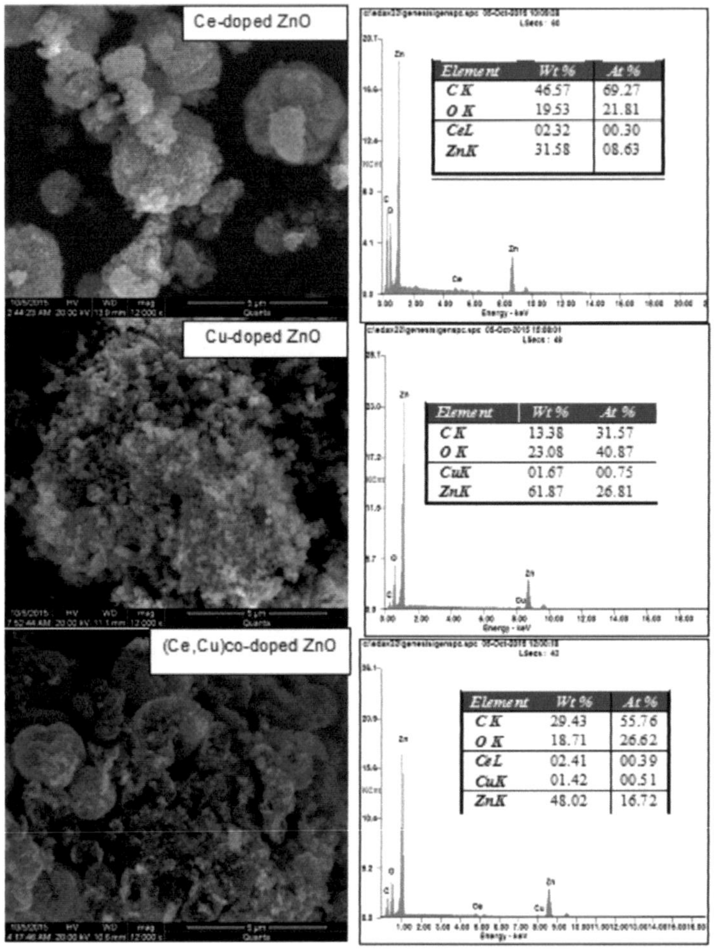

Figure 5. SEM images and EDAX results of Cu-doped ZnO, Ce-doped ZnO and (Ce,Cu)-codoped ZnO.

The normalized optical absorption spectra of all elaborated aerogels are shown in figure 6. The absorption edge is found to be at 346 nm (3.58 eV) in the undoped ZnO, 350 nm (3.54 eV) in Ce-doped ZnO, and 353 nm (3.51 eV) in Cu-doped ZnO and 352 nm (3.52 eV) in (Ce, Cu)-codoped ZnO aerogels. After the introduction of dopant elements in the ZnO lattice, the absorption edge shifts to higher wavelengths. This shift is probably due to the formation by the dopant elements of an impurity band, which overlaps with the conduction band edge and leads to the decrease of the optical gap. Muthukumaran et al. [36] reported the decreasing in the bandgap of Cu doped ZnO nanoparticles synthesized by co-precipitation method for low Cu concentration. Panda et al. [2] reported that the UV band observed at 368 nm in pure ZnO is shifted to 372 nm in cerium (1.5 %) doped ZnO synthesized through soft solution route. This band is attributed to the band edge absorption of ZnO and this band is compressed by cerium doping. Djadja et al. [37] reported the same behavior of the absorption edge in Ce-doped ZnO nanoparticles prepared by the co-precipitation method. They believed that the incorporated Ce ions into ZnO matrix lead to the formation of electron localized states into the band-gap of ZnO. These electron states locate closer to the lower edge of conduction band to form the new lowest unoccupied molecular orbital, which results in the band gap reduction. In (Ce, Cu)-codoped ZnO spectrum a shoulder at 340 nm (3.64 eV) is observed. Based on the first principle theoretical calculations carried out to study the effect of Ce and Cu dopants on the structural stability and magnetic properties of ZnO, Arul Mary et al. [16] found the formation of two new bands in both the valence and conduction bands near the Fermi energy. They suggested that the generation of these bands is due to the interactions between Cu (d), O (p) and Ce (f) electrons. Based on these data, the decrease of the optical gap can be explained by the formation of impurity bands at the bottom of the conduction band and/or at the top of the valence band.

The room temperature photoluminescence spectra of the undoped, Cu-doped, Ce-doped and (Ce, Cu) co-doped ZnO aerogels are shown in figure 7. As can be seen, the PL spectra of the undoped and Cu-doped ZnO aerogels consist of a sharp UV emission and a broad visible one. However, Ce doping and codoping leads to the quenching of the broad visible emission and the appearance of two sharp emissions centered respectively at 546 and 612 nm. In particular, Ce co-doping aerogel leads the emergence of a relatively weak and broad blue emission at 443 nm. The narrow UV emission observed at 386 nm is due to near-band edge emission of the ZnO. The visible emission observed in the PL spectra of the undoped and Cu-doped ZnO demonstrates that this emission consists of 3 bands centered at 510, 546 and 617 nm.

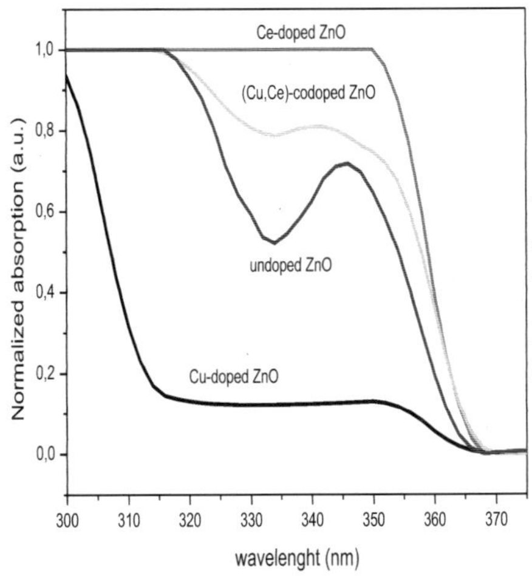

Figure 6. Absorption spectra of undoped, doped and codoped ZnO aerogels

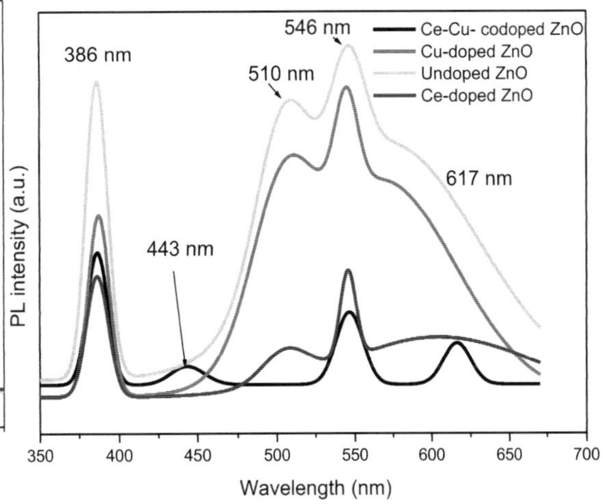

Figure 7. RT PL spectra of undoped, doped and codoped ZnO aerogels

The green emission at 510 nm is assigned to the transition related with the single ionized oxygen vacancies [37]. Many authors attribute this emission to the transitions between the singly ionized oxygen vacancies to the photoexcited holes [38]. The green emission at 546 nm observed in all the aerogels comes from the recombination of electrons in singly occupied oxygen vacancies with photoexcited holes [39]. The orange emission observed at 617 nm (2.03 eV) is usually attributed to oxygen interstitials [40].

It is remarkable that the introduction of cerium in the ZnO (Ce-doped and (Ce,Cu) codoped samples) reduces considerably the intensity of the visible emission. In particular, the green (510 nm) emission is totally quenched. The reduction in total emission upon Ce doping can be attributed to the increase in non-radiative transitions. This quenching may be due to the introduction of Ce atoms in Vzn sites. It is noted that several reports have attributed the red emission at 2 eV to V_{Zn} [41,42]. Furthermore, the decrease of the visible emission intensity after the introduction of Ce atoms can be assigned to the increase of non-radiative relaxation, which is due to the surface bound states introduced by cerium ions and the degradation of ZnO crystal quality [43,44]. In the PL spectrum of (Ce, Cu) co-doped ZnO aerogel, a new emission band is observed at 443 nm. It has been reported that this emission originates from the electron transition from shallow donor levels of oxygen vacancies to the top of valence band levels [45]. Panda et al. [12] reported a visible light absorption band at 438 nm in cerium doped ZnO synthesized via soft solution route. They attributed this emission to cerium ions. The appearance of this emission can be explained in the light of Ce^{3+} ($4f^1 5d^1 6s^1$) incorporation into ZnO there by forming empty energy levels below the conduction band [12]. In addition, to this emission can contribute the transitions between impurity energy levels and conduction band of ZnO or 4f–4f or 5d–5d transitions in Ce^{4+}states. One of the explanations for the appearance of Ce^{3+}emission in the co-doped aerogel resides in the difference of the ions radii. As Ce^{3+}ions are bigger than Zn^{+2} ions, the incorporation of trivalent Ce^{3+}into ZnO could cause a big distortion in the ZnO crystal lattice and also require association with other charge-compensating lattice defects or impurities, such as Cu^{2+} ions [46]. The appearance of this emission only with the presence of Cu atoms can be explained by the fact that Cu^{2+} ions might change the Ce^{3+} ions environment and increase the transition probabilities with the result of inducing the energy transfer from ZnO to Ce^{3+} ions [47,48].

4. Conclusion

Undoped, Ce-doped, Cu-doped and (Ce, Cu)-codoped ZnO aerogels were successfully synthesized in supercritical ethanol. [Cu]/[Zn] and [Ce]/[Zn] atomic ratios were fixed to 0.02 (2%) in all solutions. The (Ce, Cu) iso-codoping caused the changes in the crystallite size, lattice parameters and strain of ZnO lattice. XRD characterizations manifest that the aerogels have high crystallinity and shared the presence of wurtzite structure. (Ce, Cu)-codoping strongly deteriorates the crystalline quality of the ZnO aerogel. The presence of Cu^{2+} ions in the codoped aerogel favors the replacement of Zn^{2+} sites by Ce^{3+} thus dilating the cell parameters of the ZnO lattice. EDAX studies confirm the high purity of the aerogels and the incorporation of dopant atoms in ZnO lattice. FT-IR spectra confirm that the presence of Ce and Cu ions simultaneously in ZnO lattice induces a shift of Zn-O bond vibration band toward low wavenumbers. The introduction of dopant elements in the ZnO lattice shifts the absorption edge toward higher wavelengths implying a decrease in the optical band gap. The room temperature photoluminescence spectra show a narrow UV emission at 386 nm and a broad visible luminescence (490- 650 nm). (Ce, Cu) co-doping strongly decreases the visible luminescence intensity because of the scattering of the excitation radiation by surface-adsorbed dopant atoms and the non-radiative recombination process introduced by dopant ions. The deterioration of the crystallinity and the change of the Ce^{3+} ions environment with the presence of Cu atoms are responsible of the apparition of a blue emission located at 443 nm.

References

[1] Bagnall D M, Chen Y F, Zhu Z, Yao T, Koyama S, Shen M Y and Goto T 1997 *Appl. Phys. Lett.* **70** 2230

[2] Sonawane Y S, Kanade K G, Kale B B and Aiyer R C 2008 *Mater. Res. Bull.* **43** 2719

[3] Tang Z K, Wong G K L and Yu P 1998 *Appl. Phys. Lett.***72** 3270

[4] Bolboshenko V Z, Djouadi D, Kassiyan A V and Nedeoglo D D 1992 *Phys. Status Solidi (A)***133** 121

[5] Zhang X T, Liu Y C, Zhi Z Z, Zhang J Y, Lu Y M, Shen D Z, Xu W, Fan X W and Kong X G 2002 *J. Lumin.* **99** 149

[6] Peng X, Xu J, Zang H, Wang B and Wang Z 2008 *J. Lumin.***128** 297

[7] Xi C, Wang F and Hu C 2014 *J. Alloys Compd* **589** 604

[8] Fu M, Li Y, Wu S, Lu , Liu J and Dong F 2011 *Appl.Surf.Sci.***258** 1587

[9] Milenova K, Stambolova I, Blaskov V, Eliyas A, Vassilev S and Shipochka M 2013 *J.Chem. Technol. Metal.* **48** 259

[10] Samavati A, Ismail A F, Nur H, Othaman Z and Mustafa M K 2016 *Chin.. Phys.* **25** 077803

[11] Meddouri M, Hammiche L, Slimi O, Djouadi D and Chelouche A 2016 *Mater. Sci-Poland* **34** 659

[12] Panda N R, Acharya B S, Singh T B and Gartia R K 2013 *J.Lumin.***136** 369

[13] Reddy A J, Kokila M K Nagabhshana, H, Rao J L, Shvakumara C, Nagabhushana B M and Chakradhar R P S 2011 *Spectrochim. Acta Mol. Biomol. Spectrosc.* **81** 53

[14] Assadi M H N, Zhang Y B, Zheng R K and Ringer S P 2011 *Nanoscale Res. Lett.* **6** 357

[15] Amiri A E, Abid M, Lassri H and Hlil E K J 2012 *J. Supercond. Nov. Magn.* **25** 2529
[16] Arul Mary J, Judith Vijaya J, Bououdina M, John Kennedy L, Dai J H and SongY 2015 *Physica E* **66** 209
[17] Avgouropoulos G and Ioannides T 2003 *Appl. Catal. A Gen.* **244** 155
[18] Zhang Y G, Zhang G B and Wang Y X *J. Appl. Phys.* **109** 063510
[19] López-Romero S, Quiroz Jiménez M J and García-Hipólito M 2016 *World J. Condens. Matter Phys.* **6** 300
[20] Pemartin-Biernath K, Vela-González A V, Moreno-Trejo M B, Leyva-Porras C, Castañeda-Reyna I E, Juárez-Ramírez I, Solans C and Sánchez-Domínguez M 2016 *Materials* **9** 480
[21] Meddouri M, Hammiche L, Djouadi D, Chelouche A, Touam T and Boudine B 2016 *J. Sol-Gel Sci. Technol.* **80** 642
[22] Djouadi D, Meddouri M and Chelouche A 2014 *Opt. Mater.* **37** 567
[23] Meddouri M, Djouadi D, Chelouche A, Touam T and Chergui A 2014 *Eur. Phys. J-Appl. Phys.* **66** 10402
[24] Iribarren A, Hernández-Rodríguez E and Maqueira L 2014 *J. Alloys Compd* **60** 376
[25] Chen Y, Xu X L, Zhang G H, Xue H and Ma S Y 2009 *Physica B* **404** 3645
[26] Chen K J, Fang T H, Hung F Y, Ji LW, Chang S J, Young S J and Hsiao Y J 2008 *Appl.Surf. Sci.* **254** 5791
[27] Ren F, Xin R, Ge X and Leng Y 2009 *Acta Biomater.* **5** 3141
[28] C S Barret and T B Massalski 1980 Structures of Metals: Crystallographic Methods,Principles and Data, Pergamon Press, Oxford, UK, 1980.
[29] Thaweesaeng V N, Supankit S, Techidheera W and Pecharapa W 2013 *Energy Procedia* **34** 682
[30] Mohseni H, Shokrollahi H, Sharifi I and Gheisari K 2012 *J. Magn. Magn. Mater.* **324** 3741
[31] Yang J H, Lang J H, YangL L, Zhang Y J, Wang D D, Fan H G, Liu H L, Wang Y X and Gao M 2008 *J. Alloys Compd* **450** 521
[32] Kulyk B, Sahraoui B, Figà V, Turko B, Rudyk V and Kapustianyk V 2009 *J. Alloys Compd* **481** 819
[33] Gayathri S, Ghosh O S N, Sathishkumar S, Sudhakara P, Jayaramudu J, Ray S S and Viswanath A K 2015 *Appl.. Sci. Lett.* **1** 8
[34] Wang B L, Hu L L, Zhang L Yand Zhao S L 2006 *Chin. Phys. Lett.* **23** 2799
[35] Jawad I A M, Al-Hamdani A A and Hassan R M A 2016 *Inter. J. Enh. Res. Sci. Tech.Eng.* **5** 130
[36] Muthukumaran S and Gopalakrishnan R 2012 *Opt. Mater.* **34** 1946
[37] Djaja N F and Saleh R 2013 *Mater.Sci.Appl.***4** 145
[38] Sahu D, Acharya B S and Panda N R, *Int. Conf. Cond. Mat. Appl. Phys. (ICC 2015) AIP Conf. Proc.* 1728 (2016) 020165.
[39] Sahu D, Panda N R, Acharya B S and Panda A K 2014 *Ceram. Inter.* **40 B** 11041
[40] Li W, Bin Z, Xian C L and Jun Z W 2007 *Sci. China Chem.* **50** 224
[41] Chelouche A, Touam T, Tazerout M, Boudjouan F, Djouadi D and Doghmane A,2017 *J. Lumin.* **181** 448
[42] Kohan A F, Ceder G, Morgan D and Van de Walle C G 2000 *Phys. Rev. B* **61** 15019
[43] El Ghoul J, Kraini M and El Mir L *2015 J. Mater. Sci. Mater. Electron*, **26** (2015) 2555
[44] Wang A, Chen T, Lu S, Wu Z, Li Y, Chen H and Wang Y 2015 *Nanoscale Res. Lett.* **10** 75
[45] Li G R, Lu X H, Zhao W X, Su C Y, Tong Y X, Su C Y and Tong Y X 2008 *Cryst. Growth Des.* **8** 1276
[46] Chelouche A, Touam T, Djouadi D and Aksas A 2014 *Optik* **125** 5626
[47] Zhu Y J, Zhao X R, Duan, Bai X J, Sun H N and Duan W F 2013 *J. Sol-Gel Sci. Technol.* **67** 155
[48] Chen G Y, Liu H C, Somesfalean G, Sheng Y Q, Liang H J, Zhang Z G, Sun Q and Wang F P 2008 *Appl. Phys. Lett.***92** 113114

Thermally induced fading of Mn-doped YAP nanoceramics

Ya Zhydachevskyy[1,2], N Martynyuk[2], A I Popov[3], D Sugak[2], P Bilski[4], S Ubizskii[2], M Berkowski[1] and A Suchocki[1,5]

[1] Institute of Physics, Polish Academy of Sciences, Warsaw, Poland
[2] Lviv Polytechnic National University, Lviv, Ukraine
[3] Institute of Solid State Physics, University of Latvia, Riga, Latvia
[4] Institute of Nuclear Physics, Polish Academy of Sciences, Kraków, Poland
[5] Institute of Physics, University of Bydgoszcz, Bydgoszcz, Poland

E-mail: nmartynyuk@i.ua

Abstract. Thermally induced fading of the thermoluminescent (TL) detectors in the form of $YAlO_3$:Mn ceramics prepared from the nanocrystalline materials synthesized by the solution combustion method have been measured and analyzed in comparison with the single crystalline $YAlO_3$:Mn detectors studied previously. The observed differences in thermoluminescent and dosimetric properties of nanoceramic and single crystalline detectors are discussed in terms of peculiarities of the nanoceramic material.

1. Introduction

Application potential of Mn^{2+}-doped $YAlO_3$ (YAP) for thermoluminescent (TL) dosimetry of ionizing radiation has been shown previously (see [1] and references herein). For this purpose, one of two types of detectors can be used. The first type produces green emission near 530 nm (caused by Mn^{2+} ions) in the main TL peak at 200 °C, whereas the second type produces an orange emission around 640 nm in the TL peak near 350 °C.

Main features of the single crystalline $YAlO_3$:Mn^{2+} detectors are as following: high thermochemical and time stability, high resistance to radiation damage, high sensitivity to ionizing radiation, extremely wide range of linearity (from few μGy up to few kGy), high effective atomic number (Z_{eff} = 31.4) and consequently high energy response, low thermal fading of single crystalline detectors. In such a way the material is a good candidate for wide-range dose measurements, especially when tissue equivalence is not required, as well as for a purpose of the radiation quality determination if used alongside with other low-*Z* materials.

The Mn-doped YAP mainly in the form of single crystals grown by the Czochralski method has been studied previously. However, an easy, cheap and practical method is preferred for synthesis of an efficient material applicable for TL dosimetry. The solution combustion synthesis is an appropriate method for obtaining of the Mn^{2+}-doped YAP nanoceramic detectors which have thermoluminescent properties [2,3] comparable with the corresponding single crystalline material. However, the thermally induced fading of the ceramic samples has not been studied yet.

Therefore, the purpose of the present work was detailed study of thermoluminescent and dosimetric properties of YAP:Mn ceramic detectors, in particular thermal fading, and analyze the results in comparison with the single crystalline detectors of similar nominal composition studied previously.

2. Experimental

As object of investigation, YAP:Mn^{2+} detectors in the form of ceramics prepared from the YAP:Mn,Hf nanocrystalline material synthesized by the solution combustion method using urea as a fuel were used The synthesis method of the nanocrystalline YAP:Mn,Hf powders was described in details in [2]. The Hf^{4+} co-doping was made for the charge compensation of Mn^{2+}. Samples of ceramic TL detectors were produced by pressing of the nano powders into pellets of 12 mm diameter and 3 mm thickness. After sintering in an oven at 1600 °C in air, the pellets were cut into the bits of the same size as single crystalline samples studied previously in [4] – 3×3×1 mm. TL readout of the detectors was performed using the Risø TL/OSL Reader Model DA-20 with a bialkali EMI 9235QB photomultiplier with β-(^{90}Sr/^{90}Y) and α-(^{241}Am) irradiation sources, the heating rate of 2 °C/s was used.

For studying of thermally induced fading of the YAP:Mn detectors, the samples were stored in darkness (wrapped in aluminium foil) at room temperature. Before the irradiation procedure, the samples were annealed at 450 °C during half of an hour in air to erase any possible TL signal due to previous irradiation. An irradiation of the detectors similarly as storage was done in darkness.

3. Results and discussion

The studied YAP:Mn ceramic samples similarly to single crystalline YAP:Mn detectors have a main dosimetric TL peak near 200 °C (see [3] for details).

The figure 1 and figure 2 demonstrate thermally induced fading of the TL signal during dark storage of ceramic and crystalline YAP:Mn detectors at room temperature. As it is seen from the figure 1, ceramic detectors lose a half of TL signal magnitude already after 10 h of dark storage and up to 90 % of the signal after one-month dark storage. For comparison, the TL signal from crystalline YAP:Mn detectors lose no more than 20% during one-year dark storage (figure 2).

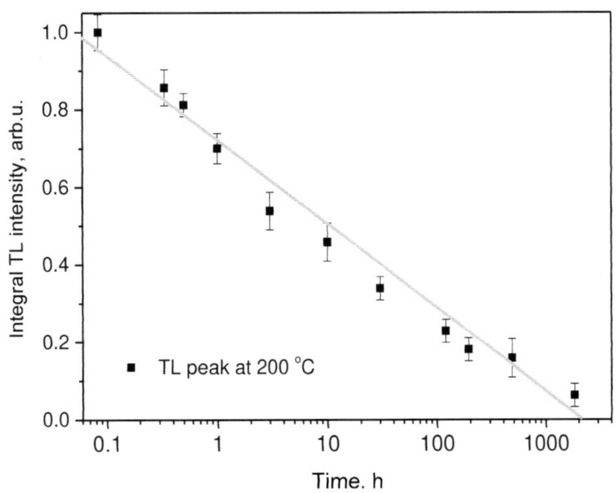

Figure 1. Thermal fading of the TL peak at 200 °C measured on ceramic YAP:Mn detectors after β-irradiation (from ^{90}Sr/^{90}Y) and dark storage at room temperature.

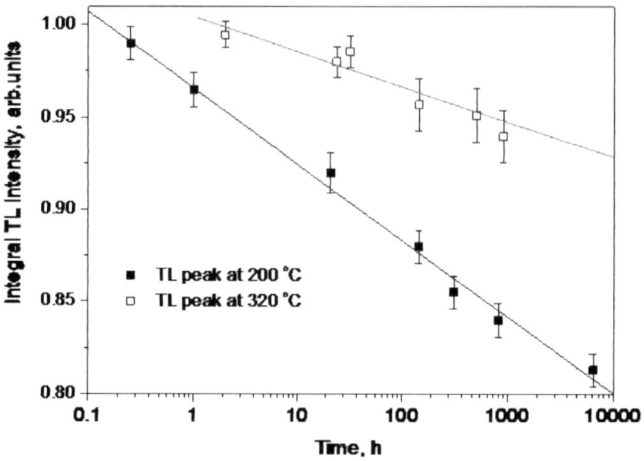

Figure 2. Thermal fading of the TL peak at 200 °C measured on single crystalline YAP:Mn detectors after γ-irradiation (D=100 Gy) and dark storage at room temperature [4].

Similar anomalous fading was also observed previously for Al_2O_3:Mg,Y ceramic detectors (see [5] and references therein). This common feature for the polycrystalline materials can be considered as typical for the oxide materials with large surface of grain boundaries and is caused most probably by recombination processes involving defects on grain boundaries.

4. Conclusions

The thermally induced fading of the nanoceramic YAP:Mn-based detectors has been studied and compared with results obtained on the single crystalline detectors.

The YAP:Mn-based ceramic TL detectors are found to have significant thermally induced fading (up to 90 %/month) anomalous for the TL peak at 200 °C, that is caused most probably by recombination processes involving defects on grain boundaries. By this reason, an applicability of nanoceramic YAP:Mn detectors in dosimetry of ionizing radiation is essentially limited when compared to the single crystalline YAP:Mn.

References

[1] Zhydachevskii Ya, Morgun A, Dubinski S, Yu Yan, Glowacki M, Ubizskii S, Chumak V, Berkowski M and Suchocki A 2016 *Radiat. Meas.* **90** 262-264.

[2] Zhydachevskii Ya, Kamińska I, Berkowski M, Twardak A, Bilski P, Ubizskii S and Suchocki A, 2014 *Opt. Mater.* **37** 125-131.

[3] Zhydachevskii Ya, Kamińska I, Berkowski M, Twardak A, Bilski P, Ubizskii S and A Suchocki 2014 *Proc. Int. Conf. on Oxide Materials for Electronic Engineering (Lviv, Ukraine)* (IEEE) p 261-262.

[4] Zhydachevskii Ya, Suchocki A, Berkowski M, Bilski P and Warchol S 2010 *Radiat. Meas.* **45** 516-518.

[5] Chen R and McKeever S W S 1997 *Theory of thermoluminescence and related phenomena* (Singapore: World Scientific).

Acknowledgements

The work was supported by the NATO Science for Peace and Security Program (Project G4649), by the Ministry of Education and Science of Ukraine (project DB/Reader, no. 0117U004443) and by the Latvian-Ukrainian Joint Research Project (LV-UA/2016/1 in Latvia, and M/78-2017 in Ukraine).

Design and analysis of non-linear properties of photonic crystal fiber with Various Doping Concentration

M. Benhaddad [1,2], F. Kerrour [1] and O. Benabbes [2]

[1]laboratoire de Modélisation des dispositifs a énergies renouvelables et nanométrique, Université Frères Mentouri de Constantine1, Constantine, Algérie
[2]laboratoire de physique mathématique et subatomique,Université Frères Mentouri de Constantine1,Constantine, Algérie

Abstract. With exceptional light guiding mechanism and deliver significantly improved performance compared to conventional optical fibers, Photonic Crystal Fibers (PCFs) are a new class of optical waveguide. In this work, a solid core photonic crystal fiber (PCF) is designed using Comsol Multiphysics software based on finite element method. In our analysis the structure introduced is a structure formed by a doped core surrounded by a cladding which is composed by elliptical air holes in silica arranged in a hexagonal array. The important optical properties like effective area and nonlinear coefficient has been studied. Each characteristic has been investigated under different doping concentration within range of wavelength 800-1600 [nm].

1. Introduction

Crystal fibers (PCFs)[1] are a new class of optical fibers which have the potential to revolutionize optical fiber technology. These fibers are essentially low-loss waveguides and consist of a core which is surrounded by a periodic array of air holes in the cladding region. This configuration has led to a number of novel properties like endless single mode (ESM) operation[2], controllable birefringence and dispersion characteristics[3,4] and a high degree of nonlinearity . The cladding of a photonic crystal fiber can be constructed with a structure similar to that found in photonic crystal. This is where the term ‘photonic crystal fiber’ originates.
There are two classes of photonic crystal fibers. solid core PCF (SC-PCF) and hollow-core PCF (HC-PCF). The term solid-core photonic crystal fiber refers to those structures that have a solid core which is usually made of silica. These fibers guide light by the phenomenon of total internal reflection. On the other hand, hollow-core photonic crystal fiber has an air hole in the core region and transmits light by photonic band gap type guidance[5]. In this work, a solid core photonic crystal fiber (PCF) is designed using Comsol Multiphysics software based on finite element method. In our analysis the structure introduced is a structure formed by a doped core surrounded by a cladding which is composed by elliptical air holes in silica arranged in a hexagonal array. The important optical properties like effective area and nonlinear coefficient has been studied. Each characteristic has been investigated under different doping concentration within range of wavelength 800-1600[nm].

2. Fiber Design

The cross section of the proposed (PCF) structure is shown in Fig(1). It is composed of elliptical air holes in the cladding arranged in a hexagonal array with lattice constant Λ, The ellipse have a major radius and a minor radius. The central core region is perturbed by a GeO_2 doped SiO_2. The properties of the (PCF) can be controlled to a large extent by varying these parameters.

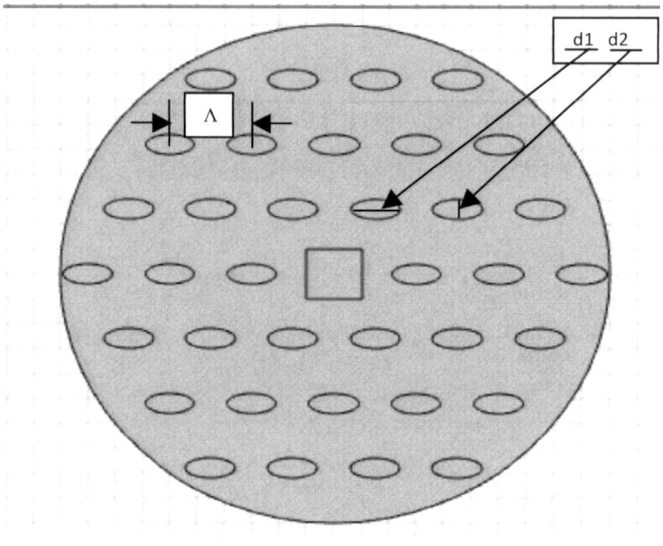

Figure 1. Geometry of solid-core photonic crystal fiber

3. Simulation Method

The finite element method (FEM) is used to analyze the various properties of (PCFs). Once the effective refractive index n_{eff} is obtained by solving an eigenvalue problem drawn from the Maxwell's equations using the (FEM), the parameter like effective area (A_{eff}) and nonlinear coefficient (γ) can be obtained.

The effective area of (PCF) is the quantitative measure of the area which a waveguide or fiber mode effectively covers in transverse dimensions. The effective area is calculated by [6]:

$$A_{eff} = \frac{\left(\iint |E|^2 dxdy\right)^2}{\iint |E|^4 dxdy} \; [m^2] \tag{1}$$

where E is the transverse component of the electric field.

The nonlinearity of (PCF) can be measured using the parameter (γ) termed as nonlinear coefficient. (PCFs) can have high nonlinearity because of its ability to confine high intensity light [8]. The nonlinear coefficient of (PCF) can be calculated by [7]:

$$\gamma = \frac{2\pi n_2}{\lambda \; A_{eff}} \quad [W^{-1}m^{-1}] \tag{2}$$

where n_2 is the nonlinear refractive index of silica material, λ is the wavelength of the light and(A_{eff}) is the effective mode area of (PCF).

4. Results and Discussions

The results obtained after the graphical modeling and simulation of proposed (PCF) are presented here.

4.1. Simulation Results

The COMSOL simulated output for the proposed design is given in Fig(2), which represents a highly confined light beam. The graphs of various (PCF) parameters are plotted within a wavelength range of 800 nm to 1600 nm.

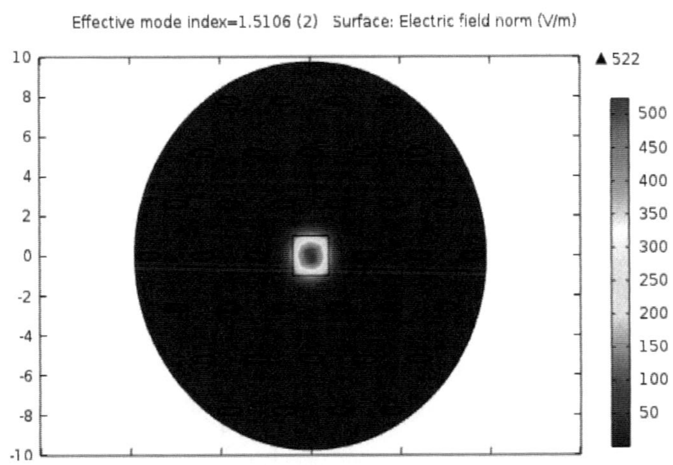

Figure 2. Electric field distribution of proposed PCF structure

4.2. Graphical Results

The variation of different parameters of proposed PCF like effective index, effective area and nonlinearity with respect to wavelength and doping concentration are analysed and plotted:

4.2.1. Effective refractive index. The variation of effective index with respect to wavelength is shown in figure(3). It shows that effective index decreases with wavelength[8,9] and increases with increases of doping concentration[10].

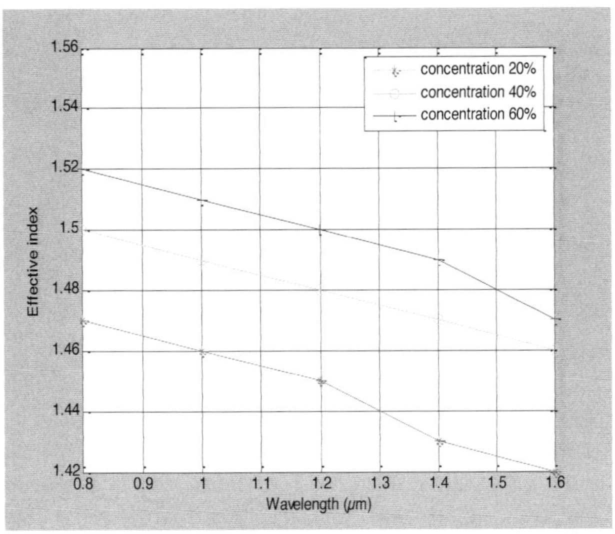

Figure 3. Plot of Effective index vs. Wavelength

4.2.2. Effective Mode Area. The variation of effective area with respect to wavelength is shown in figure (4). It shows that effective area increases with wavelength[11] and increases with the decreases of doping concentration[10]. This is because as the wavelength increases the propagating mode leaks in to cladding and so the area covered by propagating mode also increases.

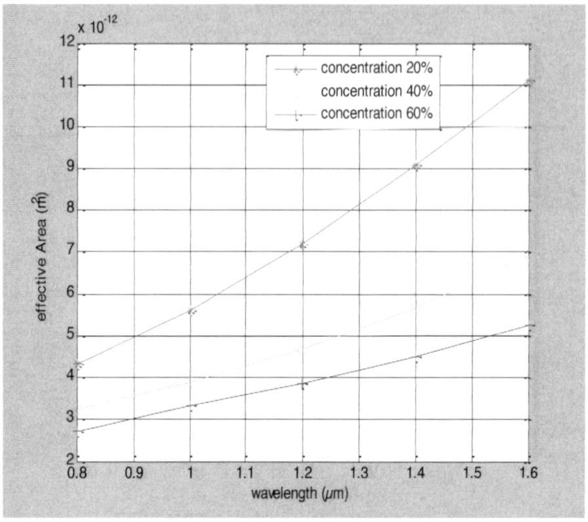

Figure 4. Plot of Effective Area vs. Wavelength

4.2.3. Nonlinear Coefficient. The variation of nonlinear coefficient with respect to wavelength is shown in figure (5). It can be seen that nonlinearity decreases with wavelength and increases with the increases of doping concentration[10]. This is because of increase in effective area. At lower wavelength, mode is well confined, reducing its effective area and as the energy is concentrated in small area, nonlinearity increases.

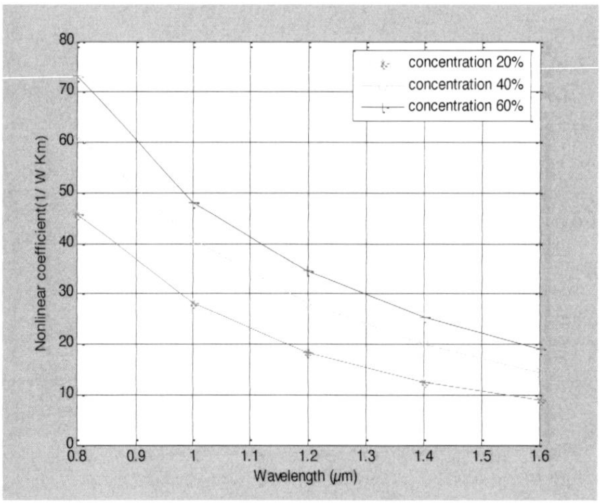

Figure 5. Plot of nonlinear coefficient vs. Wavelength

5. Conclusion. In this paper, we have modeled a solid-core photonic crystal fiber (PCF) by using full vectorial finite element method (FEM). Modal properties of the (PCF) have been investigated by calculating the effective index of the supported mode. Results for effective index ,effective area and non linear parameter are presented. COMSOL Multiphysics software has been used to calculate the effective index of the supported mode of the fiber, with Various Doping Concentration in the core of fiber.

6. References

[1] Knight J. C, Birks T. A, Russell P. St. J and Atkin D. M 1996 opt.lett p 1547-1549
[2] Birks T. A, Russell P. St. J and Knight J. C 1996 *opt.lett* p 961-963
[3] Sinha R. K and Varshney S. K 2003 *Microwave Opt Technol Lett* p 129–132

[4] Knight J. C, Arriaga J.and Birks T. A 2000 *Photonics Technology Lett* p 807-809
[5] Broderick N.G.R, Monro T.M, Bennett P.J and D.J. Richardson 1999 *Opt. Lett* p 1395-1397
[6] Pourmahyabadi M and Mohammad Nejad Sh 2009 *Iranian journal of Electrical & Electronic Engineering* No.3
[7] Agrawal G.P 2007 "Nonlinear Fiber Optics",Fourth Edition,USA p 432
[8] Gautam Prabhakar A, Akshit Peer B, Ajeet Kumar C and Vipul Rastogi D 2012 India.(*IEEE*) p1-5
[9] Sharma S and Kumar J 2013 India.(*IEEE*) p 1-4
[10] Sajjad Hossain Md, Neupane K, Shihab Bin Hafizl Md and Prasad Majumder S 2014 Bangladesh.(*IEEE*) p 500-503
[11] Anand N, Sudheer S.K and Mahadevan Pillai V.P 2013 Katakshstan.(*IEEE*) p 36-39

Effect of Processing Parameters on the Morphology of PVDF Electrospun Nanofiber

M A Zulfikar[1], I Afrianingsih[1], M Nasir[2] and A Alni[3]

[1] Analytical Chemistry Research Group, Institut Teknologi Bandung, Jl. Ganesa 10 Bandung, Indonesia
[2] Clean Technology Research Center, Indonesian Institute of Sciences, Jl. Cisitu Lama 5 Bandung, Indonesia
[3] Organic Chemistry Research Group, Institut Teknologi Bandung, Jl. Ganesa 10 Bandung, Indonesia

Abstract. Electrospinning is a process that produces continuous polymer fibers with diameters in the submicron range through the action of an external electric field imposed on a polymer solution or melt. Because of the tiny diameter in several hundreds of nanometers and the high porosity, electrospun membranes show potential applications in extensive areas such as filtration systems, biomedical tissue templates, drug delivery membranes, and so on. In the electrospinning process, some parameters such as polymer concentration, feeding rate of the polymer solution, additives, humidity, viscosity, surface tension, applied voltage, and nozzle-to ground collector distance will affect the fiber diameter and morphology. In this work, we have evaluated the effects of two processing parameters including the flow rate of the polymer solution and nozzle-to ground collector distance, on the morphology of the fibers formed. The solutions used in the electrospinning experiments were prepared using Poly(vinylidene fluoride) (PVDF). This material was dissolved in N,N-dimethylformamide (DMF) to make solutions with concentrations of 20 wt%. These solutions was electrospun using a 5 mL plastic syringe with an 8 gauge stainless needle at an applied voltage of 20.0 kV, a flow rate of 0.02-0.04 mL/min and nozzle-to ground collector distance of 12 and 15 cm. Electrospinning of PVDF polymer solution was performed in horizontal alignment having a grounded aluminum foil which serves as a collector. The nanofibers obtained were characterized by polarizing optical microscope. We find that the low flow rate of the polymer solution and nozzle-to ground collector distance are strongly correlated with the formation of bead defects in the fibers.

1. Introduction

Polymer nanofibers are an important class of nano-materials, which have attracted increasing attention in the last decade because of their high surface-to-mass (or volume) ratio and special characteristics attractive for advanced application, such as filtration and water treatment [1,2], catalyst and catalyst support [3,4], adsorbents [5-9], polymer electrolyte membrane [10], membrane distillation [11], hydrogen generator [12] and sensors [13]. Traditional methods of obtaining polymer fibers include melt spinning, spinning from solution or liquid crystalline state, and forming fibers from a gel state [14]. Nowadays, there has been increased interest in another method of fiber production, namely electrospinning, which can consistently produce fibers that are sub-micron in diameter [14].

The electrospinning process consists of the following steps: (i) a strong electric field is applied between a polymer solution contained in a nozzle and a conductive substrate; (ii) when the voltage reaches a critical value, electrostatic forces overcome the surface tension of the solution; (iii) charged droplets or jets are sprayed from the tip of the nozzle, in a dry atmosphere; and (iv) the dried droplets or jets are finally collected on the substrate to form a thin film [15]. In the electrospinning process, some parameters such as polymer concentration, flow rate of the polymer solution, additives, humidity, viscosity, surface tension, applied voltage, and nozzle-to ground collector distance will affect the fiber diameter and morphology [15].

Polyvinylidene fluoride (PVDF) is a thermoplastic polymer built on -(CH2CF2)n monomer and usually contain 59,4% fluorin and 3% hydrogen [16]. PVDF possesses excellent thermal and chemical stability, excellent mechanical strength [17-19], resistance towards UV radiation and corrosion, high dielectric permittivity and characteristic pyroelectric and piezoelectric properties [20]. PVDF is a non-flammable material with a good sliding property and high abrasion capacity. Its soluble in several polar organic solvent namely N,N-Dimethylacetamide (DMAc), N,N-Dimethylformamide (DMF) and N-Methyl-2-pyrrolidone (NMP) [17].

PVDF is an electro active polymer due to high polarizability of the C-F moiety. It has at least four main crystalline structure (α, β, γ, δ) with different polarity, i.e. α-phase is a non-polar whilst β-phase has polar properties. Each phase of PVDF can be used for specific application [15,16]. Owing to its properties, PVDF is developed as high performance membrane for application in distillation, membrane contactor and wastewater treatment [16,17].

2. Materials and method

2.1. Materials

PVDF polymer (Scientific Polymer Products, USA), N,N-dimethylformamide (DMF; Aldrich, Milwaukee, WI, USA), Electrospinning apparatus, optic microscopy (Nikon Eclips E400, Japan).

2.2. Method

The PVDF polymer was dissolved in N,N-dimethylformamide to form a 20 wt% solution. The polymer solution was transferred to a spin pack containing spinnerets (capillary tip = 0.5 mm diameter). Electrospinning was carried out by applying a high positive voltage (20 kV) to the polymer solution via the spinneret tip with flow rate ranging from 0.02 to 0.04 mL/min and nozzle-to ground collector distance of 12 and 15 cm. The electrospun fibers were collected as a thin web on a current collector plate with a suction collector. The electrospun nanofiber web was then passed under rollers at 140 ^{o}C to fabricate a dense PVDF nanofiber membrane. The morphology and structure of the PVDF nanofiber membrane were characterized by optic microscopy.

3. Results and Discussion

3.1. Effect of flow rate

The effect of flow rate on morphology of the nanofibers formed can be seen at Figure 1. The nanofibers produced under 0.02 mL/min conditions have a cylindrical morphology with many bead defects present and not uniform (Figure 1a).

As flow rate is increased, the electrospun fibers produced still have essentially a cylindrical morphology, but there is a distinct decrease in number of bead defects present in the fiber mat. At 0.04 mL/min, the morphology of the fibers produced becomes more uniform and the bead defects are lost. Formation of beads can occur when the wet polymer fibers were collected on the grounded collector with the self-repulsion force by the electric field. Without strain force, this condition caused the fibers to melt and solidify, hence formation of beads [21]. When the flow rate was increased the fibers are formed faster hence more strain force on the grounded collector and the beads do not has sufficient time to form.

The flow rate of the PVDF solution also affects the surface charge density and the electrical current of the polymer solution [22]. Increasing the flow rate will cause an increase in electrical current due to the electroactive nature of PVDF polymer. On the contrary, the surface charge density decreases as the flow rate increases, thus preventing the formation of beads defects on the fiber mat. The wavy morphology caused by the presence of beads will reduce the membrane porosity hence the ability to minimize its formation will be preferred.

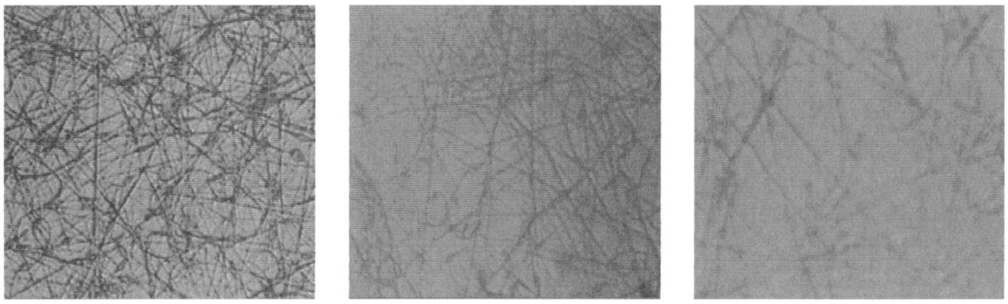

(a) (b) (c)

Figure 1. Optical micrographs of nanofiber by electrospinning from PVDF solution at different flow rate (a) 0.02 mL/min (b) 0.03 mL/min, and (c) 0.04 mL/min.

3.2. Effect of nozzle to ground collector distance

Figure 2 shows that higher nozzle to ground collector distance produce uniform fibers. This effect is common for electrospun polymers [21,23]. Nozzle to ground collector distance will affect the jet pathway and electric field strength hence the morphology of the nanofibers. Narrow nozzle and ground collector distance cause the formation of wavy fiber mat (Figure 2a). Higher distance will form more uniform fibers (Figure 2b).

The nozzle-to-ground collector distance affects the solvent evaporation and the continuous stretching due to the electrical force [24]. As the distance increases, the electrical field strength at constant applied voltage decreases; however, the evaporation time of the polymer jet to reach the ground collector increases.

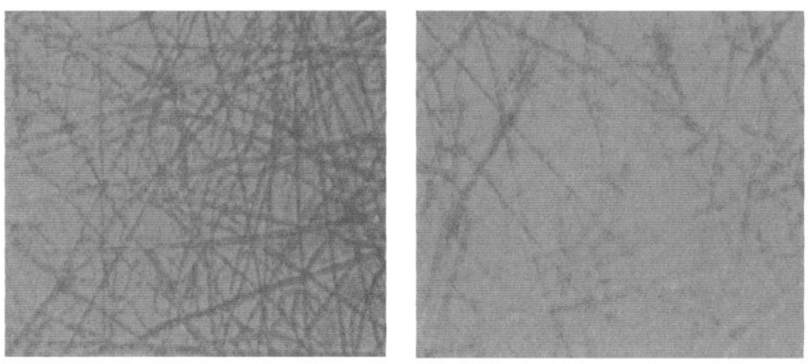

(a) (b)

Figure 2. Optical micrographs of nanofiber by electrospinning from PVDF solution at different nozzle to ground collector distance (a) 12 cm (b) 15 cm.

4. Conclusion

In the present study, the determining factors on the beads formation on fibers mat were investigated. Lower flow rate (0.02 mL/min) induces the formation of beads defects and produce fibers with cylindrical morphology. As the flow rate was increased to 0.04 ml/min, the beads defects are lost . Another parameter that influences the formation of beads defects is nozzle-to-ground collector distance. A clear difference was observed when the fibers are formed at 12 and 15 cm nozzle-to-ground collector distance. At higher nozzle to ground collector distance the fibers are more uniform with little beads formation observed on the optical micrographs images. These results provide information for preparation of high-performance PVDF nanofiber by electrospinning method.

Acknowledgements

The authors are very grateful to Research, Technology and Higher Education Ministry for their financial support through Riset Unggulan Perguruan Tinggi 2016/2017.

References

[1] Feng C, Khulbe K.C, Matsuura T, Tabe S and Ismail A.F 2013 Preparation and characterization of electro-spun nanofiber membranes and their possible applications in water treatment *Sep. Purif. Technol.* **102** 118

[2] Tian M, Qiu C, Liao Y, Chou S and Wang R 2013 Preparation of polyamide thin film composite forward osmosis membranes using electrospun polyvinylidene fluoride (PVDF) nanofibers as substrates *Sep. Purif. Technol.* **118** 727

[3] Daels N, Radoicic M, Radetic M, Hulle S.W.H.V and Clerck, K.D 2014 Functionalisation of electrospun polymer nanofibre membranes with TiO_2 nanoparticles in view of dissolved organic matter photodegradation *Sep. Purif. Technol.* **133** 282

[4] Kanjwal M.A, Barakat N.A.M and Chronakis I.S 2015 Photocatalytic degradation of dairy effluent using $AgTiO_2$ nanostructures/polyurethane nanofiber membrane *Ceram. Int.* **41** 9615

[5] Almasian A, Olya M.E and Mahmoodi N.M 2015 Synthesis of polyacrylonitrile/polyamidoamine composite nanofibers using electrospinning technique and their dye removal capacity *J. Taiwan Inst. Chem. Eng.* **49** 119

[6] Doh J.H, Kim J.H, Kim H.J, Ali R.F, Shin K and Hong Y.J 2015 Enhanced adsorption of aqueous copper(II) ions using dedoped poly-N-phenylglycine nanofibers *Chem. Eng. J.* **277** 352

[7] Hallaji H, Keshtkar A.R and Moosavian M.A 2015 A novel electrospun PVA/ZnO nanofiber adsorbent for U(VI), Cu(II) and Ni(II) removal from aqueous solution *J. Taiwan Inst. Chem. Eng.* **46** 109

[8] Sharma D.K, Li F and Wu F 2014 Electrospinning of Nafion and polyvinyl alcohol into nanofiber membranes: A facile approach to fabricate functional adsorbent for heavy metals *Colloids Surf. A: Physicochem. Eng. Aspects* **457** 236

[9] Xin Q, Fu J, Chen Z, Liu S, Yan Y, Zhang J and Xu Q 2015 Polypyrrole nanofibers as a high-efficient adsorbent for the removal of methyl orange from aqueous solution *J. Environ. Chem. Eng.* **3** 1637

[10] Kim Y.J, Ahn C.H, Lee M.B and Choi M.S 2011 Characteristics of electrospun PVDF/SiO_2 composite nanofiber membranes as polymer electrolyte *Mater. Chem. Phys.* **127** 137

[11] Liao Y, Wang R, Tian M, Qiu C and Fane A.G 2013 Fabrication of polyvinylidene fluoride (PVDF) nanofiber membranes by electro-spinning for direct contact membrane distillation *J. Membr. Sci.* **425-426** 30

[12] Chinnappan A, Kang H.C and Kim H 2011 Preparation of PVDF nanofiber composites for hydrogen generation from sodium borohydride *Energy* **36** 755

[13] Cho E, Kim C, Kook J.K, Jeong Y.I, Kim J.H, Kim Y.A, Endo M and Hwang C.H 2012 Fabrication of electrospun PVDF nanofiber membrane for Western blot with high sensitivity *J. Membr. Sci.* **389** 349

[14] Deitzel J.M, Kleinmeyer J, Harris D and Tan N.C.B 2001 The effect of processing variables on the morphology of electrospun nanofibers and textiles *Polymer* **42** 261

[15] Nasir M, Matsumoto H, Danno T, Minagawa M, Irisawa T, Shioya M and Tanioka A 2006 Control of diameter, morphology, and structure of PVDF nanofiber fabricated by electrospray deposition *J. Polym. Sci. Part B Polym. Phys.* **44**(5) 779

[16] Liu F, Hashim N.A, Liu Y.T, Abed M.R.M and Li K 2011 Progress in the production and modification of PVDF membranes *J. Membr. Sci.* **375** 1

[17] Kang G.D and Cao Y.M 2014 Application and modification of poly(vinylidene fluoride) (PVDF) membranes - A review *J. Membr. Sci.* **463** 145

[18] Ma X.H, Dong Z.Q, Zhang P.Y, Zhong Q.F and Xu Z.L 2015 Preparation and characterization of superhydrophilic PVDF electrospun nanofibrous membrane based on insitu free radical polymerization *Mater. Lett.* **15** 58

[19] Prince J.A, Singh G, Rana D, Matsuura T, Anbharasi V and Shanmugasundaram T.S 2012 Preparation and characterization of highly hydrophobic poly(vinylidene fluoride)-Clay nanocomposite nanofiber membranes (PVDF-clay NNMs) for desalination using direct contact membrane distillation *J. Membr. Sci.* **397-398** 80

[20] Mokhtaria F, Latifia M and Shamshirsaz M 2016 Electrospinning/electrospray of polyvinylidene fluoride (PVDF): piezoelectric nanofibers *J. Tex. Inst.* **107** 1037

[21] Zong X, Kim K, Fang D, Ran S, Hsiao B.S and Chu B 2002) Structure and process relationship of electrospun bioabsorbable nanofiber membranes *Polymer* **43** 4403

[22] Theron S.A, Zussman E and Yarin A.L 2004 Experimental investigation of the governing parameters in the electrospinning of polymer solutions *Polymer* **45** 2017

[23] Munir M.M. Iskandar F, Yun K.M, Okuyama K and Abdullah M 2008 Optical and electrical properties of indium tin oxide nanofibers prepared by electrospinning *Nanotechnology* **19** 145603

[24] Reneker D.H, Yarin A.L, Fong H and Koombhongse S 2000 Bending instability of electrically charged liquid jets of polymer solutions in electrospinning *J. Appl. Phys.* **87**(9I) 4531

IOP Conf. Series: Journal of Physics: Conf. Series **987** (2018) 012012 doi:10.1088/1742-6596/987/1/012012

Thermal and non-thermal explosion in metals ablation by femtosecond laser pulse: classical approach of the Two Temperature Model

Ahmed Abdelmalek[1], **Zeyneb Bedrane**[1*], **El-Hachemi Amara**[2]

[1]Physics Department, Theoretical physics Laboratory, Tlemcen University, 13000 Tlemcen, Algeria

[2]Laser Material Processing Team, Center for Advanced Technologies Development, CDTA, PO. Box 17 Baba-Hassen, 16303 Algiers, Algeria

E-mail: *zeyneb_bedrane@yahoo.fr

Abstract. We propose a classical Two Temperature Model TTMc where we consider the metal film during the irradiation like an ideal plasma. The numerical results are comparing to those finding by the existing TTM and the experimental data. In our model The cooper is taken as a target irradiated by a single laser pulse with 120 fs at 800 nm wavelength in air room. Our numerical results shown that there are a thermal and non-thermal explosion successively occurs in metal ablation by ultrashort laser pulse.

1. Introduction

Metal ablation by ultrashort laser pulse has taken a spacial attention siense 1990s. On the one hand, because the new regim of lasers rather femtosecond lasers opened a new field of indusriel-thechnology with a vaste applications developped owing to the high quality of ablation hole in contrast to those obtained by conventional (nanosecond) regim. In the other hand, this kind of laser pulse have opened a new posibilities of fundamental investigations in physics.

it has been shown that the quality of ablated holes induced by femtosecond laser pulse is much better than those produced by nanosecond or even more large pulses [1, 6], where the main charactiristic of femtosecond laser ablation of metals are: very rapid creation of plasma phase and no heat effect zone around the craters at low fluence [1, 2], and a thin layer of liquid appear at high fluence [3]. Even if all authors are in agreement with the fact that the matter ejected in liquid form is necessarily due to phase explosion whether by nanosecond pulse [1, 4] or at high fluence by femtosecond pulse [3, 5], we must investigate the physical mecanism responsible of the metal ablation by femtosecond (fs) laser pulse at low and high fluence.

Many works have been published in this field to answer this quenstion where it was proposed that the ablation mecanism is always due to phase explosion e.i. thermal explosion basing on the existing quantum Two Temperature Model (TTMq) [5, 7, 8]. But the problem is that this model does not give any information on the ablation state at low fluence (aroud ablation thershold), also the phase explosion model does not explane the absense of liquid around the crater at low fluence. In this effect, Gamaly *et al.* [9, 10] proposed a new model called Coulomb Explosion (CE) where they explained that the metal ablation mecanism at fs laser pulse consists in the ion acceleration in the electrostatic field created by hot electrons escaping from the target.

CE is very interesting, since it explains clearly the ablation mechanism of solids by ultra-short lasers, without thermal physics. Gamaly has succeeded in giving explanation for the lack of liquid phase at the surface during ablation. However the model presents an inconvenient concerning the nature of the solid material. Since 2002, Stoian *et al.* [11] demonstrated that this kind of explosion is dominant only in the case of dielectrics. In 2006, Dachraoui's *et al.* [12] calculations shown that this kind of explosion exists by evidence in all solids but at different rates: it is very important in dielectrics, moderately weak in semi-conductors, and very weak in metals because of the screening effect.

In this paper we try to contribute to resolve this problem by giving an answer to the rising question: what is the physical mechanism responsible of the ablation by ultra-short laser pulses? we propose a new classical Two Temperature Model (TTMc) which agree with the exprememtal data at low and high fluence.

2. Theoretical Model

2.1. quantum Two Temperature Model TTMq

Considering that a copper target is normally irradiated by a femtosecond laser pulse, the evolution of electron and lattice temperatures can be described by the famous one-dimensional Two Temperature Model called quantum Two Temperature Model (TTMq) [7, 13]:

$$C_e\frac{\partial T_e}{\partial t} = \frac{\partial}{\partial z}(k_e\frac{\partial T_e}{\partial z}) - G_q(T_e - T_l) + S \tag{1}$$

$$C_l\frac{\partial T_l}{\partial t} = G_q(T_e - T_l) + S \tag{2}$$

where the subscript e and l denote, respectively, electron and lattice. C is the heat capacity, k is the thermal conductivity, G_q is the electron–phonon coupling factor, S is laser heat density and t is the time. The laser beam is propagated along the z-axis.
the heat density S can be expressed as [7]:

$$S = 0.94\frac{1-R}{t_p(\delta+\delta_b)(1-e^{-d/(\delta+\delta_b)})}F.exp[-\frac{z}{(\delta+\delta_b)} - 2.77(\frac{t}{t_p})^2] \tag{3}$$

t_p is the FWHM (full width at half maximum) pulse width, $\delta = 1/\alpha$ the absorption depth, R the reflectivity, d the thickness of the sample, and F the fluence. δ_b is the ballistic range. For the parameter values of $C_e = \gamma T_e$ where $\gamma = 97\ J(m^3K^2)$ [14, 15] is the electron heat capacity constant and $C_l = 3.46\ 10^6 J(m^3K)$ [15].
α is the absorption coefficient and R is normal-incidence reflectivity. k_e is the electron thermal conductivity value, it was calculated using the general solution provided by Animisov *et al.* [13]. G_q is the electron-phonon coupling factor related to the rate of the energy exchange between the electrons and the lattice, which can be expressed as [5]:

$$G_q = G_0[\frac{A_e}{B_l}(T_e + T_l) + 1] \tag{4}$$

where $G_0 = 0.5551\ 10^{17}(W/m^3/K)$ is the G factor at room temperature [5], A_e, B_l are the material constants for cooper; $1.28\ 10^7(s^{-1}K^{-2})$ and $1.23\ 10^{11}(s^{-1}K^{-1})$, respectively [16]. we defined also the effective frequency of electron-phonon collision, where can be expressed as [10]:

$$f_{eph}(T_e) = (\frac{m_e^*}{M})^{1/2}\frac{Ji}{{}^{-}h}\frac{T}{T_D} \tag{5}$$

where m_e^* is the electron effective mass [21], $J_i = 7.72eV$, $T_D = 343K$ and $M = 63.54\ a.m.u$ are first ionization potential, Debye temperature and mass molar for copper respectively [22].

2.2. classical Two Temperature Model TTMc

We propose a new classical Two Temperature Model (TTMc) where we consider the cooper target during the irradiation like an ideal plasma (electrons +ions).
we adapt the existing TTM Eq.1 and Eq. 2 to our assumption where we propose a new G factor.

$$C_e\frac{\partial T_e}{\partial t} = \frac{\partial}{\partial z}(k_e\frac{\partial T_e}{\partial z}) - G_c(T_e - T_i) + S \tag{6}$$

$$C_i\frac{\partial T_i}{\partial t} = G_c(T_e - T_i) + S \tag{7}$$

G_c is the electron-ion coupling factor related to the rate of the energy exchange between the electrons and ions.
We define an average relaxation time $\tau_c(Te)$ for the exchange of momentum between electrons and ions such as $G = \frac{\pi^2}{6}\frac{m_2^* v_s^2 n_e}{\tau_c(Te)T_e}$ [17]. The amount $f_{ei}(Te) = 1/\tau_c(Te)$ is the collision frequency for momentum exchange which can be expressed as: $f_{ei}(Te) = n_e\sigma_1 v_e^*$, σ_1 is called the cross-section for momentum transfer between the electron and ions once ionized, given by the relation $\sigma_1 = 4\pi p_0^2 ln\Lambda$ where $p_0 = e^2/(4\pi\varepsilon_0 m_e^* v_e^*)$ [18] is called the critical impact parameter; p_0 is the distance at which the two particles should be placed so that the potential energy is equal to twice the initial kinetic energy. $v_e^* = \sqrt{(3k_B T_e)/(m_e^*)}$ is the average quadratic velocity of electrons. The amount $ln\Lambda = ln(\lambda_D/p_0)$ [6] is called the Coulomb logarithm; it is related to the plasma parameter, this factor measures the relative importance of distant collisions compared to near collisions, λ_D is the Debye length.
Initial and boundary conditions are expressed as:

$$T_e(z,0) = T_{l,i}(z,0) = T_0 \tag{8}$$

$$\frac{\partial T_e}{\partial z}|_{z=0} = \frac{\partial T_e}{\partial z}|_{z=d} = \frac{\partial T_{l,i}}{\partial z}|_{z=0} = \frac{\partial T_{l,i}}{\partial z}|_{z=d} \tag{9}$$

3. Results and discussion

Our model was simulated using MATLAB software, the results were compared to those obtained by the TTMq model and with the experimental data.

Figure 1(a) represent the evolution of electron and lattice temperatures after a femtosecond laser irradiation, modelling by TTMq model. We see that the lattice reach 90% T_C ($T_C = 7696K$ is the critical temperature of phase explosion (thermal explosion)) at ($3\ J/cm^2$) after vibrational (equilibrium) relaxation time. Because the emitted phonons carry little energy, the electron-phonon collision takes many scattering processes, and therefore several picoseconds, before the carriers and the lattice reach thermal equilibrium [19]. This fluence ($F = 3$) called the strong ablation threshold where the matter start to be ejected in liquid form during the ablation. It is

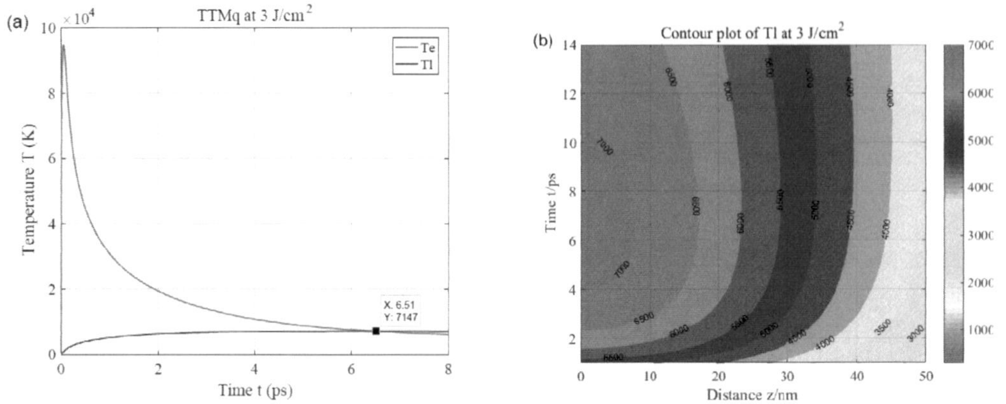

Figure 1. (a)the evolution of electron and lattice temperatures, (b) lattice temperature versus time and position in a copper sample for an absorbed fluence just above the thermal ablation threshold (3 J/cm^2)

similarly equal to Mannion's *et al.* experimental results [20] where they find $F = 3.19$ J/cm^2. However, as mentioned above, it has been observed experimentally [2, 5, 6, 20] a very clear ablation at low fluence about of $F = 0.5$ J/cm^2, where there is no trace of liquid on the surface, therefore, it has been suggested that the matter has been ejected in plasma form.

In Figure 1(b) we represent the lattice temperature versus the time and the position in a copper sample for an absorbed fluence just above the thermal ablation threshold (3 J/cm^2). It is the contour of the thermal penetration of lattice temperature during electron-phonon collision time.

Figure 2(a) represent the evolution of electrons and ions temperature after the irradiation by femtosecpnd laser, obtained by TTMc model proposed.

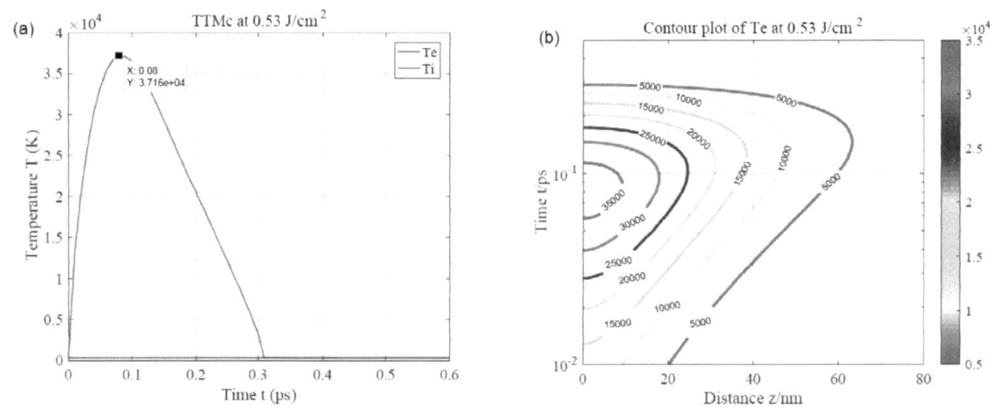

Figure 2. (a) the evolution of electrons and ions temperature, (b) electrons temperature versus time and position in a copper sample for an absorbed fluence just above gentle ablation threshold (0.53 J/cm^2)

When the metal surface irradiated by a femtosecond laser, the electrons will absorbe the laser energy and then transfer it to set of ions, but we remark that at ($0.53\ J/cm^2$), the electrons reach a classical critical temperature T_{CC} greater than or equal to E_b where $E_b = 3.125\ eV/atom$ is the binding energy [10]. So, $T_{CC} = \frac{E_b}{k_b} = 3.62 \times 10^4 K$. therefore, we suggest that ions absorbe electrons energy in momentum exchange at sub-vibrational time via electron-ion collision and stay thermally cold (Figure 2(a)), so there will be a random motion of ions which can be considered as a non-thermal liquid (*plasma*) that will escape or leave the material, so it may be called a non-thermal explosion. This critical fluence is in good agreement with the ablation threshold (also called gentle ablation) in various experimental results [5, 6, 20].

In Figure 2(b) we represent the electron temperature versus the time and the position in a copper sample for an absorbed fluence just above the non-thermal ablation threshold ($0.53\ J/cm^2$). It represents the contour of penetration of electron temperature during electron-ions collision time.

Figure 3 represents a logarithmic evolution of depth in function of fluence. We observe that our theoretical results obtained by TTMc model are in good agreement with the experimental data.

But since the thermal explosion is obviously present at high fluence, we suggest that at low fluence all the ablation inside the laser focal volume is due to non-thermal explosion, but with increasing fluence the ablation in the side of the spot laser have to be caused by thermal explosion, because electron-phonon collision frequency increase in contrast the electron-ions collision frequency decrease (Fig. 4).

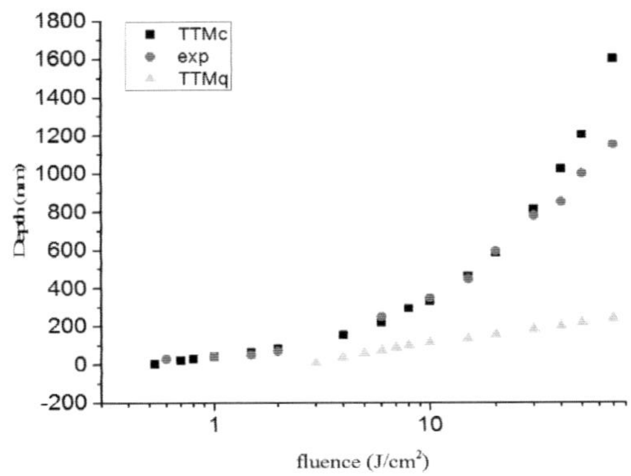

Figure 3. Simulation and experimental ablation depths. the experemental datas from Refs. [5, 20]

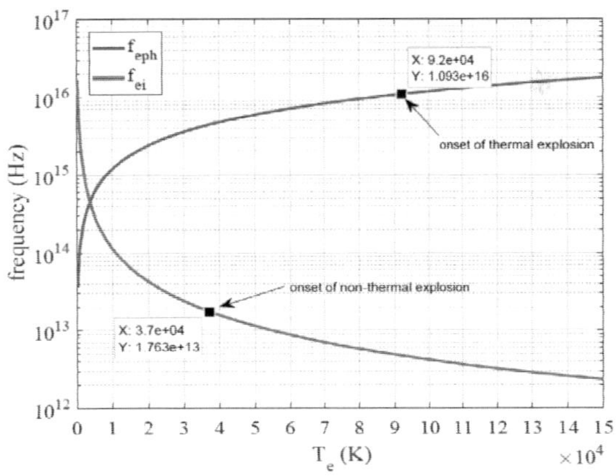

Figure 4. electon-phonon and electrons-ions colision frequency in function of electron temperature.

Finally, we can say that at low fluence all the crater volume due to the non-thermal explosion (N-TE) and the mater is ejected in a plasma form, and at high fluence, deeper region of crater due to N-TE and the inter side region due to thermal explosion (TE) and the mater successively ejected in plasma form and in liquid form.

4. Conclusion

In this paper, we highlight a recent problem in the modern physics, that one concerning the physical mechanism of the metal ablation by femtosecond laser pulse, where it has been shown that there is a large difference between the quality of the crater produced by short or ultrashort laser pulse. Therefore, we can say that there are a large difference also in the physical mechanism corresponding.

It experimentally observed that no trace of liquid around the crater during the irradiation by fs pulse at low fluence and the presence of a thin liquid layer at high fluence. The TTMq model show that at high fluence there is a thermal explosion (TE), so the matter is ejected in a liquid form, but at low fluence no information is given. Our proposed model TTMc show that there is a non-thermal explosion (N-TE) at low fluence due to the creation of a plasma phase at sub-vibrational time, during electron-ion collision when the material keep thermally cold. At high fluence there is always a non-thermal explosion (N-TE) but exist also a phase explosion (TE) because of the different rate in collision frequency value.

As a conclusion, basing to our model TTMc, metal ablation at low fluence is necessary due to a non-thermal explosion which produce a very clear ablation. In contrast, at high fluence, both non-thermal and thermal explosion successively occurs.

References

[1] Chichkov, B. N., Momma, C., Nolte, S., Von Alvensleben, F., Tnnermann, A. (1996). Femtosecond, picosecond and nanosecond laser ablation of solids. Applied Physics A, 63(2), 109-115.

[2] Nolte, S., Momma, C., Jacobs, H., Tnnermann, A., Chichkov, B. N., Wellegehausen, B., Welling, H. (1997). Ablation of metals by ultrashort laser pulses. JOSA B, 14(10), 2716-2722.

[3] Leitz, K. H., Redlingshfer, B., Reg, Y., Otto, A., Schmidt, M. (2011). Metal ablation with short and ultrashort laser pulses. Physics Procedia, 12, 230-238. ISO 690

[4] Song, K. H., Xu, X. (1998). Explosive phase transformation in excimer laser ablation. Applied surface science, 127, 111-116.

[5] Cheng, C. W., Wang, S. Y., Chang, K. P., Chen, J. K. (2016). Femtosecond laser ablation of copper at high laser fluence: Modeling and experimental comparison. Applied Surface Science, 361, 41-48.

[6] Momma, C., Nolte, S., Chichkov, B. N., Alvensleben, F. V., Tnnermann, A. (1997). Precise laser ablation with ultrashort pulses. Applied surface science, 109, 15-19. ISO 690

[7] Chowdhury, I. H., Xu, X. (2003). Heat transfer in femtosecond laser processing of metal. Numerical Heat Transfer: Part A: Applications, 44(3), 219-232.

[8] Chen, J. K., Beraun, J. E. (2003). Modelling of ultrashort laser ablation of gold films in vacuum. Journal of Optics A: Pure and Applied Optics, 5(3), 168.

[9] Gamaly, E. G., Rode, A. V., Luther-Davies, B., Tikhonchuk, V. T. (2002). Ablation of solids by femtosecond lasers: Ablation mechanism and ablation thresholds for metals and dielectrics. Physics of plasmas, 9(3), 949-957.

[10] Gamaly, E. G.., Rode, A. V., Tikhonchuk, V. T., Luther-Davies, B. (2002). Electrostatic mechanism of ablation by femtosecond lasers. Applied surface science, 197, 699-704.

[11] Stoian, R., Rosenfeld, A., Ashkenasi, D., Hertel, I. V., Bulgakova, N. M., Campbell, E. E. B. (2002). Surface charging and impulsive ion ejection during ultrashort pulsed laser ablation. Physical review letters, 88(9), 097603.

[12] Dachraoui, H., Husinsky, W., Betz, G. (2006). Ultra-short laser ablation of metals and semiconductors: evidence of ultra-fast Coulomb explosion. Applied Physics A: Materials Science Processing, 83(2), 333-336.

[13] Anisimov, S. I., Rethfeld, B. (1997, April). Theory of ultrashort laser pulse interaction with a metal. In Nonresonant Laser-Matter Interaction (NLMI-9) (Vol. 3093, pp. 192-204). International Society for Optics and Photonics.

[14] Lin, Z., Zhigilei, L. V., Celli, V. (2008). Electron-phonon coupling and electron heat capacity of metals under conditions of strong electron-phonon nonequilibrium. Physical Review B, 77(7), 075133.

[15] Furusawa, K., Takahashi, K., Kumagai, H., Midorikawa, K., Obara, M. (1999). Ablation characteristics of Au, Ag, and Cu metals using a femtosecond Ti: sapphire laser. Applied Physics A, 69(1), S359-S366.

[16] Tsibidis, G. D. (2014). Thermal response of double-layered metal films after ultrashort pulsed laser irradiation: The role of nonthermal electron dynamics. Applied Physics Letters, 104(5), 051603.
[17] Lin, Z., Zhigilei, L. V., Celli, V. (2008). Electron-phonon coupling and electron heat capacity of metals under conditions of strong electron-phonon nonequilibrium. Physical Review B, 77(7), 075133.
[18] Delcroix, J. L., Bers, A. (1966). Physique des plasmas (Vol. 2). Les Ulis (Essonne): EDP sciences.
[19] Sundaram, S. K., Mazur, E. (2002). Inducing and probing non-thermal transitions in semiconductors using femtosecond laser pulses. Nature materials, 1(4), 217-224.
[20] Mannion, P. T., Magee, J., Coyne, E., Oconnor, G. M., Glynn, T. J. (2004). The effect of damage accumulation behaviour on ablation thresholds and damage morphology in ultrafast laser micro-machining of common metals in air. Applied surface science, 233(1), 275-287.
[21] Nowak, D., Lee, M. J. (1972). Optical and Thermal Effective Masses of Copper. Physical Review B, 5(8), 2851.
[22] Kittel, C. (2005). Introduction to solid state physics. Wiley.

IOP Conf. Series: Journal of Physics: Conf. Series **987** (2018) 012013 doi:10.1088/1742-6596/987/1/012013

Characterization of Polycaprolactone and Rice Husk Silica Composite (PCL-SiO_2) by E-Spinning to Apply Supporter for Drug Release

Sinae Song,[1] Askwar Hilonga,[2] and Hee Taik Kim[3*].

[1]Department of Advanced Materials Science and Engineering, Hanyang University, Ansan, Republic of Korea
[2]Department of Material Science and Engineering, Nelson Mandela African Institution of Science and Technology, P.O. Box 447, Arusha, Tanzania
[3]Department of Fusion Chemical Engineering, Hanyang University, Ansan, Republic of Korea

khtaik@hanyang.ac.kr

Abstract. Polycaprolactone (PCL) is an interesting material to apply biomedical field owing to its biodegradability and biocompatibility which is suitable for a specific site with longer healing times. Blending the polymer with other materials has degradation property improved with the effective and economic method. This study was conducted to fabricate supporter based on Polycaprolactone and Rice husk silica (PCL-SiO2) by using electrospinning. Nano-porous silica in the composite was synthesized from rice husk having properties of economic, eco-friendly and high surface area. It drew to enhance the amount of drug loading in the carrier. Electrospinning technique is used to fabricate fibrous component by optimization condition obtained from previous mechanical properties experiments. Release experiment was carried out by the degree of dye absorbance at 544nm by ultraviolet–visible spectroscopy, the RhB in SiO2 alternative drug for modelling of drug release was released for 1 ~ 20 days at 37°C in phosphate buffer. Furthermore, the Mechanical property was confirmed by DSC, TGA. Morphology and degree of biodegradation were shown as SEM images and EDS.

1. Introduction

Nanoporous Silica have known as a materials to use widely application with high specific surface area such as cosmetics, filler, pharmaceuticals, chromatography which is derived from its hygroscopic and hydrophilic properties. The usage as a drug delivery carrier has increased in medical application because high porous structure property is able to impregnate the active ingredient onto its surface. However the challenge in this experimental is expensive cost of its precursor. In this study, silica extracted from rice husk was used to synthesize nanoporous silica with economical and biocompatible properties.

Polycaprolactone (PCL) is utilized of medical field having biodegradable, but it has weak mechanical properties. In this study, PCL composite with rice husk silica was fabricated as type of nanofiber to overcome the challenge by electronic spinning. By extension, the polymer nanofiber was prepared to apply a material for guided bone regeneration (GBR) with resembling bone structure, which the inorganic component of the nanofiber mimics the hydroxycarbonate-apatite structure while the polymer component mimics the collagen rich extracellular matrix.

2. Experiment

Materials and Pretreatment

Rice husk was obtained from Namhae where is located in southern part of the Republic of Korea. Sodium hydroxide (NaOH) and acetone used in this study are products of DAEJUNG. Hydrochloric acid (HCl) and Polycaprolactone (PCL) were purchased from SAMCHUN chemical and Aldrich respectively. Three steps were conducted to extract silica from rice husk: 1st step was washing rice husk with deionized water to remove dust and drying in an oven. The following step was leaching rich husk with 0.1 M HCl for 2 h at 90 °C then rinsing sufficiently to neutralize. Calcination of clean rice husk to eliminate inorganic impurities remained was the final step. After all these steps, Silica/Carbon powder (rice husk ash; RHA) was obtained

Nano-porous rice husk silica

The RHA was used as raw material to synthesize nano-porous silica. 500mg RHA was added in a 10 ml vial containing 9ml 1.5 M NaOH and boiled in a water bath at 90 °C for 1 h. Conversion from RHA in NaOH to sodium silicate solution was generated during boiling step. The solution was moved to a 50ml beaker and diluted until 1 M concentration with adding deionized water. The diluted sodium silicate solution was quickly titrated up to pH4 by 1.0 M HCl. Steady stirring magnetic bar in the solution was played a role to avoid partial transformation from the silicate sol state to gel upon pH during titration. The mixture was placed in an air oven at 70 °C for 5~8days, which high aging temperature was accelerated reactive silanol group in the gel formation to exist during the process of hydrolysis and condensation. Aged samples were washed with deionized water by centrifuge at 10,000 rpm for 5 min three times then drying in an air oven. The synthesized rice husk silica products were ground to powder and stored in a vacuum desiccator.

PCL-Rice husk silica nanofiber

1.17 g PCL and 20wt% rice husk silica were dissolved in 10 ml acetone at 50 °C for 10 min. The solution was delivered a constant flow rate (Q = 3 ml/min) through a metal needle (21 gauge) connected to a high-voltage power supply. The polymer solution was ejected aluminium foil on the surface of the collector from the capillary upon applying a high voltage (18kV) for 40 min. The nanofibers collected on the aluminium foil surface were obtained by peeling with tweezers.

Dye release test

Release test of dye in nano-porous silica was done by the amount of absorbance intensity at 544nm using UV-visible spectroscopy. Prior to conduct release test, a calibration curve of Rhodamine B upon different concentration (0.21mM ~ 6.38*10-6 mM) was constructed by UV absorbance at 544nm of each dye concentration. 50mg nano-porous silica sample was pressed into disks (1cm* 0.3cm) using hand pressure machine. Then nano-porous silica tablet sample put into 3ml 3.26*10-3 mM Rhodamine B solution and stayed overnight. Amount of dye loading in the tablet samples was checked through the UV absorbance intensity at 544nm of the rest solution by the calibration curve. Each tablet samples with 1 ml Phosphate buffer solution (PBS) were added into 1ml microtube with one side sealing membrane (). The microtubes put into vials containing 20 ml PBS and placed in an incubator at 37°C with 100rpm. A sampling of dye release from silica tablets was carried out on 1, 4, 7, 10 and 15 days. The Samples obtained on each day were dried under freeze-drying and re-dispersed in 5ml DW then analysed with UV-visible spectroscopy.

Polymer degradation test

PCL nanofibers with nano-porous silica samples having a different specific surface area were sized 1cm * 1cm and put into vials containing 5ml PBS. The vials were in an incubator at 37°C with 100rpm for 1 ~ 15days. Scanning Electron Microscopy (SEM; Hitachi S-4800S) and Energy Dispersive X-ray (EDX; Horiba EMAX mics) were used to confirm their degradation properties such as change morphology and new elements generated after degradation of PCL.

3. Results and discussion

Nano-porous rice husk silica

Rice husk ash (RHA), raw source to synthesize nanoporous rice husk silica, was obtained through pre-treatments three steps of rice husk (Figure 1).

Figure 1. Images of Rice Husk with each experimental steps

Metallic impurities were eliminated by acidic leaching and calcination process removed maintaining inorganic dust in rice husk to get rice husk ash having high pure silica. Twelve samples were synthesized by several factors such as the amount of RHA, level of pH and aging times, which was named R, P and A implying R to the quantity of RHA, P to pH and A to aging times; e.g. 6R4P7A means 600mg RHA, pH4 and 7 days aging. The reason why factors were chosen was that the specific surface area was likely to be influenced by the level of pH and aging time. Values of specific surface area of the samples were shown in table 1.

Table 1. Specific surface area, pore volume, pore diameter, particle size of nanoporous rice husk silica

Samples	Surface Area $_{BET}$ (m^2/g)	Pore Volume(cm^3/g)	Pore Size(nm)
3R4P1A	506.17	0.55	4.38
3R7P5A	240.05	0.87	14.61
3R9P7A	53.02	0.40	30.35
6R9P1A	131.77	0.56	17.16
6R7P7A	111.92	0.59	21.26
6R4P5A	339.60	0.85	10.06
9R7P1A	186.72	0.68	14.62
9R9P5A	35.12	0.20	23.20
9R4P7A	449.92	0.80	7.18
6R4P7A	474.61	0.22	1.89
6R4P6A	756.02	0.49	2.59
6R4P8A	44.9251	0.07	6.91

The nanoporous silica samples were impregnated with Rhodamine B dye to test dye releasing from the samples in aqueous condition at 37°C for several days. Accumulative percentage of dye on releasing time was computed by the trend line equation from the calibration curve of different concentration (1 umol ~ $3.19*10^{-5}$ umol) of Rhodamine B dye shown in figure 2.

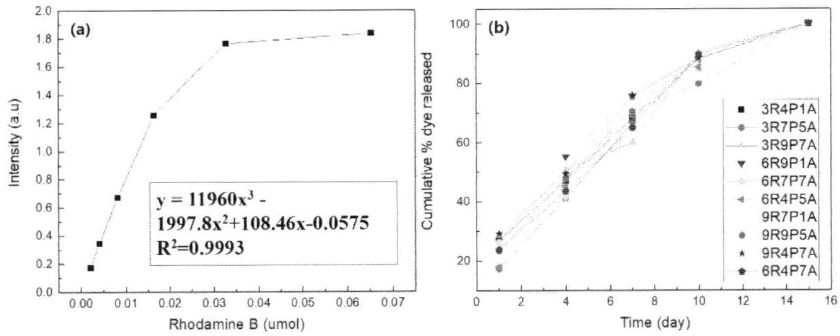

Figure 2. (a) Calibration curve depending on diverse concentration of Rhodamine B dye and trendline equation, (b) cumulative percentage of dye released from nanoporous silica

Silica having large pore size confirmed releasing randomly meanwhile having high specific surface area released the dye steadily and slowly.

Degradation test of PCL-Rice husk silica composite nanofiber

6R4P5A, 6R4P6A, and 6R4P8A were chosen to apply for degradation test of PCL-Rice husk silica nanofiber because this study was aimed to confirm dye release along the specific surface area of silica, which meant aging time was the influencing factor to the value of the specific surface area. The morphology of samples was shown in figure 3.

Figure 3. Scanning Electron Microscopy (SEM) images of nano-porous rice husk silica (insert: average particle size)

Particle sizes were ranged from 17.94 nm to 50.76 nm inducing small particles to the high specific surface area and big to a low one. Prior experimental was conducted to confirm optimized dynamic mechanical properties of PCL composites along containing rice husk silica (S_{BET}: 454.845 m^2/g). PCL-20wt% rice husk silica was resulted to be a suitable material for practical application (table 2).

Table 2. Dynamic mechanical properties of PCL-SiO_2 Composites depend on amount of silica containing from the previous study

Sample	T_g(°C)	E′ at 30°C (MPa)	E′ at 55°C (MPa)
Neat PCL	-46.98	2.38	1.11
PCL-SiO_2 10wt%	-52.46	2.43	1.14
PCL-SiO_2 20wt%	-54.38	2.48	1.44
PCL-SiO_2 30wt%	-49.19	2.16	0.93

The nanofiber samples were analyzed to characterize the other mechanical properties by thermogravimetric analysis (TGA; Dong-il SHIMADZU TGA 50) and differential scanning calorimetry (DSC; TA Q20P). Figure 4 was the result of TGA confirmed PCL-6R4P6A nanofiber with high surface area was rapidly decomposed.

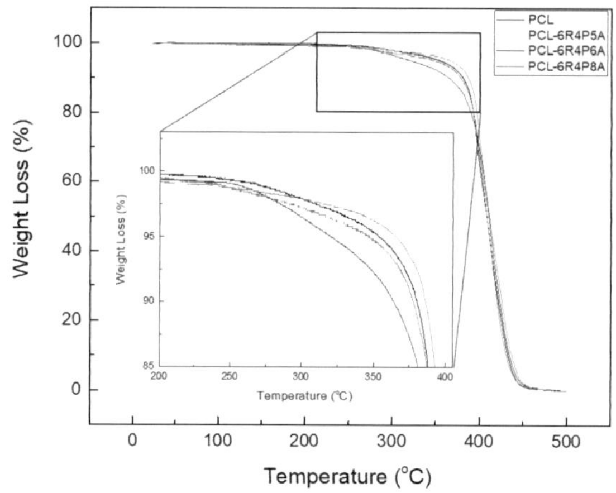

Figure 4. TGA thermal graph of PCL, PCL-6R4P5A, PCL-6R4P6A, PCL-6R4P8A

High specific surface area of nanoporous silica was also verified to induce melting point(T_m) and crystallinity (X_c) to increase shown in figure 5 and table 3 from DSC analysis.

Table 3. Thermal properties of PCL-Silica eco-composites analyzed by DSC

Samples	T_m(°C)	H_m(J/g)	X_c(%)
PCL	59.90	39.13	28.98
PCL-6R4P5A	61.10	24.03	14.24
PCL-6R4P6A	61.64	33.43	19.81
PCL-6R4P8A	59.00	28.27	16.75

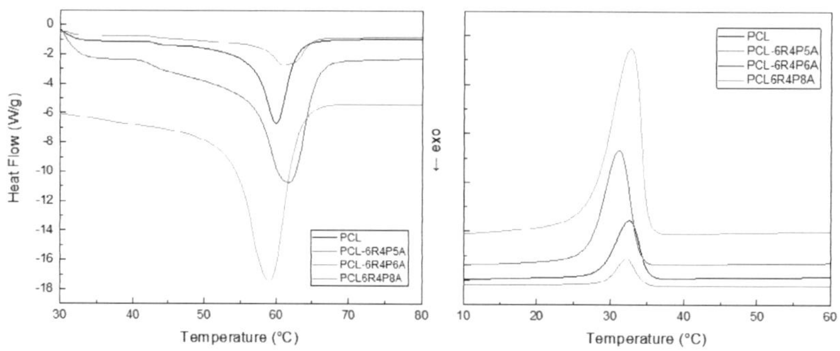

Figure 5. Differential Scanning Calorimetry (DSC) thermal analysis graphs of PCL, PCL-6R4P5A, PCL-6R4P6A, and PCL-6R4P8A

Degradation test of PCL-Rice husk silica nanofiber was progressed in phosphate buffer solution at 37°C along reaction days and the samples were extracted and analyzed to check their morphology change and Ca/P ratio by using SEM and Energy dispersive spectrometer (EDX). Increasing the Ca/P ratio was revealed along reaction days as illustrated in table 4.

IOP Conf. Series: Journal of Physics: Conf. Series **987** (2018) 012013 doi:10.1088/1742-6596/987/1/012013

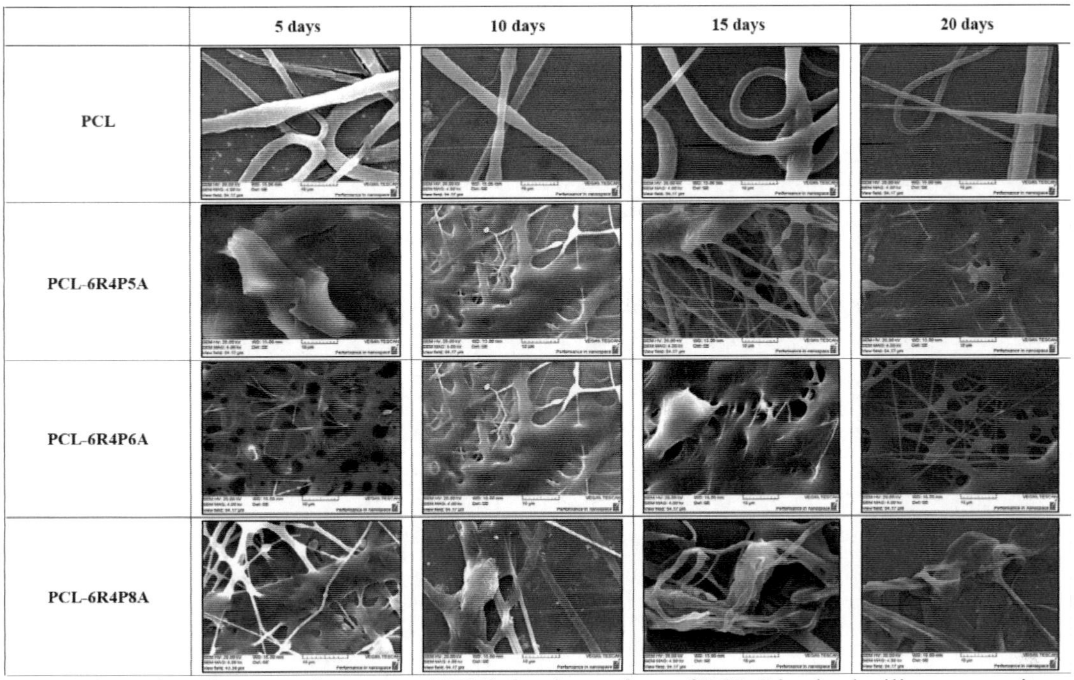

Figure 6. Scanning electron microscopy (SEM) for the surface of PCL-Rice husk silica composites after degradation from 5 to 20 days (scale bar; 10um)

Table 4. Ca/P ratio depending on diverse degradation days of PCL-Silica eco-composites

Ca/P	5 days	10 days	15 days
PCL	0.12	0.18	0.23
PCL-6R4P5A	0.16	0.18	0.59
PCL-6R4P6A	0.15	0.27	0.93
PCL-6R4P8A	0.25	0.16	0.18

The Ca/P ratio of PCL-6R4P6A nanofiber is high after 15 days having large difference comparing the other samples. Nanoporous silica having high specific surface area was indicated to accelerate PCL to decompose itself.

4. Conclusion

Eco friendly nanoporous silica samples having diverse specific surface area were prepared from rice husk which was generated as biomass. The silica samples was impregnated with rhodamine B dye as alternative of real drug and conducted releasing test along times. Rice husk silica having large pore size was released randomly meanwhile high specific surface area silica was constantly done. Subsequently several synthesized nanoporous silica was composited as nanofiber with PCL by using electro spinning. PCL-6R4P6A, nanofiber of high surface area silica with PCL, had increasing melting point and faster degradation comparing with other samples. In the future work, rice husk silica impregnated dye is composed as nanofiber with PCL and conducted dye releasing test along degradation of its polymer.

Acknowledgement
This work was supported by the Human Resources Development program(No.20174030201830) of the Korea Institute of Energy Technology Evaluation and Planning(KETEP) grant funded by the Korea government Ministry of Trade, Industry and Energy.
This subject is supported by Korea Ministry of Environment(MOE) as "Advanced Technology Program for Environmental Industry Program".

References

[1] Jin Seok C and Sung Jin A, 2015, *Korean Journal of Materials Research*, **25**, 209-13
[2] Naghizadeh F, Sultana N, Abdul Kadir M, Tengku Md Shihabudin T, Hussain R and Kamarul T, 2014, *Journal of Nanomaterials*, **2014**, 9
[3] Slowing I I, Vivero-Escoto J L, Wu C W and Lin V S Y, 2008, *Advanced Drug Delivery Reviews*, **60**, 1278-88
[4] Smitha S, Shajesh P, Aravind P R, Kumar S R, Pillai P K and Warrier K G K, 2006, *Microporous and Mesoporous Mater.*, **91**, 286-92
[5] Yang F, Both S, Yang X, Walboomers X. F. and Jansen J A, 2009, *Acta Biomaterialia*, **5**, 3295-04
[6] Yoshimoto H, ShinY M, Terai H and Vacanti J P, 2003, *Biomaterials*, **24**, 2077-82

IOP Conf. Series: Journal of Physics: Conf. Series **987** (2018) 012014 doi:10.1088/1742-6596/987/1/012014

Indium-saving effect and physical properties of transparent conductive multilayers

M Kawamura, T Kiba, Y Abe, K H Kim

Kitami Institute of Technology, Kitami, Hokkaido 090-8507 Japan

E-mail: kawamumd@mail.kitami-it.ac.jp

Abstract. Indium-free transparent conductive multilayer structures consisting of top and bottom MoO_3 layers and an Ag interlayer (MoO_3/Ag/MoO_3; MAM) are deposited onto glass substrates by vacuum evaporation. The transmittance and sheet resistance of the structures are evaluated, and the optimum structure is determined to be MAM (20/14/30 nm) as it shows the best figure of merit (FOM), which is used as the index for transparent conductive films, with a value of 6.2 $\times$ 10^{-3} Ω^{-1}. To further improve the performance of the films, we attempt to fabricate a multilayer consisting of MoO_3 and indium zinc oxide (IZO), based on previous results. The obtained IAM (30/14/50 nm) multilayer shows an FOM higher than that of the MAM, with a value of 32 $\times$ 10^{-3} Ω^{-1}. Moreover, it reduces the amount of required indium as compared with the IZO/Ag/IZO multilayer.

1. Introduction

Display devices such as organic light emitting diodes (OLEDs) and organic solar cells generally use transparent conductive films as transparent electrodes [1, 2]. The most commonly used material in these films is indium tin oxide (ITO). However, it would be favorable to reduce the use of indium because its rarity causes its price fluctuation significantly [3]. In addition, a study has highlighted inferior mechanical properties of ITO films when used on flexible substrates [4]. Indium-free or indium-saving transparent conductive films have been intensively developed; an example of such a film is a sandwich structure consisting of top and bottom metal oxides and a metal interlayer (oxide/metal/oxide). Such a multilayer can reduce sheet resistance (R_s) due to the metal interlayer as well as the use of indium even if the top and bottom oxide is ITO compared to ITO single film [5]. In general, the same substance among various oxides has been introduced into the top and bottom layers and the properties have been recently reviewed [6, 7]. Previously, we reported the optimization of layer thickness and sputtering conditions for a multilayer consisting of indium zinc oxide (IZO) and Ag layers (IZO/Ag/IZO; IAI). It was shown to exhibit excellent properties as a transparent conductive film and used 80% less indium than an IZO single layer film with the same sheet resistance value. In addition, the optimized multilayer could be used as an anode in an OLED that demonstrated excellent EL properties [8].

Among the various metal oxides, MoO_3 [9, 10] has the advantage of being not only indium free but also an excellent hole injection material in OLEDs. It has already been shown that a MoO_3/Ag/MoO_3 (MAM) electrode can lower the barrier height for hole injection into an organic layer (hole transporting layer) when compared with an ITO electrode [9].

In this study, we first prepare indium-free MAM multilayers with different thickness combinations of the MoO_3 and Ag layers. Their transmittance, sheet resistance, and figure of merit (FOM) [11] are subsequently obtained, and their structures are optimized. We then attempt to fabricate multilayers consisting of MoO_3 layers and sputtered IZO layers. We determine the optimum IZO/Ag /MoO_3 (IAM) multilayer structure. Finally, we compare the optimum MAM and IAM multilayers on FOM and indium-saving effect.

2. Experimental procedure

To begin with, we prepared the MAM structures. Commercial MoO_3 powders (99.9999% purity) and an Ag wire (99.99% purity) were deposited onto Corning Eagle XG glass substrates through vacuum evaporation after evacuation to below 4.0×10^{-4} Pa. The thickness of the deposited layer was controlled using a quartz crystal microbalance. The thickness of the Ag layers was also confirmed using X-ray fluorescence spectrometry.

To fabricate the IAM structures, the bottom IZO layer was deposited via sputtering onto a glass substrate. The deposition of a 30-nm-thick IZO layer onto a glass substrate was performed by RF sputtering using a 2-inch diameter IZO (10 wt% ZnO) target under an O_2 atmosphere, as previously reported [8]. The sample was then transferred into the vacuum evaporation chamber in which the Ag and top MoO_3 layers were deposited exactly the same conditions as they were in the MAM structure.

The sheet resistance (R_s) of the samples was measured using a four-point-probe method, and transmittance was measured in the wavelength region of 200–800 nm using a spectrophotometer. The figure of merit (FOM) is expressed as T^{10}/R_s [11], where T is the transmittance of the films and R_s is the sheet resistance. We used two values for transmittance in this calculation: the first was that at a wavelength of 550 nm because it is the approximate maximum human visual sensitivity; the second was the average transmission value in the visible range because the use of this average is also practiced in other reports. The surface morphology was observed using an atomic force microscope (AFM).

3. Results and discussion

3.1. MoO_3/Ag/MoO_3 (MAM) multilayer

The thicknesses of the top and bottom MoO_3 layers were optimized, whereas that of the Ag layer was fixed to 14 nm. The optimum IAI structure has been shown to be 30/14/30-nm thick [8]; therefore, we attempted to explore the thickness of each layer in a similar range. First, the transmission spectra of MAM, in which the top layer was fixed to 30 nm, whereas the bottom layer was varied from 15 to 30 nm were measured. As a result, all the spectra were similar, but the structure with a 20-nm-thick bottom layer could have the highest transmittance. The sheet resistance (R_s) of the structures was consistently low, with values of 3.4–3.8 Ω/sq., because it mostly depends on the sheet resistance of the Ag layer.

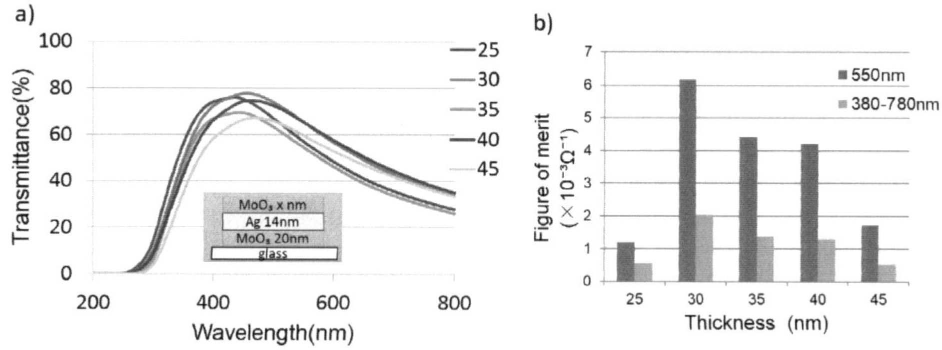

Figure 1. The properties of MoO_3(20 nm)/Ag(14 nm)/MoO_3(x nm) multilayer samples, where x is varied from 25 to 45 nm. a) the transmission spectra and b) the FOM calculated using two different transmittance values.

Figure 1(a) shows the transmission spectra of MAM, in which the bottom layer was fixed to 20 nm, whereas the top layer was varied from 25 to 45 nm. A larger change was observed in these spectra, and 30 nm could afford the highest transmittance at λ = 550 nm. The R_s of the structures was again consistently low, being in the range of 3.2 to 3.8 Ω/sq. Figure 1(b) shows the FOM values of the MAM structures as a function of the thickness of the top MoO_3 layer. Transmittances at both the 550 nm and average visible range wavelengths (380 to 780 nm) were used, and in both cases, our results show that a 30-nm-thick top layer affords the highest FOM value.

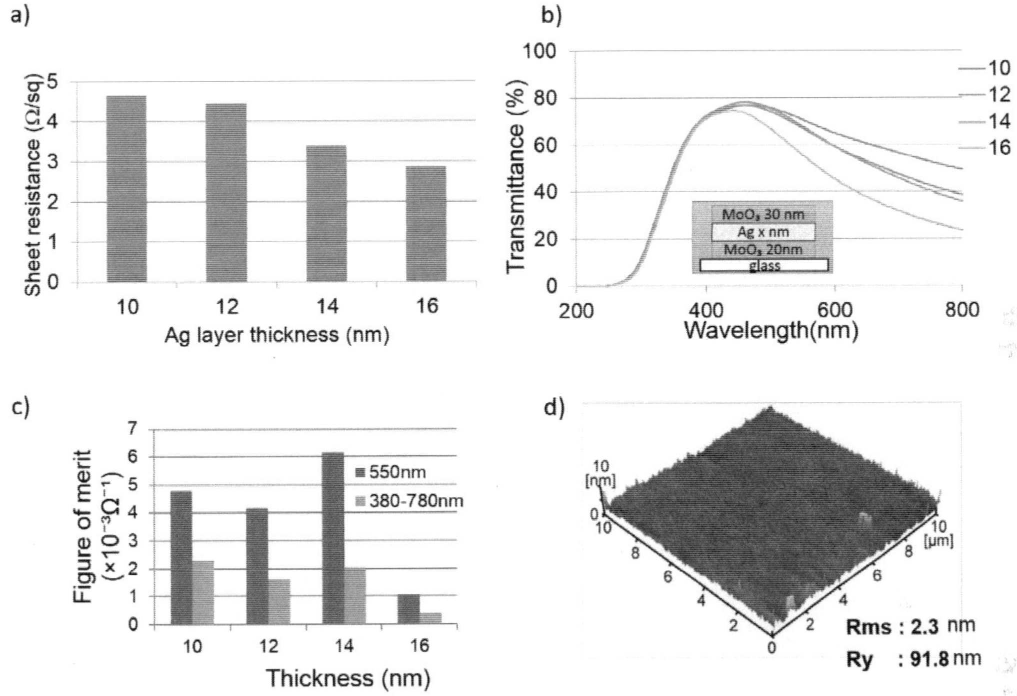

Figure 2. The properties of MoO_3(20 nm)/Ag(x nm)/MoO_3(30 nm) multilayer samples, where x is varied from 10 to 16 nm. a) the sheet resistance, b) the transmission spectra, c) the FOM calculated using two different transmittance values, and d) AFM image and roughness values of MoO_3(20 nm)/Ag(14 nm)/MoO_3(30 nm) multilayer (scanning area: 10×10μm^2).

The effect of the thickness of the Ag layer was also confirmed; by increasing the thickness from 10 to 16 nm, the sheet resistance decreased from 4.5 to 2.8 Ω/sq., as shown in Fig. 2(a). However, transmittance was considerably lower for the thicker Ag layer, and a high transmission was only obtained in a narrow region, as shown in Fig. 2(b). Consequently, the highest FOM value was obtained with 14 nm thickness when the transmittance at λ = 550 nm is used, as shown in Fig. 2(c). In conclusion, the optimum MAM structure is observed to be MAM (20/14/30 nm). This structure shows a high transmittance (at λ = 550 nm) of 67.9%, a low sheet resistance (3.4 Ω/sq.), and an FOM of $6.2 \times 10^{-3}\ \Omega^{-1}$. When the average transmittance (60.8%) was used, FOM was decreased slightly to $2.0 \times 10^{-3}\ \Omega^{-1}$. We observed that the optimum structure in our results exhibit a thickness range similar to those of the optimum MAM structures in literatures. *Nguyen et al.* reported that an MAM structure with a 35-nm-thick top MoO_3 layer, a 20-nm-thick bottom MoO_3 layer, and a 10-nm-thick middle Ag layer afforded an average transmittance of 70% [10], which is higher than our average transmittance. However, the R_s and FOM of this optimum structure were not shown. *Naka et al.* reported that a combination of MoO_3 layers between 20 and 50 nm and an Ag layer between 5 and 15 nm led to high transmittance (70% at

λ = 550 nm) and low sheet resistance (5.8 Ω/sq.) [9]; therefore, we obtained similar FOM values in this study.

Finally, the surface morphology of the MAM multilayer was also investigated. AFM image and root mean square roughness and height difference values of the MAM (20/14/30 nm) are shown in Fig. 3(d). It is found that the surface roughness is very small, however, some spikes are also observed therefore relatively large Ry value is measured.

3.2. IZO/Ag/MoO$_3$ (IAM) multilayer

The FOM value of MAM is found to be much lower than that of the IAI multilayer [8]. We attempted to improve the FOM by combining a MoO_3 layer and an IZO layer in the structure. A 14-nm-thick Ag layer and a MoO_3 top layer (with thicknesses between 20 and 60 nm) were deposited onto a 30-nm-thick sputtered IZO film. Figure 3(a) shows the transmittance spectra of all the samples. The different MoO_3 layer thicknesses resulted in varying positions of the peaks, but apparently the sample consisting of a 50-nm-thick MoO_3 layer shows a transmittance (at λ = 550 nm) of 79.2%, which is high. The sheet resistance of all the samples remained constant because the thickness of the Ag layer remained constant. Figure 3(b) shows the FOM of all the samples based on the two transmittance values. In both cases, the IAM(30/14/50 nm) sample showed the highest values, and a value of 32 $\times 10^{-3}$ Ω^{-1} was achieved. We expect that the IAM electrode could be successfully used as an anode in OLEDs.

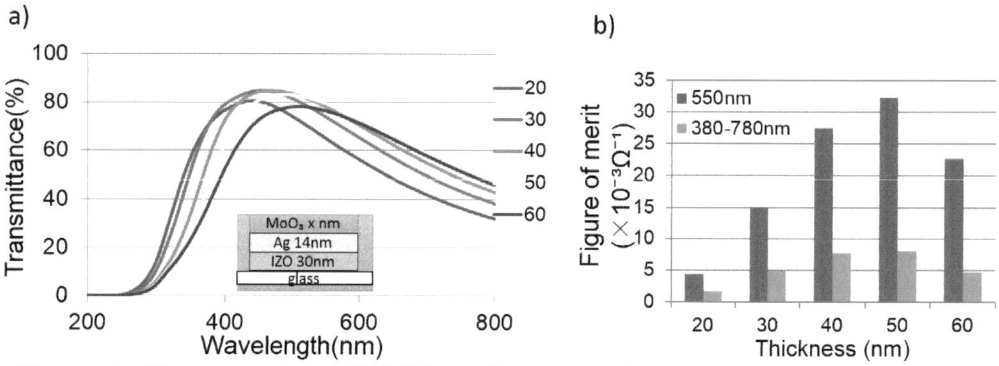

Figure 3. The properties of IZO(30 nm)/Ag(14 nm)/MoO_3(x nm) multilayer samples, where x is varied from 20 to 60 nm. a) the transmission spectra and b) the FOM calculated using two different transmittance values.

Table 1 summarizes results of representative film structures concerning about FOM values and thicknesses of IZO layer. A higher FOM value can be obtained by using the IAM structure instead of the MAM structure, although the FOM value is still lower than that of the IAI structures [8]. On the other hand, the MAM is indium-free and the IAM consisting of hybrid oxides could reduce the thickness of the IZO layer, which implies significantly less indium is used, compared to IAI. Both MAM and IAM are superior to a 70-nm-thick IZO single film which has a low FOM value due to its high electrical resistivity (3.0×10^{-4} Ωcm) and sheet resistance (42 Ω/sq.).

There is a report describing a hybrid structure of MoO_3 and In_2O_3 [12], where post-annealing of the multilayers was required to obtain a low sheet resistance. The as-deposited film exhibited a high sheet resistance of approximately 50 Ω/sq.; then, annealing at 120 °C lowered the value to 8.2 Ω/sq. The transmittance at 550 nm in the study was stated to be 85%; therefore, its FOM was estimated as 24×10^{-3} Ω^{-1}. However, our IAM multilayer achieved a higher FOM without any annealing, with a similar indium-saving level.

Table 1. The FOM values and IZO layer thicknesses of representative samples.

Structures	FOM ($\times 10^{-3}\ \Omega^{-1}$)	Thickness of IZO layer (nm)
MAM(20/14/30)	6.2	0
IAM(30/14/50)	32	30
IAI(30/14/30) [8]	56	60
IZO single film	2.3	70

Acknowledgment
The authors deeply thank Mr. Shogo Takehara for his assistance on sample preparation.

4. Conclusion

Indium-free, transparent conductive multilayers are fabricated, and the optimum structure is found to be MAM (20/14/30 nm). The FOM value is observed to be lower than that for a previously studied IAI multilayer; therefore, the fabrication of a hybrid oxide multilayer consisting of MoO_3 and IZO is attempted for the first time. Following this, IAM (30/14/50 nm) structure exhibits the highest FOM value, which is still lower than that of the IAI multilayer, but it is important because less indium can be used in display devices.

References

[1] Hosono H 2007 *Thin Solid Films* **515** 6000
[2] Granqvist C G 2007 *Sol. Energy Mater. Sol. Cells* **91** 1529
[3] Hoel C A, Mason T O, Gaillard J F and Poeppelmeier K R 2010 *Chem. Mater.* **22** 3569
[4] Leterrier Y, Medico L, Demarco F, Manson J A E, Betz U and Escola M F 2004 *Thin Solid Films* **460** 156
[5] Guillen C and Herrero J 2009 *Opt. Commun.* **282** 574
[6] Guillen C and Herrero J 2011 *Thin Solid Films* **520** 1
[7] Cattin L, Bernede J C and Morsli M 2013 *Phys. Status Solidi A* **210** 1047
[8] Chiba T, Kudo Y, Kawamura M, Abe Y and Kim K H 2015 *Vacuum* **121** 320
[9] Banzai K, Naka S and Okada H 2015 *Jpn. J. Appl. Phys.* **54** 054101
[10] Nguyen D T, Vedraine S, Cattin L, Torchio P, Morsli M, Flory F and Bernede J C 2012 *J. Appl. Phys.* **112** 063505
[11] Haacke G 1976 *J. Appl. Phys.* **47** 4086
[12] Varnamkhasti M G and Shahriari E 2015 *Appl. Phys. B* **120** 517

IOP Conf. Series: Journal of Physics: Conf. Series **987** (2018) 012015 doi:10.1088/1742-6596/987/1/012015

Rapid thermal process by RF heating of nano-graphene layer/silicon substrate structure: Heat explosion theory approach

M Sinder[1], J Pelleg[1], V Meerovich[2], V Sokolovsky[2]

[1]Materials Engineering Department, Ben Gurion University of the Negev, Beer Sheva, 84105, Israel

[2]Physics Department, Ben Gurion University of the Negev, Beer Sheva, 84105, Israel

E-mail: micha@bgu.ac.il

Abstract. RF heating kinetics of a nano-graphene layer/silicon substrate structure is analyzed theoretically as a function of the thickness and sheet resistance of the graphene layer, the dimensions and thermal parameters of the structure, as well as of cooling conditions and of the amplitude and frequency of the applied RF magnetic field. It is shown that two regimes of the heating can be realized. The first one is characterized by heating of the structure up to a finite temperature determined by equilibrium between the dissipated loss power caused by induced eddy-currents and the heat transfer to environment. The second regime corresponds to a fast unlimited temperature increase (heat explosion). The criterions of realization of these regimes are presented in the analytical form. Using the criterions and literature data, it is shown the possibility of the heat explosion regime for a graphene layer/silicon substrate structure at RF heating.

1. Introduction

There is a number of investigations devoted to the induction heating of a conductor film/ silicon substrate structure by a radio-frequency (RF) magnetic field perpendicular to the substrate surface [1-3]. The heat explosion theory [4] was used to describe the heating process. Results of the theory well agree with the experiments with Ti and Ta-Ti thin films (85-220 nm thickness) deposited on Si (111) and on Si (100) substrates [1-3]. Application of the RF heating allows one to prepare these structures with required properties. One drawback of using this technology is pollution of silicon substrates due to the fast diffusion of metallic impurities.

Here we propose a new rapid thermal process (RTP) which is based on RF heating of a nano-graphene layer deposited on the back side of a silicon substrate. The characteristic property of the method is that no additional heater is required, despite the fact that thickness of the graphene layer/Si substrate structure is less than a skin depth. In contrast to metals, the diffusion of carbon in silicone is very slow. The novel RTP can be successfully applied in technological processes of fabrication of films and layers on silicon substrates as well as to study graphene transformations.

IOP Conf. Series: Journal of Physics: Conf. Series **987** (2018) 012015 doi:10.1088/1742-6596/987/1/012015

2. Model

We consider a structure of a thin conductor film on a silicon substrate in a uniform RF magnetic field normal to its surface (Fig. 1). For simplification, we assume that the length of the structure is much larger than its width, i.e. the specimen has the form of a strip.

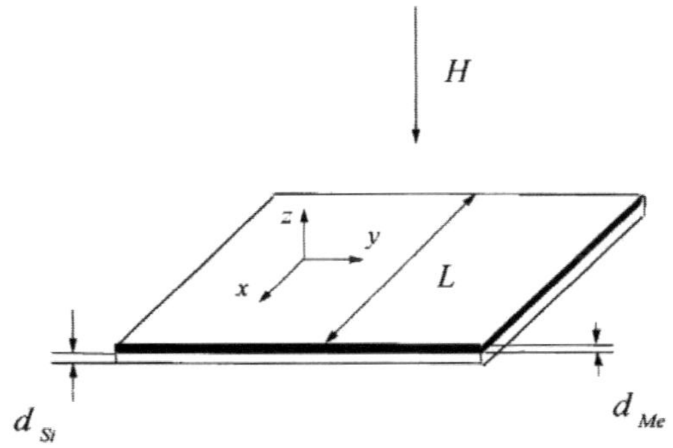

Fig. 1. Configuration of a conducting layer/silicon substrate structure in magnetic field.

The expressions for the heat generation W (averaged eddy current loss power per unit area of the structure surface [1]) and heat transfer Q from the surface (the Stefan-Boltzmann law) are

$$W = K(\mu_0 H\omega L)^2(d_{Si}/\rho_{Si} + d_c/\rho_c) \quad (1)$$

$$Q = \sigma T^4 - \sigma T^4_{room} \quad (2)$$

where K is the specimen shape coefficient, e.g. equalled 1/24 for a long strip, μ_0 is the vacuum magnetic permeability, H and ω are the amplitude and the angular frequency of the external magnetic field, L is the characteristic size of the specimen (strip width), σ is the Stefan-Boltzmann constant, d_{Si}, d_c and ρ_{Si}, ρ_c are the thicknesses and resistivities of the Si substrate and conducting layer, respectively.

The thermal state of the structure is approximately described by the following equation

$$CdT/dt = W - Q \quad (3)$$

where T and C are the averaged temperature and total heat capacity per a unit area of the structure, respectively.

3. Analysis of RF heating

In Fig.2 presents the typical dependences of the heat generation W and heat transfer Q on the structure temperature. If the thickness of the conducting layer is sufficiently small such as d_{c1} (curve 1), the system is heated up to a finite temperature T_S corresponding to point S. The thermal state corresponding to this point is stable. Point G is another stationary state but is unstable. With increasing the conducting layer thickness the stationary points move toward and at a critical thickness d_{c2} (curve 2) these points merge. No stationary state exists for a conducting layer with thickness larger than the critical thickness d_{c2}: temperature increases unrestrictedly (curve 3). In this case, heating of the conducting layer/silicon substrate structure obeys the explosion law.

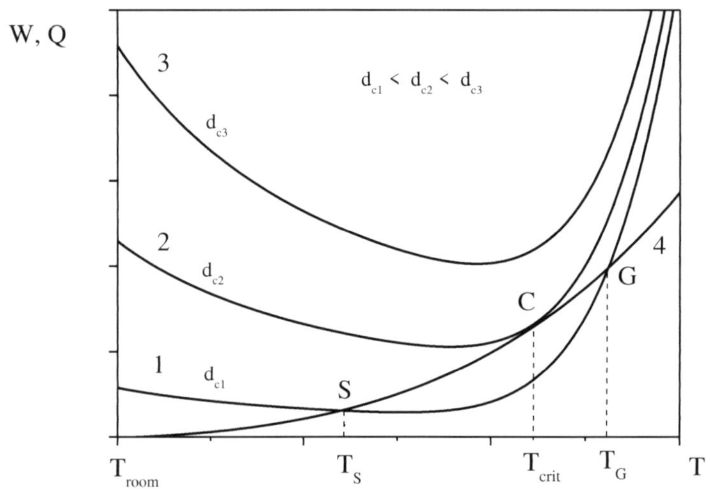

Fig. 2. Heat generation W and heat transfer Q temperature diagrams. Heat generation for different thicknesses of graphene layer is shown by curves 1, 2 and 3, whereas curve 4 represents the heat transfer.

The main result of the qualitative analysis is: there are two essentially different heating regimes of the structure: a) the stationary regime with a constant temperature; b) the non-stationary heat explosion regime with an unlimited increase of the structure temperature.

The conditions for realization of the regimes can be represented as [2, 3]:

a) the stationary regime with a constant temperature is achieved at

$$R_\square/(H\omega L)^2 > A, \qquad (4)$$

b) the heat explosion regime is observed at

$$R_\square/(H\omega L)^2 < A, \qquad (5)$$

where $R_\square = \rho_c/d_c$ is the sheet resistance of a conducting layer.
Here A is constant that can be calculated from the parameters of the structure [3]:

$$A = \mu_0\gamma_{cr}K\beta_0/(\sigma\Theta^4) \qquad (6)$$

where $\Theta = k_B T_1^2/E_a$; $\beta_0 = T_{room}/\Theta$, $\beta_1 = T_1/\Theta$; $\gamma_{cr} \cong (1+\beta_1)/[\beta_1^2(\beta_1^4 - 4\beta_1^3 - \beta_0^4)]$. Here the characteristic temperature T_1 is determined as a temperature at which the loss power in the conducting layer and the substrate are equaled $d_{Si}/\rho_{Si}(T_1) = d_c/\rho_c\,(T_1)$. Activation energy is determined by the expression $\rho_{Si} = \rho_{Si0}\exp(+E_a/k_B T)$. A detailed description of the criterions Eqs. (4) – (6) is presented in [3].

The obtained results can be applied to analysis of thermal state of structure with any conducting layer. In the next section the theoretical prediction will be compared with experimental results with Ta-Ti layer, and then we will discuss the explosion regime for monolayer graphene/silicon structures.

4. Experiment with Ta-Ti film/Si substrate structure

Experimental realization of the explosion regime was performed using the Ta – Ti film/Si substrate structure [2]. Substituting our experimental parameters, for which the heat explosion was realized H = 12000 A/m, f = 13.5 MHz, L = 0.005 m, $R_{\square}$ = 12.2 – 23.4 Ω/□ , in to (6) we obtain

$$A_{theor} = 18\mathrm{x}10^{-19}\ (\Omega/\square)(\mathrm{sec/A})^2.$$

This value is in reasonable agreement with the critical parameter estimated from the experimental data

$$A_{exper} = (4.7\text{-}9.0)\mathrm{x}10^{-19}\ (\Omega/\square)(\mathrm{sec/A})^2$$

In [3] it was shown that the heating in the explosion regime can be separated in two stages. In the first stage the structure is slow heated, while in the second stage the sample is fast heated, theoretically up to an unlimited value. The theoretical predictions are well confirmed by experiment with the Ta-Ti film/silicon substrate structures, parameter $R_{\square}/(H\omega L)^2$ of which is less than the critical value (Fig. 3). The detailed description of the experiments is presented in [2].

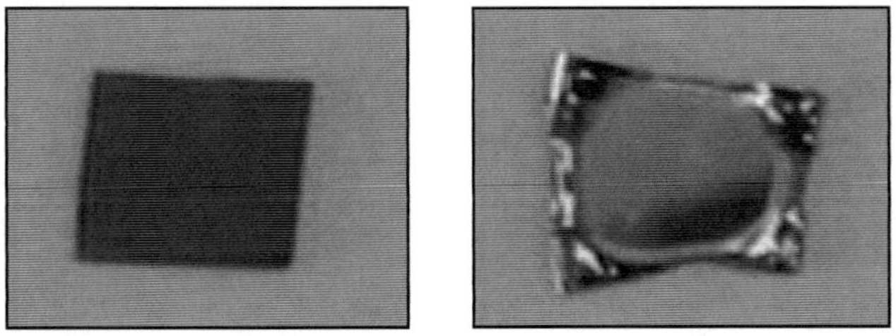

a) b)

Fig. 3. Ta-Ti film/ Si substrate structures after RF heating in the heat explosion regime, parameter $R_{\square}/(H\omega L)^2$ less than the critical value: a) no melting; magnetic field was switched off before a fast unlimited heating; b) melting at substrate edges; to prevent the sample destruction the field was switched off in the initial stage of the fast unlimited heating [2].

5. Evaluation of heat explosion in nano-graphene layer/Si substrate structure

The sheet resistance values of two monolayer graphene films were measured as 26.97 Ω/□ and 61.7 Ω/□ [5]. These experimental values are close to the sheet resistance of $R_{\square}$ = 12.2 – 23.4 Ω/□ in our experiments which confirmed the explosion regime. Note that variation of the frequency or/and amplitude of the RF field, or structure sizes allows one to achieve the heat explosion regime for conducting layer with the sheet resistance varying in a wide range. This points to a real opportunity to realize the heat explosion regime for a graphene layer/silicon substrate structure.

6. Summary

The RF heating of the conducting layer/silicon substrate structure was analysed. It is shown that two regimes of the RF heating of these structures can be realized: the first one is characterized by heating of the structure up to a finite temperature; the second regime corresponds to a fast unlimited temperature increase (heat explosion). The criterions of realization of these regimes are presented in analytical form. Using these criterions and the experimental data for the sheet resistance of the monolayer graphene it was shown the possibility of the heat explosion regime for graphene layer/silicon substrate structures at RF heating.

References

[1] Sinder M, Pelleg J, Meerovich V and Sokolovsky V Mat. Sci. and Eng. A 2001 **302**, 31
[2] Pelleg J, Rosenberg S and Sinder M Acta Mater. 2011 **59** 4283
[3] Sinder M, Pelleg J, Meerovich V and Sokolovsky V 2013 *PIERS Proceedings* (Stockholm) p 1292
[4] Zeldovich Ya B, Barenblatt G I, Librovich V B, Makhviladze G M 1985 *Mathematical theory of combustion and explosion* (Consultants Bureau, New York)
[5] Hao L, Gallop J, Goniszewski S, Klein N Appl. Phys. Lett. 2013 **103** 123103

IOP Conf. Series: Journal of Physics: Conf. Series **987** (2018) 012016 doi:10.1088/1742-6596/987/1/012016

Electrical properties of thin films deposited from TMS/O_2 in Microwave Multipolar Plasma reactor

M Kihel[1], S Sahli[1], Y Fermi[1], P Raynaud[2], M Benhaddad[3]

[1]Université Frères Mentouri de Constantine, Laboratoire Microsystèmes et Instrumentation. Faculté sciences de la technologie, Route de Ain El Bey, Constantine 25017, Algérie
[2]Université Paul Sabatier, LAPLACE, CNRS, 118 Route de Narbonne, 31062 Toulouse cedex ¯ France
[3]Université Frères Mentouri de Constantine, Département de physique, Route de Ain El Bey, Constantine 25017, Algérie

Abstract. Thin films have been deposited from pure Tetramethylsilane and a mixture of Tetramethylsilane and oxygen (TMS/O_2). The addition of oxygen proportion to Tetramethylsilane vapors leads to the change in film structure which varies from organic to inorganic character close to SiOx-like film [1]. The electrical characterisation using Metal-Insulating-Metal structure permits the study of current-voltage (I (V)) curves behaviours. The results suggest that the carrier transport in the deposited films is limited by a space charge conduction mechanism.

1. Introduction

Dielectric layers deposited by PECVD technique have been widely employed in the design of ultra large scale integration (ULSI) devices as insulator between metal interlayers [2]. Reduction of dimensions in ULSI technology requires insulator materials with high quality such as: a low dielectric constant [2] and weak electrical current. SiO_x films remains the most insulator material used in microelectronic technology. Moreover, a lot of works have been carried out to deposit SiO_x-like films at low temperature using a large variety of precursors such as organosilicon compounds [3-4]. The electrical properties of SiO_x-like films elaborated from organosilicon compounds mixed with high proportion of oxygen (O_2, N_2O) [5] have been found depending on the deposited parameters [6-9]. The effect of the films structure on the electrical properties according to the deposited conditions needs further investigation.

In this work, we have reported the electrical properties evolution of thin films deposited from pure vapor of Tetramethylsilane (TMS) and TMS/oxygen mixture.

2. Experimental details

The thin films were elaborated at ambient temperature in a distributed electronic cyclotron resonance (RCER) plasma reactor of cylindrical geometry. This reactor is equipped with eight magnets arranged alternately with eight antennas cooled by a water circulation. It has been described elsewhere in a several reported works [10][11]. The microwave plasma frequency excitation was 2.45 GHz and the working power explored in this work has been set to 400 W. The eight antennas radiate the same power value transmitted through a waveguide equipped with a high impedance adapter and a power splitter. The films thickness varying in the range of 563-89,6 nm, were deposited from pure TMS vapors and mixed with oxygen proportions on metalized glass substrates. MIM (Metal-Insulator-Metal) structures were obtained by deposing aluminum electrodes on these films using vacuum evaporation technique. Electrical measurements were carried out using a Keithley 6512 electrometer.

3. Results and discussion

The evolution of the charging current as function of the polarization time is shown in Figure 1. Two regimes are observed: the electric current decreases rapidly for short times (transient current regime) and then slowly until reaching a quasi-constant value when the time polarization is important (permanent regime). It˘s well known that the first regime is due to a polarization effect and the second regime is due to a conduction mechanism which may be controlled by an interface effect and/or by bulk effect.

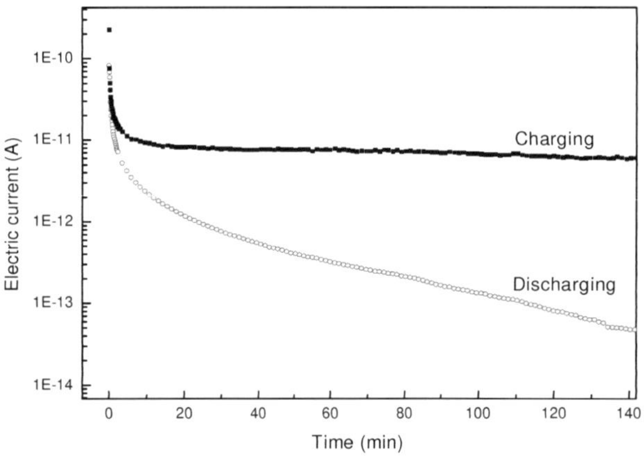

Figure 1. Variation of the charging and discharging current for an applied electric field of about $3.11*10^4$ V/cm as a function of the polarization time for films deposited from plasma created in 1 mtorr of TMS vapours and discharge power of 400 watts.

The discharge current represented in figure 1 takes place when the applied electric field is disconnected and the structure MIM electrodes was immediately short-circuited. It can be seen that the charging and the discharging currents are superposition in case of short polarization times. Beyond measuring time of about 20 sec charging and discharging curves start to diverges rapidly (They shift one from the other by two decades for a polarization time of about 3 hours). The current superposition observed for the first regime suggests a presence of dipoles in the films [12]. However, for large applied electric field values, it has been observed that the charging and discharging currents are very distinct even for short times (Figure 2).

Figure 3 shows the evolution of the quasi steady electrical current as a function of the applied electric field square root. This curve shows a nonlinear variation making weak the probability of the presence in the deposited films of Schottky and/or Pool˘Frenkel charge transport mechanisms [13,12].

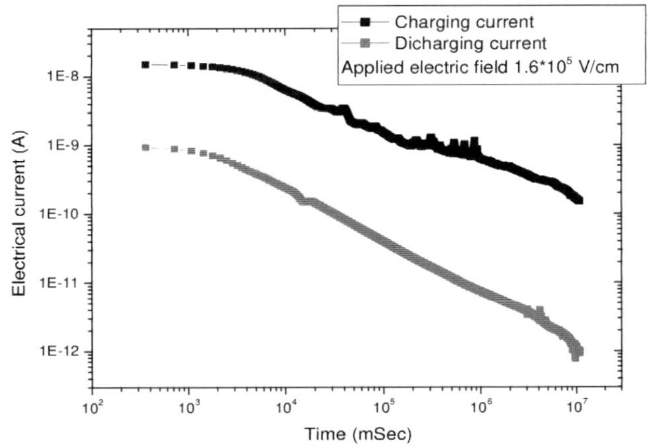

Figure 2. . Charging and discharging current evolution as a function of the polarization time for a film deposited from pure TMS vapors.

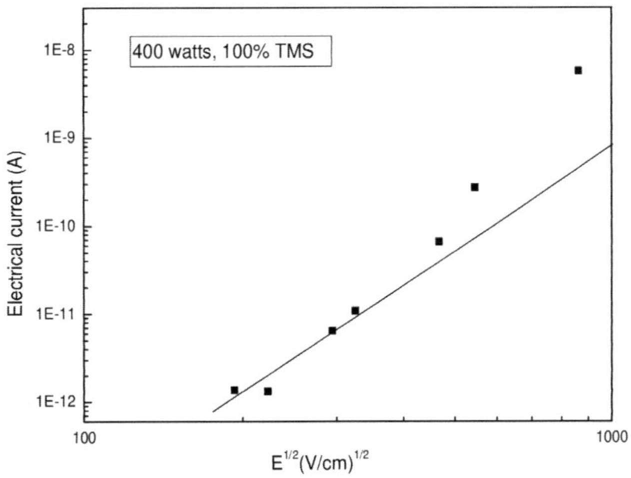

Figure 3. Variation of the electric current as a function of the square root of the applied electric field.

However, as it is shown in figure 4, the electrical current varies with a slope value close to 1 A.cm/V for low values of the applied electric field and with a slope varying from 2 to 3 A.cm/V for the medium and high applied electric field, respectively, suggesting by the way the existence of a space charge conduction mechanism effect in the elaborated films. Indeed, the presence of such conduction mechanism is characterized by an I(V) bilogharitmic curve with a linear variation of the current for low applied voltage values which can be described by the equation 1 (ohmic conduction) :

$$J = qn\mu\frac{V}{d} \tag{1}$$

where q is the electron charge, n the carrier density, μ the mobility of the electrons in the films, d the films thickness and V the applied voltage. This ohmic conduction is followed by a quadratic variation of the current given by the equation 2 [14,15].

$$J = \frac{9}{8}\mu\varepsilon_0\varepsilon_r\frac{V^2}{d^3} \tag{2}$$

with ε_0 is the vacuum permittivity ($8,85.10^{-12}$ F/m) and ε_r the films dielectric constant.

On the other hand, it is important to notice that the oxygen introduction in the precursor mixture induces a decrease of the recorded current (reaching more than a decade when the oxygen rate in the precursor is about 90%). The oxygen addition to the TMS vapors improves the electrical properties of the films probably by deceasing defects trapped in the film structure. FTIR results presented elsewhere show that the films produced by plasma using only TMS vapors consist mainly of -CH_3 and Si-CH_3 groups [1] and the oxygen addition to this vapors leads to the deposition of less organic films, quasi-inorganic when the oxygen ratio in the mixture precursor is over 90% [1].

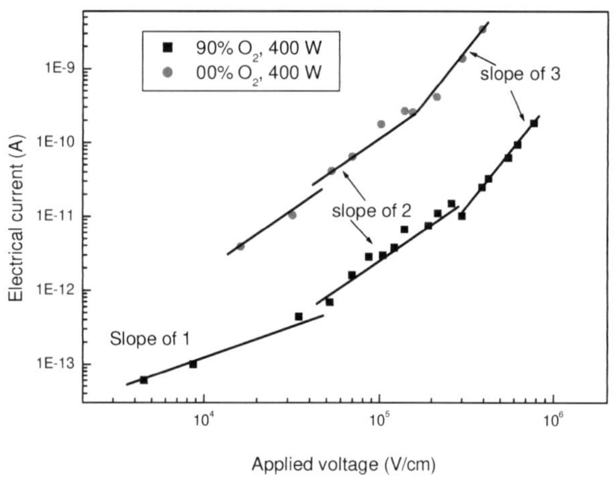

Figure 4. Evolution of the isochrones currents as a function of the applied electric field in films produced in a plasma of 400 watts created in a mixture of 10%TMS and 90% O_2.

4. Conclusion

In this study two kinds of films have been deposited, organic films elaborated using only TMS vapors (a-Si:CH) and inorganic films (SiO_x like films) using a high oxygen ratio in the TMS/O_2 mixture [1]. The electrical characterization results suggested that conduction is dipolar (for transient regime) and limited by space charge without traps (for permanent regime). The addition of high oxygen proportion could improve the electrical properties by reducing probably defects in the film. In this case the deposited films have inorganic structure close to the thermal oxide (SiO_x like films). Although the recorded I(V) characteristic is consistent with the presence of a space charge mechanism in the films. However to confirm this result, the explored voltage range must be raised at least three decades. Moreover, it will be interesting to carry out a variation study of the electric current as a function of the films thickness.

References

[1] M. Kihel, S. Sahli, A. Zenasni, P. Raynaud, Y. Segui, 2014 *Vacuum* **107** 264
[2] Yi-Lung Cheng, Chiao-Wei Haung,Wei-Jie Hung, Chung-Ren Sun,Wen-Hsi Lee, 2016 *Sur. & Coat. Tech.* **308** 182
[3] Lu Huang, Qi Jin, Xingling Qu, Jing Jin,Weiguang Yang, LinjunWang, Weimin Shi, 2017 *Sur. & Coat. Tech.* **320** 82
[4] J. Schäfer, J. Hnilica, J. πperka, A. Quade, V. Kudrle, R. Foest, J. Vodák, L. Zajʃπková, 2016

Sur. & Coat. Tech. **295** 112

[5] Xing ling Qu, Jing Jin, Qi Jin, Lu Huang, Wenqing Zhu, Weimin Shi, 2017 *Mat. Sc. in Semiconductor Processing* **71** 54

[6] Y. Nishi, T. Funai, H. Izawa, T. Fujimoto, H. Morimoto and M Ishii. 1992 *Jpn. J. Appl. Phys.* **31** 4570

[7] A. Zenasni, P. Raynaud, S. Sahli, S. Rebiai, Y. Segui. 2003 *Sur. & Coat. Tech.* **162** 174 -175

[8] Y. Inoue and O. Takai. 1996 *Plasma Sources Sci. Technol.* **5** 339¯343

[9] A. Grill and V. Patel. 1999 *J. Appl. Phycs.* **85, (6)** 3314

[10] A. Bellel, S. Sahli, Z. Ziari, P. Raynaud, Y. Segui, D. Escaich, 2006 *Surf & Coatings Technol*;**201** 129

[11] J. Pelletier, Y. Arnal, B. Petit, C. Pomot, M. Pichot, 1986 *J. Phys. D: App. Phys.* **19/5** 795

[12] M. Lamhamdi, J. Guastavino, L. Boudou, Y. Segui, P. Pons, L. Bouscayrol and R. Plana, 2006 *Microelect. Rel* **46** 1700

[13] D. K. Gupta, K. Doughtly and R. S. Brockley, 1980 *J. Phys. D: Appl. Phys.* **13 (11)** 2101

[14] T.C. Chang, S.T. Yan , P.T. Liu , Z.W. Lin , H. Aoki, S.M. Sze, 2004 *Th. Sol Films.* **516** 447-448

[15] J. J. O`Dwyer, `The theory of electrical conduction and breakdown in solid dielectrics_, 1973 pp 105-135 Clearendon press, Oxford

IOP Conf. Series: Journal of Physics: Conf. Series **987** (2018) 012017 doi:10.1088/1742-6596/987/1/012017

A built-in sensor with carbon nanotubes coated by Ag clusters for deformation monitoring of glass fibre/epoxy composites

P. Slobodian[1],*, P. Riha[2], J. Matyas[1], R. Olejnik[1], S. Lloret Pertegás[3], R. Schledjewski[3] M. Kovar[1]

[1]Centre of Polymer Systems, University Institute, Tomas Bata University, Trida T. Bati 5678, 760 01 Zlin, Czech Republic
[2]Institute of Hydrodynamics, Academy of Sciences, Pod Patankou 5, 166 12 Prague 6, Czech Republic
[3]Christian Doppler Laboratory for High Efficient Composite processing, Department of Polymer Engineering and Science, Montanuniversitaet Leoben, Otto Gloeckel-Strasse 2, A-8700 Leoben, Austria

*Corresponding author: slobodian@utb.cz (P. Slobodian)

Abstract. A multiwalled carbon nanotube network embedded in a polyurethane membrane was integrated into a glass fibre reinforced epoxy composite by means of vacuum infusion to become a part of the composite and has been serving for a strain self-sensing functionality. Besides the pristine nanotubes also nanotubes with Ag nanoparticles attached to their surfaces were used to increase strain sensing. Moreover, the design of the carbon nanotube/polyurethane sensor allowed formation of network micro-sized cracks which increased its reversible electrical resistance resulted in an enhancement of strain sensing. The resistance sensitivity, quantified by a gauge factor, increased more than hundredfold in case of a pre-strained sensor with Ag decorated nanotubes in comparison with the sensor with pristine nanotubes.

1. Introduction

Strain sensing composite materials have attracted considerable attention for their unique characteristics exceeding conventionally applied sensors. Between different solutions and various types of transducers available for these applications, piezo-resistive strain sensors are among the most investigated ones [1-6]. They are usually based on conductive polymer composites (CPCs) prepared by embedding of electrically conductive fillers as carbon nanotubes (CNTs) into a polymeric matrix. Here the carbon nanotube films are usually deposited on thicker flexible polymeric substrate which acts as a mechanical support and transfer mechanical stimulus to carbon strain sensitive layer [1-6]. The electric resistance response of a CNT film/polymer composite to extension is determined by the local resistance in nanotube contacts and the straightening and buckling of nanotubes, which affects the number of contacts between nanotubes [1]. It was found that the resistance change at deformation above approx. 1% is due to the crack formation in carbon nanotube layer [11-13]. To modify the nanotube contacts itself in the course of strain sensitivity increase, different chemical functionalization

IOP Conf. Series: Journal of Physics: Conf. Series **987** (2018) 012017 doi:10.1088/1742-6596/987/1/012017

or plasma oxidation of carbon nanotubes are used [5,7]. Moreover, decorating CNTs with Ag nanoparticles enhance also the strain sensing [8-10]. Finally, there were published results that show that increase in the sensitivity of CNT film/polymer composite to applied strain can be significantly increased by composite pre-straining method [11-13].
In the present paper the combination of MWCNTs functionalization by Ag nanoparticles (AgNPs) together with a controlled pre-straining are used to prepare a strain sensor with enhanced sensitivity to applied strain. It composes of electrically conductive entangled network of Ag-decorated multiwalled carbon nanotubes supported by polyurethane filtering membrane integrated into a glass fibre/epoxy resin composite by vacuum infusion technique.

2. Experimental methods

Materials
Purified MWCNTs produced by chemical vapour deposition of acetylene were supplied by Sun Nanotech Co. Ltd., China. According to the supplier, the nanotube diameter is 10-30 nm, the length 1-10 μm, the purity ~ 90% and the volume resistivity 0.12 Ωcm. MWCNTs decorated by Ag clusters were prepared by sonication of nanotube dispersion in Bandelin thermostatic ultrasonic bath for 30 min. Then 80 ml of 0.1 M silver nitrate ($AgNO_3$) and 20 ml of 1 wt. % of sodium salt of citric acid was added to dispersion and boiled in a glass reactor connected with the reflux condenser for 1 hour. Polyurethane (TPU), Desmopan 385S was used for fabrication of filtering membrane by technology of electrospinning supplied by Bayer MaterialScience (ultimate tensile strength 40 MPa and strain at break 450 %). As an epoxy matrix EPIKOTETM as resin and EPIKURETM RIMH 1366 as curing agents mixed in weight ratio 100:30 were used. Woven glass textiles were supplied by Lange+Ritter (03G080LT.VR080). The textile had a plain weave structure and weight 80g/m^2.

Polyurethane/MWCNT sensor integrated into glass fibre/epoxy composite
Aqueous dispersion of MWCNTs/AgNPs was prepared by sonication in an apparatus UZ Sonopuls HD 2070 for 15 minutes at room temperature. The nanotube concentration in the suspension was 0.3 wt.%. Dispersion contained also surfactants, namely sodium dodecyl sulfate and 1-pentanol with concentration 0.1M and 0.14M, respectively. Moreover, NaOH aqueous solution was added to adjust pH to 10. For making an entangled MWCNTs/AgNPs network, a porous polyurethane membrane and a vacuum filtration method was used. 30 ml of homogenized dispersion was filtered through the funnel of diameter 90 mm. Polyurethane filtering membrane was prepared by technology of electrospinning of TPU solution in dimethyl formamide/methyl isobutyl ketone solution, 3:1. This semi product was integrated by epoxy vacuum infiltration according schematic illustration in Fig. 1., and cured during 24 hours at room temperature. The size of tested MWCNTs layers integrated into epoxy composites was approx. 50x10 mm. Cu wires were fixed to the sensor by a silver-colloid electro-conductive paint Dotite D-550 (SPI Supplies) and dried before vacuum infusion process.

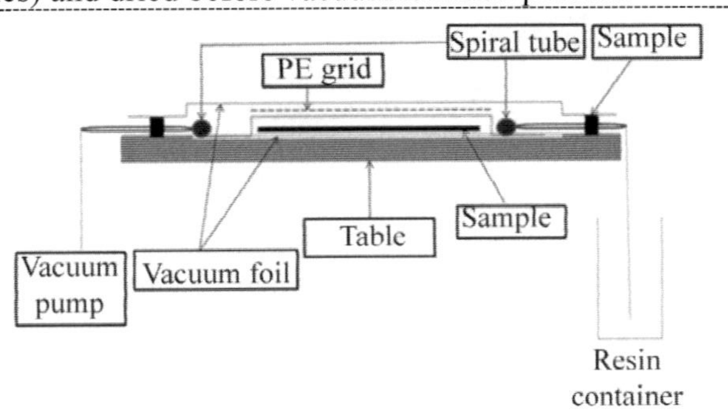

Fig. 1 The schematic illustration of fabrication process of epoxy/glass fibre composite with integrated MWCNT/TPU strain sensitive layer.

Tests and measurements

The morphology of the MWCNTs and MWCNTs decorated with Ag nanoparticles were investigated by means of TEM analysis using the microscope JEOL JEM 2010 at the accelerating voltage of 160 kV. The structure of MWCNT films was analysed by the scanning electron microscope (SEM) NOVA NanoSEM 450. Elongation and relaxation of a glass fibre/epoxy composite with integrated strain sensor was measured during the composite bending over a cylindrical surface. Seven different cylindrical surfaces with diameters from 70 to 315 mm were used. The pre-strained composites were prepared by bending around cylindrical surface of diameter 37 mm. The achieved strain ε was calculated according to Euler-Bernoulli beam bending theory, $\varepsilon = y/\rho$, where y denotes the sensor distance from neutral axis and ρ denotes radius of neutral axis, according schematics illustration in Fig. 2.

Detection ability of applied strain by integrated MWCNT layer was tested by its resistance change with deformation. The sensor resistance was monitored during bending with help of the two-point technique measured lengthwise using Multiplex datalogger 34980A with storage of data once per second.

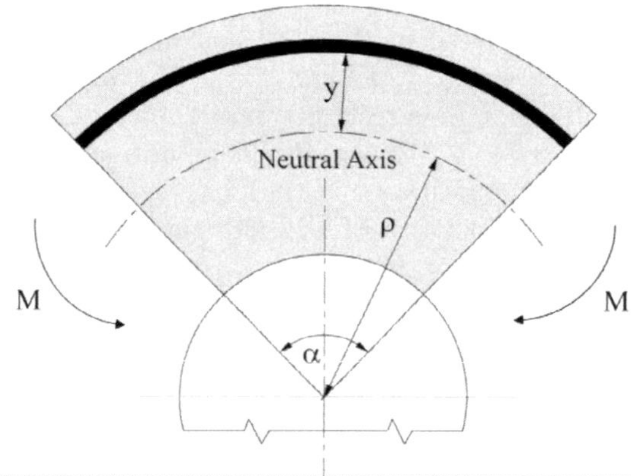

Fig. 2 The schematic of bending of glass fibre/epoxy composite over a cylindrical surface.

3. Results and Discussion

High-resolution transmission electron microscopy (HRTEM) micrographs in Fig. 3a show the structure of one individual MWCNT which consist of about 15 rolled layers of graphene. Fig. 3b is TEM analysis of MWCNT with deposited Ag clusters. The size of Ag clusters on MWCNT surface was in the range ~3-10 nm.

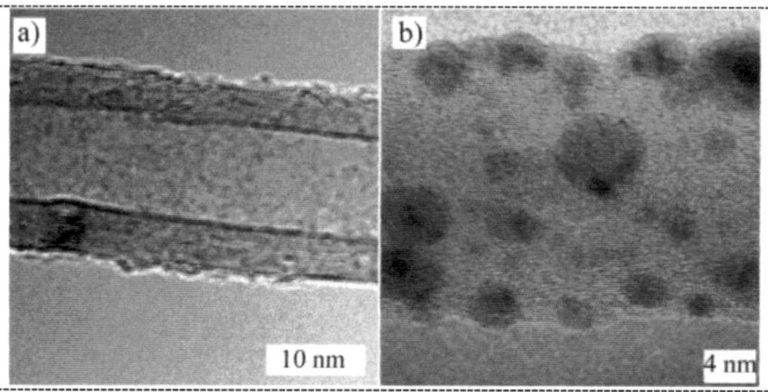

Fig. 3 a) HRTEM micrograph of the structure of individual nanotube; b) TEM micrographs of individual nanotube with deposited Ag clusters.

IOP Conf. Series: Journal of Physics: Conf. Series **987** (2018) 012017 doi:10.1088/1742-6596/987/1/012017

The scanning electron microscope (SEM) analysis of PU non-woven filtering membrane is presented in Fig. 4 a). Individual nanotubes are entrapped by filtering membrane in the course of MWCNTs aqueous dispersion vacuum filtration. The nanotubes infiltrate into membrane pores which are finally blocked and a filtering cake made of pure network of entangled MWCNTs is created. SEM micrograph in Fig. 2 b) shows the upper surface of MWCNTs network and Fig. 2 b) of network from MWCNTs/AgNPs. Both networks are porous and electrically conductive structures created by entangled nanotubes with electrical conductive junctions between individual tubes.

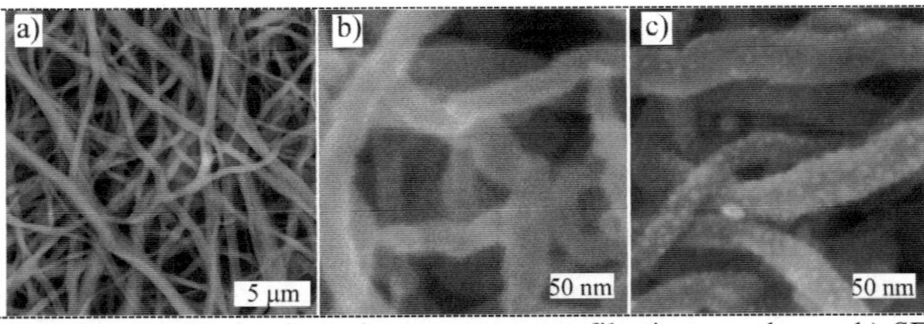

Fig. 4 a) SEM micrograph of polyurethane non-woven filtering membrane, b) SEM image of the upper surface of entangled MWCNTs and MWCNTs/AgNPs networks.

Final epoxy composite with integrated MWCNTs/PU composite sensor consisted of six layers of glass fibre textile, then a stripe of MWCNTs/PU composite sensor and finally one covering layer of glass textile, Fig. 5a.This configuration assured that the sensor was placed above the neutral axis of the glass fibre/epoxy composite. Hence, when the glass fibre/epoxy composite was bent, the sensor underwent an extension and it measured a corresponding electrical resistance change of a nanotube film.

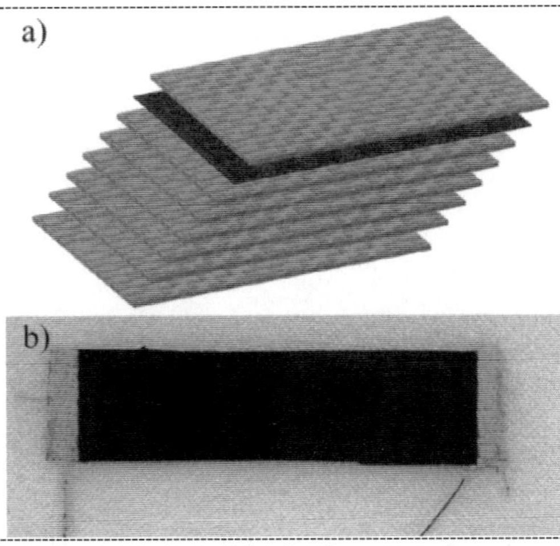

Fig. 5 The glass fibre/epoxy composite with the integrated sensor consisted of six layers of glass fibre textile, a strip MWCNT/PU sensor and finally one covering layer of glass textile; b) The inbuilt MWCNT/PU sensor with attached silver electrodes and Cu wiring.

The resistance change of the sensor made of pristine or Ag decorated MWCNTs integrated into the glass fibre/epoxy composite was tested in the course of the composite elongation by bending over the rigid cylinders of different diameters. The maximum strain of the sensor located above the neutral axis and calculated in the middle of MWCNTs layer was 0.48 %. The results of five cycles of stepwise elongation/relaxation of the composite are shown in Fig. 6 for not pre-strained and pre-strained samples. Part a) the results for the sensor with pristine nanotubes show firstly a residual strain after the sample relaxation which was not observed in tests of the integrated sensor with MWCNTs/AgNPs.

Also in comparison with data in Parts b) the resistance change is relatively small. On the other hand, the sensor with Ag decorated nanotubes did not have any residual strain and the resistance change in comparison with the sensor with pristine nanotubes substantially increased which is favourable for the effective measurement of deformation of the glass fibre/epoxy composite. A fundamental parameter of the strain sensor is its sensitivity to strain expressed usually by means of the gauge factor (GF). GF is the ratio of the fractional change in electrical resistance to the fractional change in length, or strain ε, $GF=(\Delta R/R_0)/(\Delta L/L)=\Delta R/R/\varepsilon$, where $(\Delta R/R_0)=(R-R_0)/R_0$. R_0 represents initial resistance (undeformed sensor) and R in deformed state. The sensor sensitivity to strain measured by the gauge factor was as high as 500 for the sensor with Ag decorated MWCNTs while metallic foil sensors usually have a gauge factor slightly over 2.

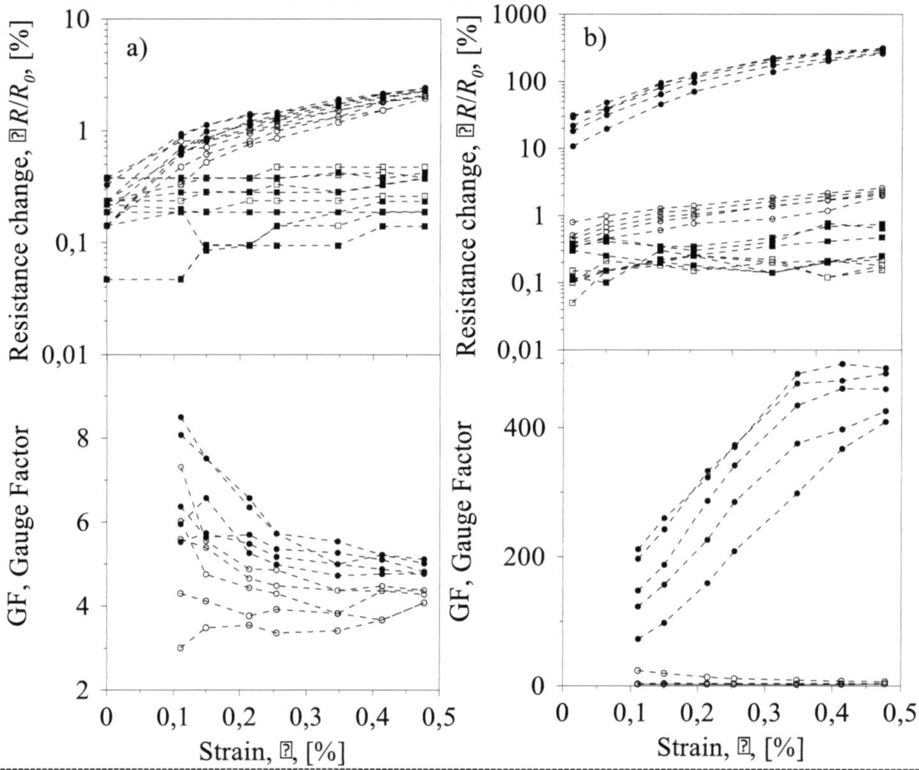

Fig. 6 The glass fibre/epoxy composites with the integrated sensors subjected to five elongation/relaxation cycles. The resistance change as well as the gauge factor corresponding to respective resistance data in the course of elongation and relaxation is denoted by circles and squares, respectively. The results for not pre-strained sensor are denoted by open symbols and the results for the pre-strained sample by 0.94 % strain are denoted by filled symbols. a) The sensor with pristine MWCNTs, b) the sensor with MWCNTs/AgNPs.

4. Conclusion

MWCNTs layers embedded in epoxy composites reinforced by glass fibres are proposed as a novel material that can be used for composite itself deformation detection. Further, it has been proven that modification of pure MWCNT with Ag clusters together with application of well-defined prestrain stimulation significantly increase composite sensitivity to applied strain defined by GF. Sensitivity for sensor with pure MWCNTs reaches GF values 6-8. When comparing with GF for composite with MWCNTs/AgNPs layer coming up to 500. This is a substantial increase ranks the polyurethane sensor integrated into the glass fibre/epoxy composite among strain gauges with the highest electromechanical sensitivity.

Acknowledgments
This work was supported by the Ministry of Education, Youth and Sports of the Czech Republic, under the project Czech-Austrian mobility (program KONTAKT II) project -7AMB16AT033 and by the OeAD, the Austrian Agency for International Mobility and Cooperation in Education Research in collaboration with the Thomas Bata University in Zlín. This work was also supported by the Ministry of Education, Youth and Sports of the Czech Republic – Program NPU I (LO1504) and with the support of Operational Program Research and Development for Innovations co-funded by the European Regional Development Fund (ERDF) and national budget of the Czech Republic, within the framework of the project CPS-strengthening research capacity (reg. number: CZ.1.05/2.1.00/19.0409). This project was supported by the internal grant of TBU in Zlin No. IGA/CPS/2017/002 funded from the resources of the Specific University Research. P. R. would like to acknowledge financial support from the Fund of the Institute of Hydrodynamics AV0Z20600510.

References
[1] M.D. Rein, O. Breuer, H.D. Wagner. Sensors and sensitivity: Carbon nanotube buckypaper films as strain sensing devices. Compos Sci Technol 2011;71:373-381.
[2] Y. Kim, Y. Kim, C. Lee, S Kwon. Thin Polysilicon Gauge for Strain Measurement of Structural Elements. IEEE Sens. J. 2010, 10, 1320–1327.
[3] T. Yamada, Y. Hayamizu, Y. Yamamoto, Y. Yomogida, A. Izadi-Najafabadi, D. Futaba, K. Hata. A stretchable carbon nanotube strain sensor for human-motion detection. Nat. Nanotechnol 6, 2011, 296–301.
[4] P. Slobodian, P. Riha, P. Saha. A highly-deformable composite composed of an entangled network of electrically-conductive carbon-nanotubes embedded in elastic polyurethane. Carbon 2012;50:3446-3453.
[5] P. Slobodian, P. Riha, R. Olejnik, U. Cvelbar, P. Saha. Enhancing effect of KMnO4 oxidation of carbon nanotubes network embedded in elastic polyurethane on overall electro-mechanical properties of composite. Compos. Sci. Technol. 2013;81:54–60.
[6] M. Hin, J. Oh, M. Lima, M. Kozlov, S. Kim, R. Baughman. Elastomeric conductive composites based on carbon nanotube forests. Adv. Mater. 2010, 22, 2663–2667.
[7] R. Benlikaya, P. Slobodian, P. Riha, Enhanced strain-dependent electrical resistance of polyurethane composites with embedded oxidized multiwalled carbon nanotube networks, J. Nanomater. (2013) Art. No. 327597.
[8] S.J. Zhang, G. Yao, F.Y. Liao, M, Z.L. Huang, K. Y. Li, Y. Lin, Highly stretchable, sensitive, and flexible strain sensors based on silver nanoparticles/carbon nanotubes composites, J. Alloy. Compd. 652 (2015) 48-54.
[9] H.S. Luo, X.D. Zhoua, Y.Y. Ma, G.B. Yi, X.L. Cheng, Y. Zhu, X.H. Zu, N.J. Zhang, B.H. Huang, L.F. Yu, Shape memory-based tunable resistivity of polymer composites, Appl. Surf. Sci. 363 (2016) 59-65.
[10] F. Xin, L. Li, Decoration of carbon nanotubes with silver nanoparticles for advanced CNT/polymer nanocomposites, Composites: Part A 42 (2011) 961–967.
[11] P. Slobodian, P. Riha, R. Olejnik, J. Matyas, M. Machovsky, Pre-strain stimulation of electro-mechanical sensitivity of carbon nanotube network/polyurethane composites, IEEE Sens. J. 16 (2016), 5898-5903.
[12] P. T. Hoang, N. Salazar, T. N. Porkka, K. Joshi, T. Liu, T. J. Dickens, Z. Yu, Engineering crack formation in carbon nanotube-silver nanoparticle composite films for sensitive and durable piezoresistive sensors, Nanoscale Res. Lett. 11 (2016) 422-429.
[13] Y. W. Choi, D. Kang, P. V. Pikhitsa, T. Lee, S.M. Kim, G. Lee, D. Tahk, M. Choi, Ultra-sensitive Pressure sensor based on guided straight mechanical cracks, Sci. Rep. 7 (2017) 40116; doi: 10.1038/srep40116 (2017).

IOP Conf. Series: Journal of Physics: Conf. Series **987** (2018) 012018 doi:10.1088/1742-6596/987/1/012018

Fabrication and characterization of Aerogel-Polydimethyl siloxane (PDMS) Insulation Film

Yeoung ah Noh[1], Sinae Song[1], and Hee Taik Kim[1,2*]

[1] Department of Advanced Materials Science and Engineering, Hanyang University, 55 Hanyangdeahak-ro, 15588, Korea l

[2] Department of Fusion Chemical Engineering, Hanyang University, 55 Hanyangdeahak-ro, 15588, Korea

khtaik@hanyang.ac.kr

Abstract. The building has a large impact on the space heating demand and the indoor environment is affected by climate or daylight. Hence, silica aerogel has generally used as a film to reduce the coefficient of the window in the building. Silica aerogel is a suitable material to apply for insulation material with lower thermal conductivity than that of air to save interior energy. However expensive precursor and drying process were the main issue of the silica aerogel synthesis and practical usage. We attempt to fabricate aerogel insulation film for energy saving through the economic process under ambient pressure. Silica aerogel was synthesized from rice husk ash, which was an agricultural waste to be able to recycle. Taguchi design was used to optimize the parameters (amount of rice husk ash, pH, aging time) controlling the surface area of silica aerogel. The silica aerogel is prepared by sol-gel processing through acidic treatment and aging. The silica aerogel was obtained by modification of silica hydrogel surface and dry at ambient pressure. Finally, aerogel film was respectively fabricated by the different content of aerogel in polydimethylsiloxane (PDMS). Silica aerogel obtained 21 – 24nm average particle size was analyzed by SEM and silica aerogel with high surface area (832.26 m^2/g), pore size (3.30nm) was characterized by BET. Then silica Aerogel – PDMS insulation film with thermal conductivity (0.002 W/mK) was analyzed by thermal wave system. The study demonstrates an eco-friendly and low-cost route toward silica – PDMS insulation film with low thermal conductivity (0.002 W/mK).

1. Introduction

The Building window is affected by the demand for space heating energy. The outside temperature is directly transmitted through the window. If this phenomenon persists, it is difficult to maintain the inside temperature. Therefore, energy is continuously consumed to maintain the indoor temperature. Because of this phenomenon, Insulation is needed to reduce the demand for such heating. Typical materials for insulation include glass wool(0.028-0.043W/mK), mineral wool(0.033-0.038W/mK), silica(0.030-0.004W/mK), and perlite(0.030-0.101W/mK). Therefore, silica aerogel is a suitable material to apply for insulation to save interior energy[1]. Silica aerogels are porous materials consisting more than 90% air and 10% solid silica with a cross-linked network structure. Silica aerogel have a large surface area, high porosity, low bulk density, low thermal conductivity. However expensive precursor and drying process were the main issue of the silica aerogel synthesis. In this study, we synthesized the silica aerogel from rice husk ash, which was an agricultural waste to be able to recycle using the Taguchi method [2]. Taguchi method is applied in determining to optimize the parameters controlling the surface area of silica aerogel on the fewer experimental results than those required to study the effect of overall input variables [3]. The silica aerogel is prepared by sol-gel processing through acidic treatment and aging. The silica aerogel was obtained by modification of silica hydrogel surface and dry at ambient pressure. To fabricate Aerogel Polydimethylsiloxane (PDMS) Insulation Film, prepared silica aerogel is combined Polydimethylsiloxane (PDMS). Then, we have investigated the specific surface area of silica aerogels through Brunauer-Emmett-Teller and analyzed the thermal conductivity of the film using thermal wave system.

2. Materials and methods

2.1 Materials

The rice husk used by raw material was obtained in Geoje, Korea. Ethyl alcohol 99.9%, Sodium hydroxide beads 97% and n-Hexane 95% were purchased from DAEJUNG. Hydrochloric acid 36% and Sylgard 184A (Polydimethylsiloxane), Sylgard 184B were prepared by SAMCHUN Sewang hitech silicone respectively.

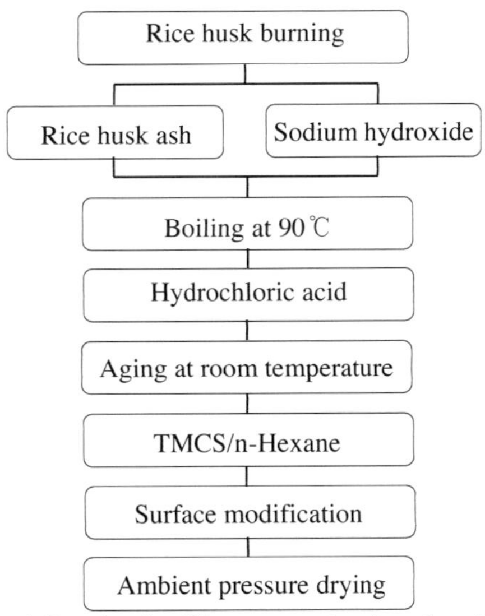

Scheme 1 Synthesis silica aerogel using the rice husk ash

2.2 Preparation of silica aerogel using the rice husk ash

Taguchi design was used to optimize the parameters (amount of rice husk ash, pH, aging time) controlling the surface area of silica aerogel[4]. The rice husk was calcinated at 600℃ for 4 hours to remove the organic matters. Then Calcinated rice husk ash (RHA) and Sodium hydroxide solution was heating 3, 6, 9g of RHA with 90ml of 1M sodium hydroxide solution at 90 ℃ for 2hour to produce sodium silicate solution, immediately the mixture was filtered to remove the contaminant. For the sol-gel polymerization, hydrochloric acid was added to the filtrated silicate solution until the pH value of the 4, 7, 9. After gelation, The prepared sample was aged for 1, 5, 7days respectively. The aged gel was washed with water for 12 hours at room temperature. The hydrogel was changed into Ethanol to remove the moisture remaining in the hydrogel pore[5]. Then, The silica gel was immersed in n-Hexane and Tetramethylchloricacid (TMCS) mixture (10:1

Volume) for 24 hours at room temperature [6]. Finally the surface modified silica gel was dried at 60, 80, 120 and 150℃ for 3, 3, 2 and 2hours as shown in scheme1.

Table 1 Synthesis condition of silica aerogel using the Taguchi method

Run no.	RHA (g)	pH	Aging time (days)
1	3	4	1
2	3	7	5
3	3	9	7
4	6	4	5
5	6	7	7
6	6	9	1
7	9	4	7
8	9	7	1
9	9	9	5

		Level		
Code	Factor	1	2	3
A	RHA (g)	3	6	9
B	pH	4	7	9
C	Aging time (day)	1	5	7

2.3 Selection of levels and factors

The rice husk ash is important in synthesis of silica aerogel. The hydrolysis and condensation reactions of the silica aerogel are different depending on the aging time as well as the pH value. The factors are amount of rice husk ash, pH value, and aging time. We chose the each parameters shown in table1: amount of rice husk 3, 6, 9g; pH value 4, 7, 9; aging time 1, 5, 7days to confirm the surface area of silica aerogel.

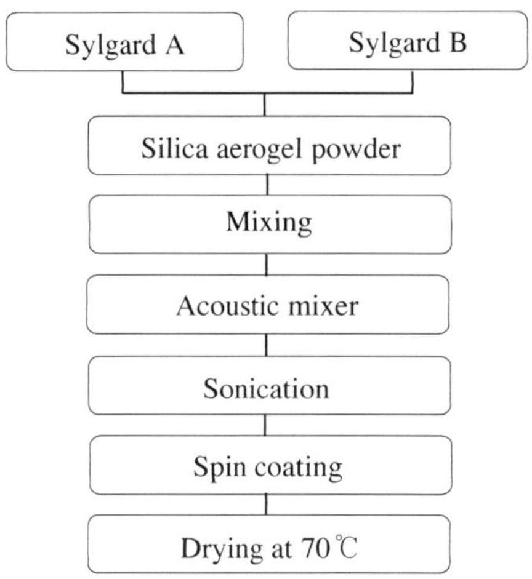

Scheme 2 A flow chart showing fabrication of Aerogel-Polydimethylsiloxane insulation film

2.4 Fabrication of Aerogel-Polydimethylsiloxane(PDMS) insulation film

The Aerogel-Polydimethylsiloxane(PDMS) insulation film was respectively fabricated by the different content of silica aerogel in polydimethylsiloxane(PDMS). Silica aerogel and polydimethylsiloxane (PDMS) were mixed by Acoustic mixer middle mode for 5 minute. Then, sonication for 15 minute and spin coating about 2ml aerogel-PDMS solution. Finally, Dry the sample of coated aerogel-PDMS solution at 70℃ as shown in scheme2.

3. Results and discussion

Table 2 The surface area and S/N ratio of silica aerogel by each factors

Run no.	Surface area (m^2/g)	Pore Volume (cm^3/g)	Pore size (nm)	S/N ratio
1	803	1.30	5.51	58.10
2	253	1.71	2.30	48.05
3	405	1.98	16.72	52.14
4	649	0.61	3.53	56.24
5	226	0.91	17.82	47.08
6	380	1.30	11.59	51.60
7	657	0.70	4.07	56.35
8	348	1.94	20.79	50.85
9	358	1.71	18.92	51.08

3.1 Determination of optimum conditions

We synthesized silica aerogel with a high specific surface area using the Taguchi method. The specific surface area of silica aerogel synthesized under different conditions was measured via Brunauer-Emmett-Teller (Tristar-2030). The table 2 shows the specific surface area, pore volume and pore size of each sample. After that, the value of specific surface area was used to calculate signal-to-noise ratio (S/N). In the Taguchi method, the Signal and noise ratio (S/N) shows the values required for each output characteristic. In this study, The bigger characteristic (specific surface area) is better when the ideal target value is infinity [7]. The signal-to-noise ratio (S/N) is defined as Eq. (1).

$$\text{S/N} = -10\log\left(\frac{1}{n}\sum_{i=1}^{n}\frac{1}{y_i^2}\right) \quad (1)$$

If the characteristic value is smaller, it is good when ideal target is zero. The Signal and noise ratio (S/N) is calculated by Eq. (2)

$$\text{S/N} = -10\log\left(\frac{1}{n}\sum_{i=1}^{n}y_i^2\right) \quad (2)$$

When the bigger characteristic is better, the signal-to-noise ratio (S/N) is calculated by Eq. (3)

$$\text{S/N} = -10\log(MSD) \quad (3)$$

The MSD is the mean square deviation from the target value of the quality characteristic.

Where:

$$\text{MSD} = \left(\frac{1}{y_1^2}+\frac{1}{y_2^2}+\frac{1}{y_3^2}+\cdots\right)/\text{n} \quad (4)$$

In MSD, y_1 and y_2 are measured results value, the specific surface area corresponds to this experiment.

The signal-to-noise ratio (S/N) values calculated by substituting the data values of the experiment into the equation1 of the bigger characteristics are shown in Table2.

Table3 Separation of S / N ratio by specific surface area

Average of S/N ratio				
		Level		
Code	Factor	1	2	3
A	RHA (g)	52.76	51.64	52.74
B	pH	56.90	48.66	51.60
C	Aging time (day)	53.51	51.79	51.95

And then, the signal-to-noise ratio (S/N) values obtained in the previous step are separated into the respective levels in order to determine what are the important factors among the variables affecting the characteristic values in Table3. For example, signal-to-noise ratio (S/N) values of 1 level on B factor is 58.10, 56.24, 56.35 and the average of signal-to-noise ratio (S/N) is 56.90. After calculating the average value of each signal-to-noise ratio (S/N), its value is indicated in Figure 1 [8]. The B (value of pH) factor with the longest in level 1-3 is the most influential factor in the specific surface area. Therefore, we can confirm the optimal condition of each factor, which is the one with the higher S / N ratio from figure1.

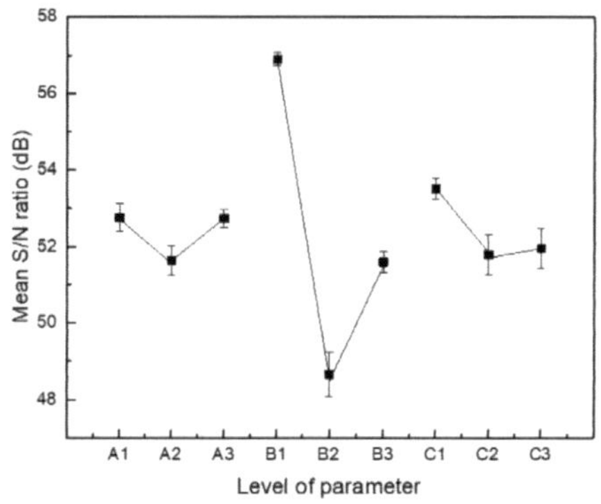

Figure1 The average of signal-to-noise ratio (S/N) ratio graph for specific surface area

For example, in the case of the B factor, the optimum condition of the B factor is B1 since the signal-to-noise ratio (S/N) ratio of B1 is larger than B2. Consequently, the optimal condition for each factor is A1B1C1. It is necessary to confirm whether the predicted results match the experimental results after the signal-to-noise ratio (S/N) ratio analysis for each factor is completed the optimal conditions. The predicted signal-to-noise ratio (S/N) is defined as Eq. (5) predicted signal-to-noise ratio (S/N)= $\bar{T} + (Optimal\ value\ of\ factor\ A - \bar{T})$
$+(Optimal\ value\ of\ factor\ B - \bar{T}) + (Optimal\ value\ of\ factor\ C - \bar{T})$ (5)
Where:
$\bar{T} = \sum_{i=1}^{n} \frac{S/N_i}{n}$ (6)
It was confirmed through Eq. (5) that the predicted value of the signal-to-noise ratio (S/N) ratio was 58.39. The silica aerogel was synthesized again under the optimum conditions. The signal-to-noise (S/N) ratio of synthesized again silica aerogel calculated to be 58.42 as shown in table 4.

Table 4 The result of optimal condition on specific surface area

Sample no.	Optimal condition	Surface area (m^2/g)	S/N ratio
Predicted value	A1B1C1	803	58.39
Experimental value	A1B1C1	834	58.42

As a result, it was confirmed that the predicted value and the result value of the signal-to-noise ratio (S/N) ratio were almost the same value. The surface of silica aerogel powder analyzed by scanning electron microscope (SEM) as shown in figure 2.

IOP Conf. Series: Journal of Physics: Conf. Series **987** (2018) 012018 doi:10.1088/1742-6596/987/1/012018

Figure 2 SEM image of silica aerogel powder under the optimized conditions (A1B1C1)

3.2 Thermal properties of Aerogel-Polydimethyl siloxane (PDMS) Insulation Film

Table5 Density, heat capacity, and thermal conductivity of insulation films containing different mass% of silica aerogel powder

	PDMS	PDMS_0.1wt% silica aerogel	PDMS_0.2wt% silica aerogel	PDMS_0.4wt% silica aerogel
Density (g/cm^3)	1.095	1.064	1.042	1.041
Heat capacity (J/g°C)	1.011	0.012	0.019	0.020
Thermal conductivity (W/mK)	0.150	0.065	0.048	0.002

We fabricated an insulation film using high specific surface area silica aerogel prepared by the ta -guchi method. The density and heat capacity of the sample should be analysed to measure the thermal conductivity [9]. Therefore, we measured the density and the heat capacity of the sample containing different amounts of silica [10]. As a result of the measurement, it was confirmed that the density decreases as the silica aerogel content increases in the insulation film. The heat capacity of the insulation film was also confirmed to increase with increasing silica aerogel content. The thermal conductivity of the film without silica aerogel (0 wt%) was measured to be 0.150 W/mK. And thermal conductivity of insulation film containing different silica aerogel contents (0.1, 0.2, 0.4 wt%) was measured by 0.065, 0.048, 0.002 W/mK. As a result of measuring the thermal conductivity, it was confirmed that the thermal conductivity value decreased as the silica aerogel content increased.

4. Conclusions

In this study, we attempt to develop economical and eco-friendly synthesis process of hydrophobic silica aerogel. The silica aerogel synthesized by the rice husk ash which was an agricultural waste to be able to recycle using Taguchi method. The silica aerogel is prepared by sol-gel processing using acidic treatment. It was dried at ambient pressure after modification of silica hydrogel surface. As a result, the silica aerogel having a high specific surface area (834 m^2/g) was synthesized under optimized conditions (amount of rice husk ash:3g, value of pH:4, aging time: 1day). The silica aerogel obtained 21 – 24nm average particle size was analyzed by scanning electronic microscope (SEM). silica aerogel with high surface area (832.26 m^2/g), pore size (3.30nm) was characterized by Brunauer-Emmett-Teller (BET). Aerogel-Polydimethylsiloxane (PDMS) Insulation Film was respectively fabricated by the different content of optimized condition silica aerogel in polydimethylsiloxane (PDMS) film. The silica Aerogel – PDMS insulation film with low thermal conductivity (0.002 W/mK) was analyzed by thermal wave system. We are fabricating a heat insulating film with the same effect as a commercially the available insulating. Therefore, we are going to carry out experiments to make insulation films to enable mass production in the future.

5. Reference

[1] CHA, J., KIM, S., PARK, K.-W., LEE, D. R., JO, J.-H. & KIM, S. 2014. Improvement of window thermal performance using aerogel insulation film for building energy saving. Journal of Thermal Analysis and Calorimetry, 116, 219-224.

[2] TADJARODI, A., HAGHVERDI, M. & MOHAMMADI, V. 2012. Preparation and characterization of nano-porous silica aerogel from rice husk ash by drying at atmospheric pressure. Materials Research Bulletin, 47, 2584-2589.

[3] TADAYON, F., MOTAHAR, S. & HOSSEINI, M. 2012. Application of Taguchi method for optimizing the adsorption of lead ions on nanocomposite silica aerogel activated carbon. Academic Research International, 2, 42.

[4] CHING, Y., LIAN, H., LEO, S., WANG, S., YAMAUCHI, Y. & KEVIN, C. 2011 Controlling Particle Size and Structural Properties of Mesoporous Silica Nanoparticles Using the Taguchi Method, Journal of Physical Chemistry. C, 27, 13158-13165.

[5] PRADIP, B., KIM, J., HILONGA, A. & KIM, H. Production of low-density sodium silicate-based hydrophobic silica aerogel beads by a novel fast gelation process and ambient pressure drying process, Journal of Solid State Sciences, 2010, 12, 911–918.

[6] DOURBASH, A., MOTAHARI, S. & OMRANPOUR, H. Effect of water content on properties of one-step catalyzed silica aerogels via ambient pressure drying, Journal of Non-Crystalline Solids, 2014, 405, 135–140.

[7] ZAINAL, N., SHUKOR, S. & RAZAK, K. Applying the Taguchi Method to Optimise the Size of Silica Nanoparticles Entrapped with Rifampicin for a Drug Delivery System, Journal of Engineering Science, 2015, 11, 9–16.

[8] YURKMEN, I., GUL, RUSTEM., CELIK, C. & DEMIRBOGA, R. Determination by the Taghchi method of optimum conditions for mechanical properties of high strength concrete with admixtures of silica fume and blast furnace slag, Journal of Civil Engineering and Environmental System, 2003, 20, 105–118.

[9] CHO, M. & HONG, S. A study on insulation fire proof materials using silica aerogels, Journal of the Korea Academia-Industrial cooperation Society, 2015, 16, 6816–6822.

[10] KIM, G. & HYUN, S. Effect of mixing on thermal and mechanical properties of aerogel-PVB composites, Journal of Materials Science 2003, 38, 1961–1966.

IOP Conf. Series: Journal of Physics: Conf. Series **987** (2018) 012019 doi:10.1088/1742-6596/987/1/012019

Shift of semimetal-semiconductor bond direction on "0 1 1" to "1 1 1" Bismuth quazi-two- dimension system

Ahmad Yazdani*, Sajad Hamreh

Department of Physics, Tarbiat Modares University, P.O. Box 14115-175, Tehran, Iran

*yazdania@modares.ac.ir

Abstract. The electronic structure of the nanocrystallines and quasi-two-dimensional systems strongly impressed by the thermodynamic- behavior mainly due to excess of hidden surface free energy. Therefore, the stability of crystalline structure's change could be related to band-offset of bond rupturing of atomic displacements. whereas for the electronic-structure of "Bi" it seams the competition of L.S and bond exchange should be effectively dominated. Besides all of the characters behave spatial like strong sensitive oxidation here it is supposed that strong correlated electronic structure in the absence of oxygen is resulted on direction of redistribution of surface chemical bond formation before any reconstructive structure. Where

- The metallic direction of electronic structure "0 1 1" is changed to "1 1 1" semiconductor direction.

- the effect of L.S is more evident on the local density of state while it is not observable around the fermi level.

- Strong effect of spin-orbit interaction on splitting of the valance to nearly conduction band around the fermi level is more evident.

1. Introduction

The coming of spintronics has inspired the investigation of the impacts of spin orbit interactions (SOI) on the electronic structure of solids. SOI prompts the spin Hall effect [1,2], which has been seen in GaAs [3], [4]. It is additionally observed [5] that in graphene the SOI leads the quantum spin Hall (QSH) effect. The finding of topological insulators (TIs) speaks to a standout amongst the most critical advances in physical science during recent years, they differentiate from normal insulator or semiconductors by insulating bulk but gapless surface or edge states [6-9]. The protected metallic surface or edge state is robust to the imperfection and non-magnetic doping, that make them suitable as dissipationless transport channels that are promising for operation in spintronics and quantum calculations [10].

Bismuth selenide and bismuth telluride are the hot topics of the past decade [11-15]. These materials exhibited certain topological properties that were of great interest. The band structure of this material is such that it has a negative band at the Γ point, indicating this in the number of odd Fermi contours. This shows the topological states are protected.

Bi_2Se_3 has a layered structure consisting of five layers containing bismuth and selenide atoms. 111 direction of these compound is parallel to these QL(s). A bilayer of Bi can be considered as a buckled honeycomb structure (Fig1), while unlike graphene, the effect of the spin-orbit is very strong, which represents a two-dimensional topological system. Amazingly, a bulk crystal of 111 Bi can be considered as a stack of Bi bilayers that are pitifully bonded to each other.

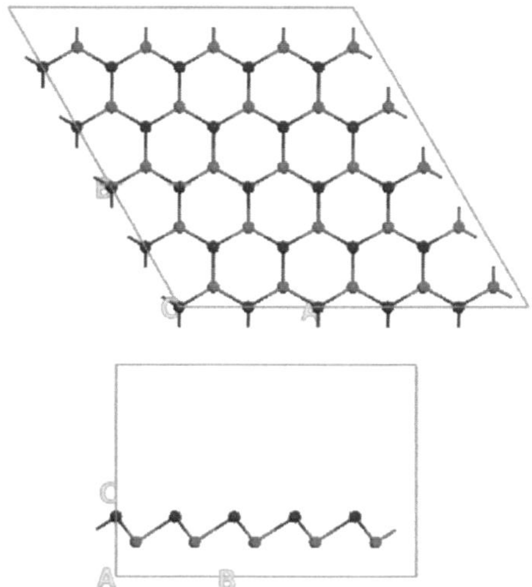

Figure 1 Top: top view of bismuth bilayer, Down: side view of bismuth bilayer

2. Methods

The electronic properties of this paper have been computed with the all-electron FHI-aims code [16] at the level of GGA in the Perdew- Burke- Ernzerhof parameterization [17]. The reciprocal space integration was performed over an 8x8x1 Monkhorst-Pack grid, after performing convergence tests on the total energy and the forces. The basis set and numerical grids are of high quality as defined by the tight settings. To treat relativistic impacts the atomic zeroth-arrange proper estimation was enforced. For relaxation of structure BFGS (Broyden-Fletcher-Goldfarb-Shanno) algorithm [18] was used until the maximum force on each atom was less than 10^{-3} eV $Å^{-1}$. Eventually spin–orbit interactions have been accomplished. We here used a post-processed, non-self-consistent, second-variational approach to include spin-orbit coupling.

Phonon band structure was calculated with the frozen-phonon approach for a fully relaxed 4x4x4 supercells using the Phonopy code [19] in combination with FHI-aims. A displacement of 0.01 Å was applied to each atom. For the phonon DOS used the Γ point sampling, while the full phonon band structure was generated using the Γ, M and K high-symmetry points in the brillouin zone. For achieving fast convergence, we specified n-max-pulay n and charge-mix-param v for Pulay mixing algorithm. For all phonon calculations we choose 10 for n and 0.6 for v. N determines the number of past iterations μ - k (k = 1, · · ·, n) to be mixed with the Kohn-Sham output density of iteration μ. The value v for the keyword charge-mix-param specify an extra (system-dependent) linear factor that is multiplied with the output density change of the Pulay mixer.

IOP Conf. Series: Journal of Physics: Conf. Series **987** (2018) 012019 doi:10.1088/1742-6596/987/1/012019

3. Results and discussion

For Bi (1 1 1) this is the normal outcome on the off chance that one perspectives the crystal as a feebly bonded stack of bilayers. For this situation, the surface ought to carry on like the bulk, as it is the situation for most layered materials. For Bi (1 1 0), be that as it may, the nonappearance of any reproduction is startling and in sharp difference with the circumstance on most semiconductor surfaces. Calculated density of states and band structures for 111 and 110 directions are shown in Fig 2 and Fig 3. As can be seen, strong spin-orbit effect breaks band degeneracy, reduces the gap in (111) direction, and induce a narrow bandgap in (110) direction. The bismuth interlayer separations can be c compared with those in bulk bismuth along the (111) direction, where a long (2.35 A°) and a short (1.59 A°) remove substitute relating to the strength of the interlayer bonds, which are of van der Waals and covalent type, respectively (Fig.4). The bonding strength between bismuth atoms is identified with the charge density, and in simple molecules, this can be described by 'bond-order'.

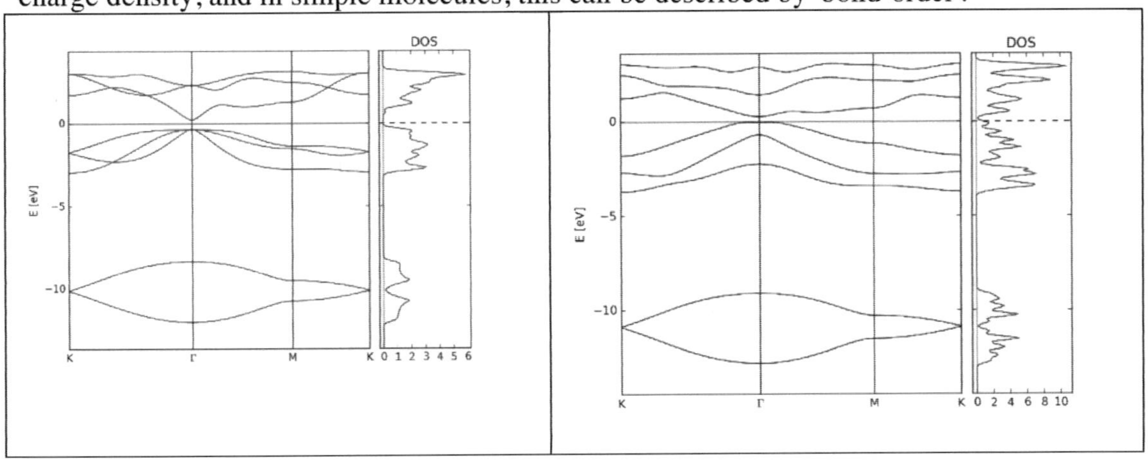

Fig2. (111) left: without spin-orbit right: with spin-orbit

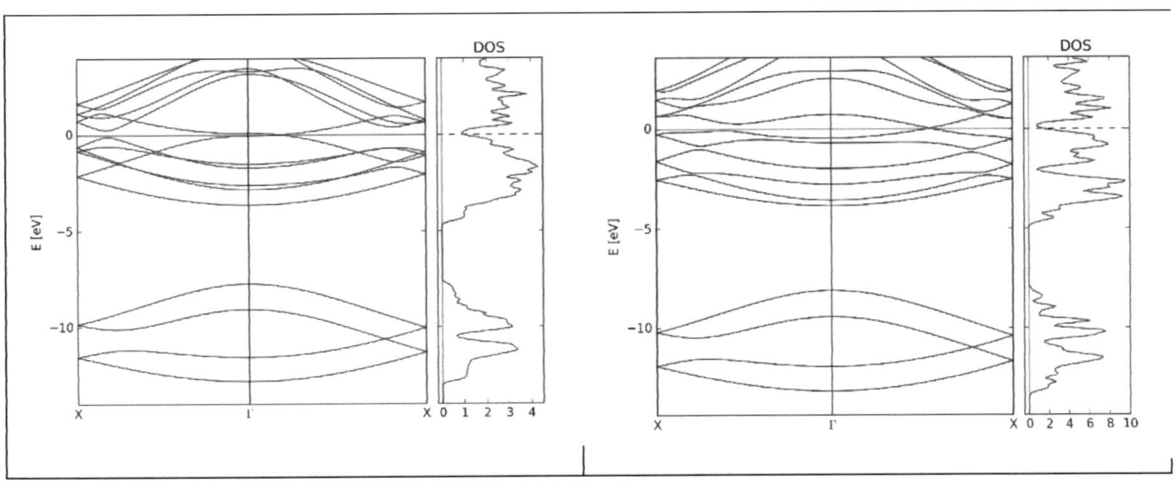

Fig3 (110) left: without spin-orbit right: with spin-orbit

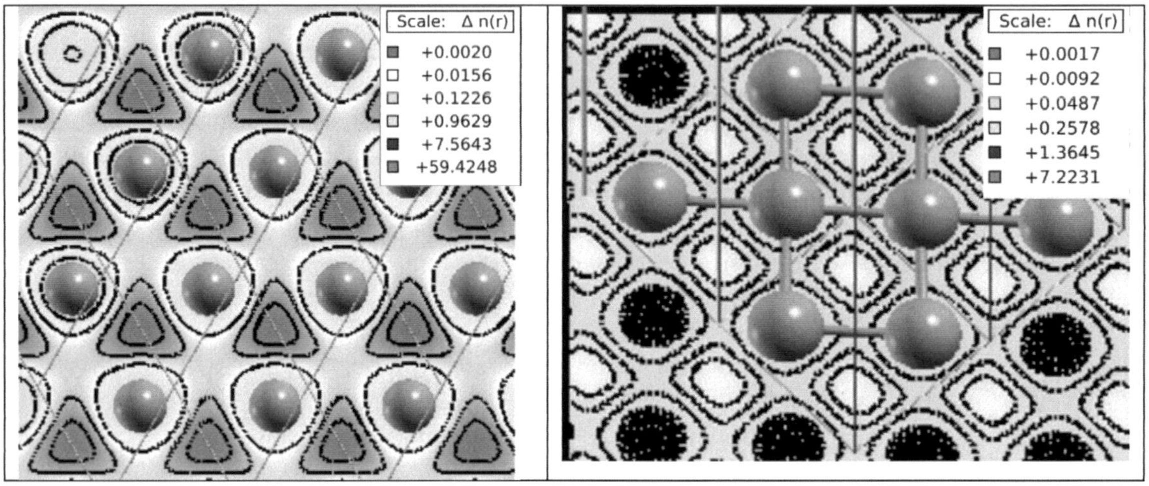

Fig4 left: (111) charge density right: 110 charge density

References

[1] S. M. N. N. S.-C. Zhang, "Dissipationless Quantum Spin Current at Room Temperature," *Science (80-.).*, vol. 31, p. 1348, 2003.

[2] J. Sinova, D. Culcer, Q. Niu, N. A. Sinitsyn, T. Jungwirth, and A. H. MacDonald, "Universal intrinsic spin Hall effect," *Phys. Rev. Lett.*, vol. 92, no. 12, pp. 126603–1, 2004.

[3] Y. K. Kato, "Observation of the Spin Hall Effect in Semiconductors," *Science (80-.).*, vol. 306, no. 5703, pp. 1910–1913, 2004.

[4] J. Wunderlich, B. Kaestner, J. Sinova, and T. Jungwirth, "Experimental observation of the spin-hall effect in a two-dimensional spin-orbit coupled semiconductor system," *Phys. Rev. Lett.*, vol. 94, no. 4, 2005.

[5] C. L. Kane and E. Mele, "Quantum Spin Hall Effect in Graphene," *Phys. Rev. Lett.*, vol. 95, no. 22, pp. 1–4, 2005.

[6] K. von Klitzing, "Developments in the quantum Hall effect.," *Philos. Trans. A. Math. Phys. Eng. Sci.*, vol. 363, no. 1834, pp. 2203–19, Sep. 2005.

[7] C. L. Kane and E. J. Mele, "Z_{2} Topological Order and the Quantum Spin Hall Effect," *Phys. Rev. Lett.*, vol. 95, no. 14, p. 146802, Nov. 2005.

[8] L. Fu and C. Kane, "Topological insulators with inversion symmetry," *Phys. Rev. B*, vol. 76, no. 4, p. 45302, Jul. 2007.

[9] L. Fu, C. Kane, and E. Mele, "Topological Insulators in Three Dimensions," *Phys. Rev. Lett.*, vol. 98, no. 10, p. 106803, Mar. 2007.

[10] B. A. Bernevig, T. L. Hughes, and S.-C. Zhang, "Quantum Spin Hall Effect and Topological Phase Transition in HgTe Quantum Wells," *Science (80-.).*, vol. 314, no. 5806, pp. 1757–1761, Dec. 2006.

[11] D. Kim *et al.*, "Surface conduction of topological Dirac electrons in bulk insulating Bi2Se3," *Nat. Phys.*, vol. 8, no. 6, pp. 460–464, Apr. 2012.

[12] M. V. Costache *et al.*, "Fingerprints of Inelastic Transport at the Surface of the Topological Insulator Bi2Se3: Role of Electron-Phonon Coupling," *Phys. Rev. Lett.*, vol. 112, no. 8, p. 86601, Feb. 2014.

[13] H. Zhang, C. Liu, X. Qi, X. Dai, Z. Fang, and S. Zhang, "Topological insulators in Bi2Se3, Bi2Te3 and Sb2Te3 with a single Dirac cone on the surface," *Nat. Phys.*, vol. 5, no. 6, pp. 438–

442, May 2009.
[14] Y. Wang *et al.*, "Topological surface states originated spin-orbit torques in Bi2Se3," *Phys. Rev. Lett.*, vol. 114, no. 25, pp. 1–5, 2015.
[15] M. V. Kuznetsov *et al.*, "Atomic structure of Bi2Se3 and Bi2Te3 (111) surfaces probed by photoelectron diffraction and holography," *Physical Review B - Condensed Matter and Materials Physics*, vol. 91, no. 8. 2015.
[16] V. Blum *et al.*, "Ab initio molecular simulations with numeric atom-centered orbitals," *Comput. Phys. Commun.*, vol. 180, no. 11, pp. 2175–2196, 2009.
[17] J. Perdew, K. Burke, and M. Ernzerhof, "Generalized Gradient Approximation Made Simple.," *Phys. Rev. Lett.*, vol. 77, no. 18, pp. 3865–3868, Oct. 1996.
[18] J. D. Head and M. C. Zerner, "A Broyden-Fletcher-Goldfarb-Shanno optimization procedure for molecular geometries," *Chem. Phys. Lett.*, vol. 122, no. 3, pp. 264–270, 1985.
[19] A. Togo, F. Oba, and I. Tanaka, "First-principles calculations of the ferroelastic transition between rutile-type and {CaCl$_2$}-type {SiO$_2$} at high pressures," *Phys.\ Rev.\ B*, vol. 78, p. 134106, 2008.

IOP Conf. Series: Journal of Physics: Conf. Series **987** (2018) 012020 doi:10.1088/1742-6596/987/1/012020

Kinetics of the melting front movement in process of centrifugal induction surfacing of powder material with nanoscale modificaters

I Sasnouski[1] and A Kurylionak[2]

[1]Senior researcher, United Institute of Mechanical Engineering of National Academy of Sciences of Belarus, 12 Academicheskaja Str., Minsk 220072, Republic of Belarus
[2]Scientific worker, United Institute of Mechanical Engineering of National Academy of Sciences of Belarus, 12 Academicheskaja Str., Minsk 220072, Republic of Belarus

E-mail: sos3@tut.by

Abstract. For solving the problem of improving the powder coatings modified by nanostructure components obtained by induction surfacing method tribological characteristics it is necessary to study the kinetics of the powdered layer melting and define the minimum time of melting.

For powdered layer predetermined temperature maintenance at sintering mode stage it is required to determine the temperature difference through blank thickness of the for one hundred-day of the define the warm-up swing on of the stocking up by solving the thermal conductivity stationary problem for quill (hollow) cylinder with internal heat source. Herewith, since in practice thickness of the cylinder wall is much less then its diameter and the temperature difference is comparatively small, the thermal conductivity dependence upon the temperature can be treated as negligible.

As it was shown by our previous studies, in the induction heating process under powdered material centrifugal surfacing (i.e. before achieving the melting temperature) the temperature distribution in powdered layer thickness may be considered even. Hereinafter, considering the blank part induction heating process quasi-stationarity under F_o big values, it is possible to consider its internal surface heating as developing with constant velocity.

As a result of development the melting front movement mathematical model in a powdered material with nanostructure modifiers the minimum surfacing time is defined. It allows to minimize negative impact of thermal influence on formation of applied coating structure, to raise productivity of the process, to lower power inputs and to ensure saving of nonferrous and high alloys by reducing the allowance for machining.

The difference of developed mathematical model of melting front movement from previously known is that the surface temperature from which the heat transfer occures is a variable and varies with a time after the linear law.

1. Main positions and admissions

The kinetics of the modified by nanostructure components powdered layer melting under centrifugal induction surfacing renders significant influence upon physico-mechanical and working characteristics of protective coatings obtained. This is connected with the fact that most of fused materials are thermally sensitive i.e. increase the surfacing temperature and time of the thermal influence, as a rule, renders the negative influence on powder layer microstructure, including modified

IOP Conf. Series: Journal of Physics: Conf. Series **987** (2018) 012020 doi:10.1088/1742-6596/987/1/012020

nanostructure components. For solving the problem of improving the powder coatings modified by nanostructure components obtained by induction surfacing method tribological characteristics it is necessary to study the kinetics of the powdered layer melting and define the minimum time of melting. At the point, the studies devoted to decision of this problems are fulfilled by developing the mathematical model of phase transition front (the front of melting) motion process under conditions of the centrifugal induction surfacing.

The decision of the problem with aggregate conditions material change is one of the most complex mathematical physics sections. This is connected with the fact that similar problems must be formulated as problems of conjugation the two temperature fields in presence of specific boundary condition at moving boundary. The Main difficulty of the problem solving is a fact that that it belongs to the class of the nonlinear problems, i.e. to problems with nonlinear boundary condition. Such a problems, as a rule, may be solved only by means of approximate methods. One of the the most efficient methods of solving is an approximate method, designed by academician L.S. Leybenzon, and developed also by academician A.V. Lykov. Essence of this method is concluded in approach after which the functions of the temperature distribution in melted and not melted layers are selected so that they satisfy the boundary condition. Matched in such a way functions are to be substituted into condition of conjugation at the boundary of the phases and after that one solve the obtained differential equation for the front of the phase transition coordinate ξ.

As it was shown by our previous studies [1-2], in the induction heating process under powdered material centrifugal surfacing (i.e. before achieving the melting temperature) the temperature distribution in powdered layer thickness may be considered even. Hereinafter, considering the blank part induction heating process quasi-stationarity under Fo big values, it is possible to consider its internal surface heating as developing with constant velocity.

On the base of these admissions problem of powdered layer melting is reduced to the following. There is unlimited hollow cylinder at the temperature $T_0 = T_m$ (T_m is a powder melting temperature). The cylinder external surface temperature increases linearly. Melted layer with thickness R-η = ξ, where η is a distance from axis of the cylinder to the layer boundary, is formed beginning from the cylinder surface.

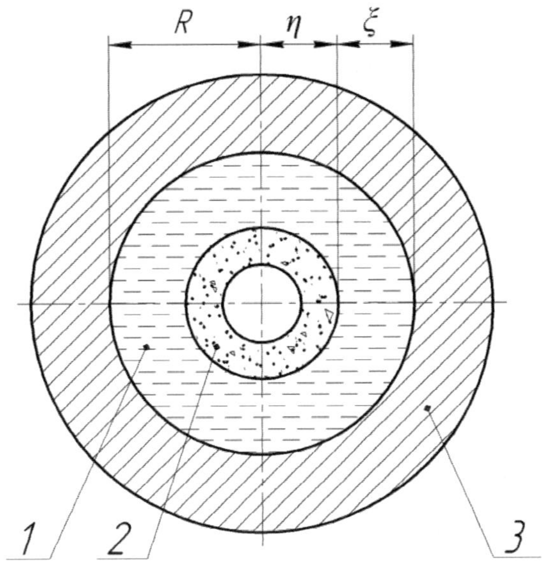

1 – melted zone; 2 – not melted zone; 3 –blank.

Fig 1 – a boundary conditions under centrifugal induction surfacing.

The temperature of not melted layer is expected uniform in all cylinder and equal to melting temperature.

The known methods of solving the heating problems for induction sintering processes and powdered coatings surfacing processes reveal the following main limitations:

- the blank surface radiant heat exchange is not taken into account and it reduces greatly the calculation accuracy;
- used for solving the problems operational methods and finite-difference method bring about exceedingly bulky mathematical expressions, requiring for calculations the use of calculating machinery, and not suitable to engineering practice;
- there are no based on the known decisions recommendations for the technological processes optimization.

The analysis of the centrifugal induction surfacing technologies main features allows to draw a conclusion that the most important stage of the technological process, defining the physico-mechanical and exploitative characteristics of two-layer products obtained, is isothermal soaking at the powdered material surfacing temperature. The main engineering problem at this stage is a powdered layer average temperature (the surfacing temperature) maintenance during certain time interval (surfacing time). Here the thermal regimes on previous stages of heating do not render essential influence upon the process of powdered layer sintering to substrate kinetics.

The powdered layer thickness, as a rule, has far less substrate thickness and its surface curvature radius. Powdered layer free internal surface and two-layer cylinder butt ends can be considered as heat-insulated. Consequently, powdered layer thermal resistance can be considered as negligible, and temperature distribution in its thickness can be treated as uniform and equal to the substrate inner surface temperature. Herewith, beginning from certain Fourier criterion value ($Fo \approx 0{,}305$) quasisteady thermal conduction mode occures, when the temperature distribution shape through the substrate changes not, but the temperature in each point grows with time retaining constant relative velocity. Upon achieving the powdered material melting temperature by internal surface of the cylindrical blank the temperature of this surfaces is support to be constant by heating source (the high frequency current generator) power management. Here the temperature distribution through blank thickness is stationary, but power, released by internal sources, is equal to heat flux from the blank external surface due to convective and radiant heat exchange with ambient medium.

For powdered layer predetermined temperature maintenance at sintering mode stage it is required to determine the temperature difference through blank thickness of the for one hundred-day of the define the warm-up swing on of the stocking up by solving the thermal conductivity stationary problem for quill (hollow) cylinder with internal heat source. Herewith, since in practice thickness of the cylinder wall is much less then its diameter and the temperature difference is comparatively small (less then 1 K), the thermal conductivity dependence upon the temperature can be treated as negligible.

2. Mathematical model development

On the grounds of mentioned above considerations the mathematical problem is formulated according to following expressions:

$$\frac{\partial T_1(r,\tau)}{\partial \tau} = a_1\left(\frac{\partial^2 T_1(r,\tau)}{\partial r^2} + \frac{1}{r}\frac{\partial T_1(r,\tau)}{\partial r}\right) \quad \left(\tau > 0; R > r > \eta; \eta = f(\tau)\right); \tag{1}$$

$$T_2(r,\tau) = T_0 = T_m = const; \tag{2}$$

$$T(\eta,\tau) = T_m = const; \tag{3}$$

$$T(R,\tau) = T_m - b\tau; \tag{4}$$

$$\lambda_1 \frac{\partial T_1(\eta,\tau)}{\partial r} = \rho\gamma_2 \frac{d\eta}{d\tau}, \tag{5}$$

Here:
T_1 – temperature of the liquid phase (melt), K;
T_2 – temperature of not melted powder layer, K;
λ_1 – liquid phase thermal conductivity factor;
λ_2 – powdered ambience thermal conductivity factor;
R – blank internal surface radius, m;
ρ – filler material density, kg/m^3;
b – rate of heating, K/s;
γ – filler material specific heat of melting, J/kg;
η – a distance from axis of the rotation to the melting boundary, m;
τ – time, s.

In accordance with method of L.S. Leybenzon take distribution of temperature in melted zone under the law of temperature distribution in hollow cylinder in stationary state, i.e.

$$T_1(r,\tau) = -\frac{b\tau \ln r + (T_m + b\tau)\ln \eta - T_m \ln R}{\ln \frac{R}{\eta}}. \quad (6)$$

The solution (6) satisfies the boundary conditions and differential equation (1). Substituting decision (6) into boundary condition (5), obtain

$$-\lambda_1 \frac{b\tau}{\eta \ln \frac{R}{\eta}} = \rho\gamma_2 \frac{d\eta}{d\tau}. \quad (7)$$

After this equations integration we have expression:

$$\eta^2 \ln \frac{R}{\eta} - \frac{1}{2}\left(R^2 - \eta^2\right) = -\frac{\lambda_1 b}{\rho\gamma_2}\tau^2. \quad (8)$$

The transcendental algebraic equation (8) defines relationship between the melting front coordinate η and time τ. Its decision with respect to η can be enough simply realized by means of the applied mathematical program packages "Matlab", "Math-CAD", "Mathematics" and others.

On the base of the relationships obtained it is possible to define the powdered layer full melting time τ_m. Under conditions $\tau = \tau_m$, $\eta = R_0$ it will be equal:

$$\tau_m = \frac{\rho\gamma_2}{b\lambda_1}\left[\frac{1}{2}\left(R^2 - R_0^2\right) - R_0^2 \ln \frac{R}{R_0}\right]. \quad (9)$$

Thereby, as a result of analysis of the technological scheme and features of powdered coating centrifugal induction surfacing, modified by nanostructure components, mathematical model of the phase transition front moving in surfacing material is designed. This has allowed, in turn, to define the minimal time, providing melting the powdered layer during deposited coating formation.

References

[1] Kashitsyn L, Sosnovsky I and Gafo Yu 2005 The engineering method of calculating the thermal modes of coating centrifugal sintering consisting from metallic powder by use of the radio frequency current internal inductor *Transactions of the Polocky state university.* The Series «Applied sciences» No **6** pp 98-102.

[2] I.Sosnovsky, Yu.Gafo, L.Kashitsyn, A.Sosnovsky Control and stabilization of technological modes of powder materials to centrifugal induction sintering surface on cylindrical parts internal surface. *Euro PM 2007 Powder Metallurgy* Congress 8 Exhibitions 14-17 October 2007 (Toulouse, France) Vol **2** P. 117-119.

IOP Conf. Series: Journal of Physics: Conf. Series **987** (2018) 012021 doi:10.1088/1742-6596/987/1/012021

Stretchable and flexible thermoelectric polymer composites

P. Slobodian[1], P. Riha[2], J. Matyas[1], R. Olejnik[1]

[1]Centre of Polymer Systems, University Institute, Tomas Bata University, Trida T. Bati 5678, 76001 Zlin, Czech Republic
[2]The Czech Academy of Sciences, Institute of Hydrodynamics, Pod Patankou 5, 16612 Prague 6, Czech Republic

Corresponding author's e-mail: P. Slobodian (slobodian@cps.utb.cz)

Abstract. Polymer composites were manufactured from pristine and oxidized multi-walled carbon nanotubes and ethylene-octene copolymer. The composites had thermoelectric properties and exhibit thermoelectric effect, that is, the conversion of temperature differences into electricity. Fourier transform infrared spectroscopy and X-ray photoelectron spectroscopy of the multi-walled carbon nanotubes in ethylene-octene copolymer matrix showed that the oxidation with HNO_3 or $KMnO_4$ enhanced its p-type electrical conductivity and that the thermoelectric power increase was proportional to the formation of new oxygen-containing functional groups on the surface of carbon nanotubes.

1. Introduction

Thermoelectric polymer composites as principal parts of thermoelectric devices are capable to convert thermal energy into electricity when there is a different temperature between the hot and cold junctions of two dissimilar conductive composites. In spite of their lower thermoelectric efficiency in terms of the figure of merit, the properties as elasticity, shape formation, light weight, low thermal conductivity, easy processing and low manufacturing cost are advantages in many technological and ecological applications.

The aim of this study is to specify the effect of pristine and oxidized multi-walled carbon nanotubes (MWCNTs) in ethylene-octene copolymer matrix on their thermoelectric power. The thermoelectric efficiency of the composites was calculated on basis of the measured electrical and thermal conductivities and the induced electric voltage.

2. Experimental

The purified multi-walled carbon nanotubes produced by acetylene chemical vapor deposition method were purchased from Sun Nanotech Co. Ltd., China with purity >90% and electrical resistivity 0.12 Ωcm. The ethylene-octene copolymer (EOC) with 45 wt% of octene content (ENGAGE 8842) was purchased from Dow Chemicals. The density of EOC was 0.8595 gcm^{-3}.

$KMnO_4$ oxidation: Oxidized MWCNTs were prepared in a glass reactor with a reflux condenser filled with 250 ml of 0.5 M H_2SO_4, into which 5 g of potassium permanganate ($KMnO_4$) as oxidizing agent and 2 g of MWCNTs were added. The dispersion was sonicated at 85 °C for 15 h using thermostatic ultrasonic bath (Bandelin Electronic DT 103H).

HNO_3 oxidation: 2 g of MWCNTs was added to 250 ml of HNO_3 (concentrated) and heated at 140 °C for 2 h. After that the dispersion was cooled and filtered. The sediment was washed by deionized water and dried at 40 °C for 24 h.

The X-ray photoelectron spectroscopy (XPS) signals were measured to get information on functional groups attached onto the nanotube surface. The XPS signals from MWCNT, MWCNT($KMnO_4$) and MWCNT(HNO_3) network surfaces were recorded using a Thermo Scientific K-Alpha XPS system (Thermo Fisher Scientific, UK) equipped with a micro-focused, monochromatic Al K⟨ X-ray source (1486.6 eV). An X-ray beam of 400 mm size was used at 6 mA x 12 kV [1]. The spectra were acquired in the constant analyzer energy mode with pass energy of 200 eV for the survey. Narrow regions were collected using the snapshot acquisition mode (150 eV pass energy) enabling rapid collection of data (5 s per region). The narrow region data was post-processed using Jansson's algorithm to remove the analyzer point spread function what resulted in improved resolution of the spectra for peak deconvolution [2].

The thermoelectric power of a material is a measure of the magnitude of an induced voltage in response to a temperature difference across that material, as induced by the thermoelectric effect. The thermoelectric power measurement was carried out for all the samples using a set-up illustrated in Fig. 1a. A circular composite sample (diameter 20 mm, thickness 2 mm) was placed between two copper electrodes. The ends of each Cu electrode were immersed in a thermostatic silicone oil baths set to different temperature. The temperature at the copper/composite interfaces was measured by a digital thermometer. The arising thermoelectric current was measured with a Multiplex datalogger 34980A.

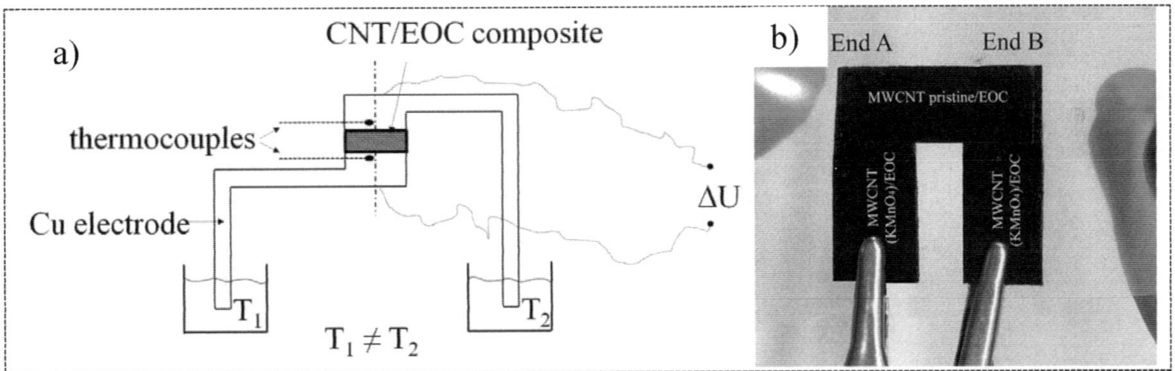

Fig. 1 a) Schematic diagram of the set up for the measurement of voltage induced in the composite in response to a temperature difference. B) Energy harvesting button assembled from MWCNT(pristine)/EOC and MWCNT(KMnO4)/EOC thermoelements.

To demonstrate a possible use of conductive MWCNT/EOC composites as thermoelectrical material, the energy harvesting from temperature gradient was examined. The arrangement of the thermoelectric generator as so called energy harvesting button, which was fabricated from three conductive thermoelements, namely, one MWCNT pristine/EOC and two MWCNT(KMnO4)/EOC composites, is shown in Fig. 1b. The composite stripes were stuck on PET foil while the composite overlaps holder together by proper stickiness of composites. The composite generated electricity was monitored by means of the Multiplex datalogger 34980A while a temperature gradient was induced by a finger touching.

3. Results

The x-ray photoelectron spectroscopy was performed on the MWCNT specimens to get information on the functional groups attached onto the nanotube surface. The surface composition (in atomic %) were determined by considering the integrated peak areas of C1 and the respective sensitivity factors. The fractional concentration of a particular element A was computed using,

$$\%A = (I_A/s_A)/\Sigma(I_n/s_n)x100\% \quad , \tag{1}$$

where I_A denotes the signal intensity of element A, I_n integrated peak areas, s_A and s_n the Scofield sensitivity factors corrected for the analyser transmission.

The main binding energy peak (284.5 eV) in XPS spectra of MWCNTs was assigned to the C1s-sp2, while the other ones were assigned to C-O (286.2 eV), C=O (287.1 eV), O-C=O (288.6-289 eV) and C1s-π-π* (291.1-291.5 eV). XPS data in Fig. 2a show that all MWCNTs have C-OH, C=O and O-C=O groups on their surface and that MWCNT($KMnO_4$) and MWCNT(HNO_3) and MWCNTs have maximum percentage of O-C=O, C-O and C=O, respectively. The XPS analysis was also a useful and simple method for evaluating the content variation of amorphous carbon in MWCNT samples by analyzing the sp^3/sp^2 carbon ratio. According to our XPS results the carbon ratios are 2.50, 3.50 and 1.69 for MWCNTs, MWCNT(HNO_3) and MWCNT($KMnO_4$), respectively. The reason why the total oxygen amount for MWCNT($KMnO_4$) is less than in case of HNO_3 oxidant could be due to removing away more amount of the oxygen species originated from amorphous carbon structures during the acidified $KMnO_4$ oxidation process. The decreased number of total groups involving oxygen increased the contact resisitance in MWCNT junctions of the network nanotube structure, Fig. 2b.

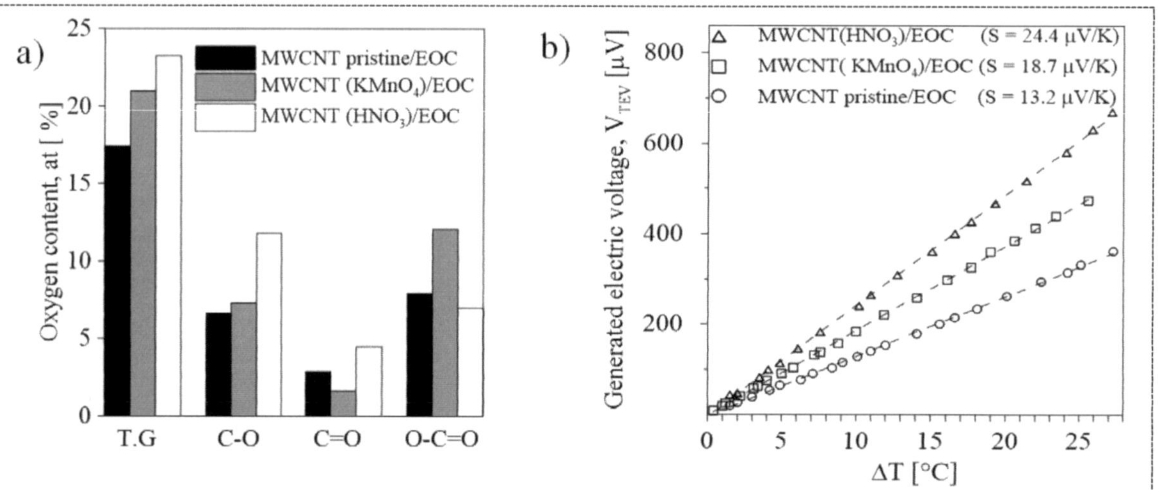

Fig. 2 a) Oxygen content (at %) on pristine and oxidized MWCNTs by different reagents. b) Generated voltage vs. temperature difference for all investigated composites (filler concentration 30 wt.%) together with the indicated thermoelectric power $S = V_{TEV}/\Delta T$ in μV/K at 22 °C.

The values of induced voltage in response to a temperature difference across the measured sample of investigated thermoelectric composites are shown in Fig. 2b. The slope of the linear temperature-difference ΔT dependence on resulting voltage V_{TEV} defines the thermoelectric power $S = V_{TEV}/\Delta T$, an essential measure evaluating the potential thermoelectric performance. The corresponding values are 13.2, 18.7 and 24.4 μV/K for MWCNTs, MWCNT($KMnO_4$) and MWCNT(HNO_3) composites, respectively. However, the enhancement of thermoelectric power by oxidation is accompanied by suppression of the electrical conductivity. Hence, the use of MWCNT(HNO_3)/EOC composite with a high thermoelectric power is only one important factor for the efficient performance of a thermoelectric generator. No less important step is careful design to enhance also the electrical conductivity σ of the composite by adjustment of the nanotube concentration which increases the thermoelectric power factor $S^2\sigma$. The power factor is important value since it is directly related to the usable power attainable from the thermoelectric material.

The arrangement of the thermocouple for energy harvesting fabricated from two conductive thermoelements (MWCNTpristine/EOC and MWCNT($KMnO_4$)/EOC composites) operates even in a small temperature gradient induced by a touch of finger (Fig. 1b). The system takes in heat from the finger and outputs electricity. Time dependence of generated voltage is shown in Fig. 3a. In practice, the energy harvesting button may use programmable thermal energy source to control electronic devices which need a small amount of electrical energy in a large time scale. The advantage of the composite energy harvesting button is its flexibility and a high level of possible integration which may be useful to remove heat in a network of electronic devices.

Thermoelectric device (thermopile) built up of two conductive thermoelements and the schematic diagram showing the material junctions in series alternating hot/cold temperatures with each junction is shown in Fig. 3b.

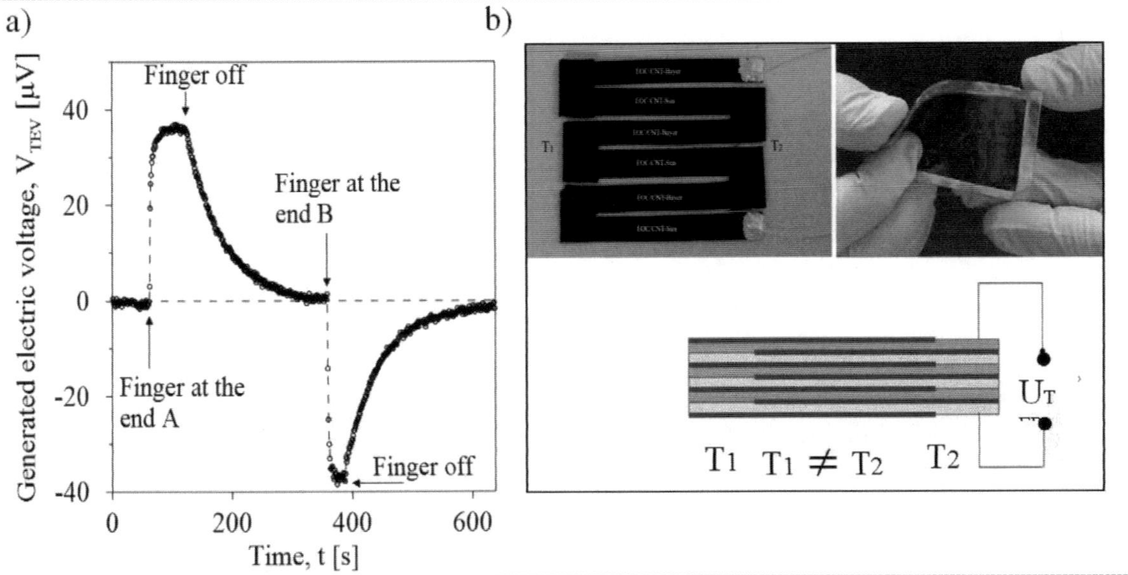

Fig. 3 a) Time-dependent generation of voltage after finger touching at A or B ends of the energy harvesting button assembled from MWCNT(pristine)/EOC and MWCNT($KMnO_4$)/EOC composites (see Fig. 1b). b) Thermoelectric device (thermopile) built up of MWCNT/EOC flexible thermoelements and the schematic diagram showing the mat erial junctions in series alternating hot/cold temperatures with each junction.

4. Conclusions

The thermoelectric power, which specifies the voltage produced per degree temperature difference, increased of 42 % for $KMnO_4$ oxidized and 85% for HNO_3 oxidize MWCNTs with respect to the thermoelectric power of the composite with pristine nanotubes [9]. The thermoelectric power increase is proportional to formation of the new oxygen-containing functional groups on the surface of carbon nanotubes.

References

[1] R. Benlikaya, P. Slobodian, P. Riha, Enhanced strain-dependent electrical resistance of polyurethane composites with embedded oxidized multiwalled carbon nanotube networks, J.

Nanomat (2013) 10 article ID 327597.
[2] P.A. Jansson (Ed.), Deconvolution of Spectra and Images, Academic Press, San Diego, 1997, pp. 119-134.
[3] P. Slobodian, P. Riha, P. Saha, A highly-deformable composite composed of an entangled network of electrically-conductive carbon-nanotubes embedded in elastic polyurethane, Carbon 50 (2012) 3446e3453.

Using graphene/styrene-isoprene-styrene copolymer composite thin film as a flexible microstrip antenna for the detection of heptane vapors

Robert Olejnik[1], Jiri Matyas[1], Petr Slobodian[1]*, Pavel Riha[2]

[1]Centre of Polymer Systems, University Institute, Tomas Bata University, Trida T. Bati 5678, 760 01 Zlin, Czech Republic
[2]The Academy of Sciences, Institute of Hydrodynamics, Pod Patankou 5, 166 12 Prague 6, Czech Republic

*slobodian@utb.cz

Abstract. Most portable devices, such as mobile phones or tablets, use antennas made of copper. This paper demonstrates the possible use of antenna constructed from electrically conductive polymer composite materials for use in those applications. The method of preparation and the properties of the graphene/styrene-isoprene-styrene copolymer as flexible microstrip antenna are described in this contribution. Graphene/styrene-isoprene-styrene copolymer toluene solution was prepared by means of ultrasound and the PET substrate was dip coated to reach a fine thin film. The main advantages of using PET as a substrate are low weight and flexibility. The final size of the flexible microstrip antenna was 10 × 25 mm with thickness of 0.48 mm (PET substrate 0.25 mm) with a weight of 0.110 g. The resulting antenna operates at a frequency of 1.8 GHz and gain –40.02 dB.

1. Introduction

Nowadays, electronic devices use mainly copper based microstrip antenna for communication between each portable device in a network [1-3]. In many fields of science one of the most promising materials is graphene. There are ever increasing possibilities for the use of graphene in electronics. The main reasons for employing this material are their characteristics such are good conductivity and flexibility achieved by necessary adjustments, and a great potential for miniaturization. These composite materials can perform several functions simultaneously, they can be applied to a variety of products and active surfaces not only in the confines of ICT. This contribution is focused on the use of graphene in a passive antenna, which consists of a layer of graphene/styrene-isoprene-styrene copolymer, applied to a PET substrate.

The aim of this research is to produce a layer with the lowest possible resistivity, which ensures that the material used for the construction of the antenna is highly suitable. The specimen produced for the antenna contains graphene particles. In addition, the final price plays an important part when eventually incorporated into the applications and devices.

2. Experimental methods

Graphene/styrene-isoprene-styrene copolymer was used as a polymer with conductive filler base for the preparation of a conductive antenna layer. The flexibility and the mechanical properties of synthetic rubbers combined with the processing abilities of plastics, makes a great material. Low hardness and low viscosity are the primary advantages for easy thermoplastic processing as well as being of a soluble state. Styrene-isoprene-styrene copolymer was dissolved in toluene to reach a 16% wt polymer solution. Graphene particles sized 50 µm were treated using sonication in toluene by means of the sonication horn UZ Sonopuls HD 2070 kit for 15 minutes to disperse the graphene particles. The obtained dispersion was mixed together with a styrene-isoprene-styrene copolymer solution and stirred for two hours. The resulting solution contained 25% wt of graphene particles. The PET substrate was dipped into the graphene/styrene-isoprene-styrene copolymer and toluene solution for 10 s to attain the resulting active layer. The composite was kept at room temperature for 48 hours in order to dry completely [4-6].

The PET substrate strip (10 × 25 mm) was coated using the dip coating method. Figure 1 demonstrates the process of dip coating the PET substrate by graphene and styrene-isoprene-styrene copolymer solution. First step: Dip the specimen perpendicularly into the solution in order to dip coat it. Second step: Coating PET substrate by prepared copolymer dispersion. Third step: Gently raise the specimen vertically upwards. Fourth step: The film thickness is possible to control by the numbers of dipping of specimen and also on the speed of removal of the specimen.

The prepared microstrip antenna was measured by frequency analyzer for the measure of the resonant frequency. The antenna was located to the close chamber with was filled by saturated heptane vapors. The parameters of sample such are resonant frequency and gain was continuously measured in time. The resulting curves are visible in the figure 3.

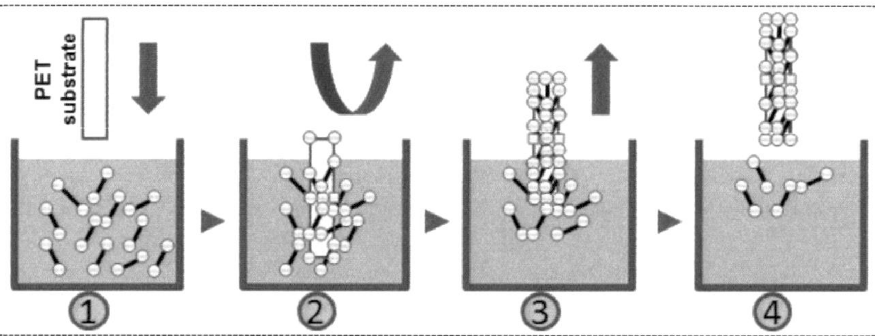

Figure 1. Preparation of electrically conductive layer using the dip coating method.

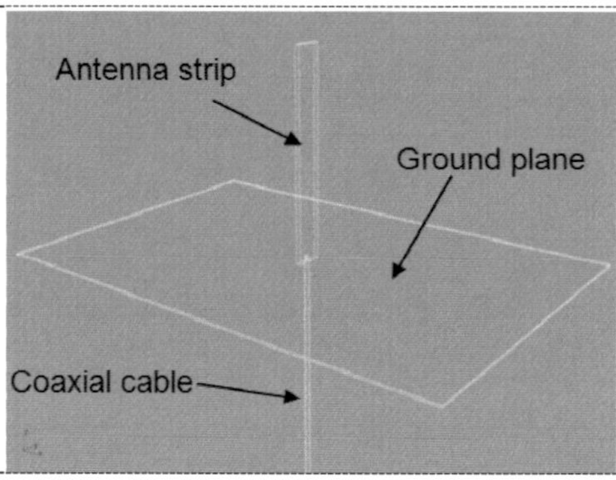

Figure 2. Construction of the antenna with antenna strip (coated PET), ground plane and coaxial cable.

3. Results

All measurements of the prepared antenna were performed in an anechoic chamber using the N9912A FieldFox Handheld RF spectrum analyzer with a measuring range between 2 MHz to 4 GHz. By means of this spectrum analyzer the parameter S11 was measured; this parameter determines the best frequencies of impedance matching for the antenna. The dimensions of the coated PET substrate are 10 × 25 mm. The total weight of the coated antenna is 0.402 g. The final size of flexible micro strip antenna was 5 x 50 mm with thickness of 0.110 g (PET substrate 0.25 mm). The ground plane of the antenna is made from FR-4 copper substrate sized 65 × 75 mm (Figure 2). All measured values (Figure 3) show that the PET substrate and graphene (styrene-isoprene-styrene copolymer) in combination with the dip coating technology proved to be very promising materials. By combining these materials a very efficient design of the microstrip antenna was achieved. Micrographs (Figure 4) obtained by scanning electron microscopy (FEI Nova NanoSEM 450) showed graphene (styrene-isoprene-styrene copolymer) conductive structure on the PET substrate.

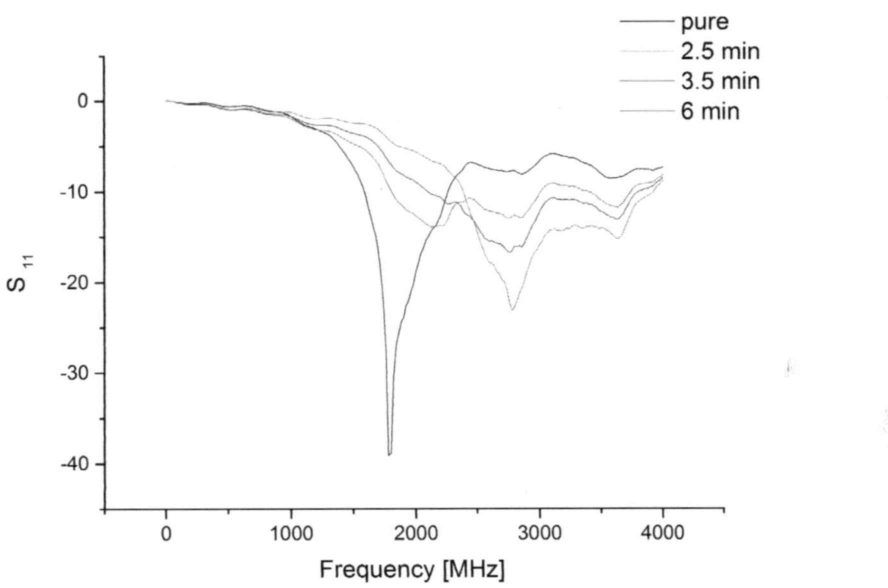

Figure 3. This graph represents the course of reflection coefficient "r" depending on the frequency in dB. (S11) for Heptane 1.8 GHz (-40.02 dB).

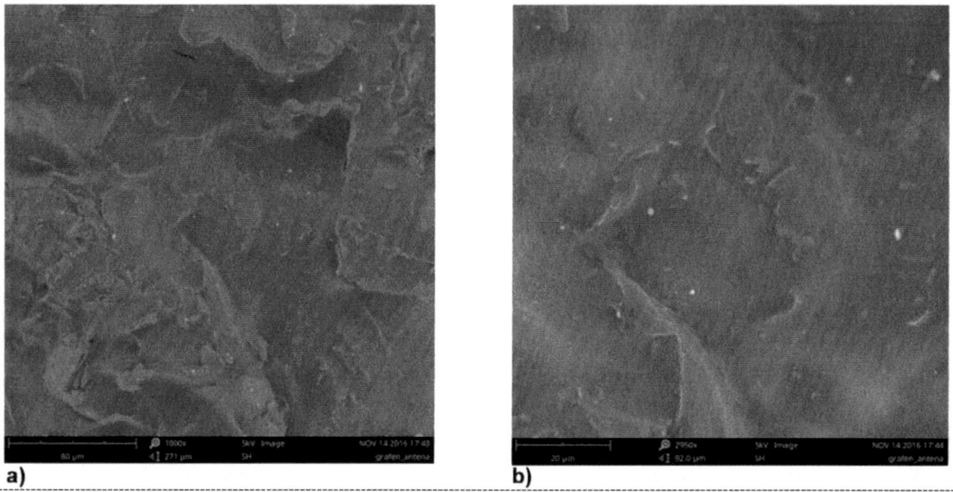

Figure 4. (a) SEM microscopy of the conductive structure on the PET substrate (scale of 80 μm), (b) SEM microscopy of the conductive structure on the PET substrate (scale of 20 μm).

4. Conclusion

Graphene particles and styrene-isoprene-styrene copolymer were used to prepare a conductive antenna layer. The antenna works at frequency 1.8 GHz (–40.02 dB) and can be used in the field of telecommunications. The main advantages are its light weight and the possibility for it to be incorporated into mobile phone covers. Additionally, the antennas could be applied to plastic covers used for portable telecommunication devices and also wearable electronics. The carbon graphene and styrene-isoprene-styrene copolymer network as an antenna could replace more commonly used materials in the future.

All portable telecommunication devices, such as mobile phones, used for communication purposes having antennas made of copper foils and strips, could be replaced in the future by the developed antenna. The majority of widespread technologies, e.g. LTE, GSM, UMTS operate at frequencies for which it is possible to produce a customized antenna using carbon nanotubes and styrene-isoprene-styrene copolymer deposited on the PET substrate because the PET substrate is frequently used for the next generation of the flexible antenna [7-15].

Acknowledgments

This work was supported by the Ministry of Education, Youth and Sports of the Czech Republic – Program NPU I (LO1504) and with the support of Operational Program Research and Development for Innovations co-funded by the European Regional Development Fund (ERDF) and national budget of the Czech Republic, within the framework of the project CPS-strengthening research capacity (reg. number: CZ.1.05/2.1.00/19.0409). This project was supported by the internal grant of TBU in Zlin No. IGA/CPS/2016/002 funded from the resources of the Specific University Research.

References

[1] K. Gupta, R. Garg, I. Bahl, P. Bhartia, Microstrip Lines and Slotlines, second ed., Artech House, Norwood, MA, 1996.
[2] R. Garg, P. Bhartia, I. Bahl, A. Ittipiboon, Microstrip Antenna Design Handbook, first ed., Artech House, Norwood, MA, 2001.
[3] M.C. Dang, T.M.D. Dang, E. Fribourg-Blanc, Adv. Nat. Sci. Nanosci. Nanotechnol. 4 (2013)

IOP Conf. Series: Journal of Physics: Conf. Series **987** (2018) 012022 doi:10.1088/1742-6596/987/1/012022

015009.
[4] M. Maiti, S. Sadhu, A. K. Bhowmick. J. Appl. Polym. Sci. 2006, 101(1), 603-610
[5] P. Slobodian, P. Riha, R. Olejnik, M. Kovar, P. Svoboda. J. Nanomater.2013, 1-7
[6] R. Theravalappi, P. Svoboda, J. Vilcakova, S. Poongavalappil, P. Slobodian, D. Svobodova. Mater. Design. 2014, 60, 458-467
[7] L. Yang., A. Rida, M. Tentzeris, Design and Development of RFID and RFID-Enabled Sensors on Flexible Low Cost Substrates, first ed., Morgan & Claypool Publishers, San Rafael, CA, 2009.A. Georgiadis, A. Collado, S. Kim, H. Lee, M. Tentzeris. M. IEEE MTT-S Int. (2012) 1-3.
[8] H.K. Yoon, Y. Yoon, H. Kim, C.H. Lee, IET Microw Antenna P. 5 (2011) 1463-1470.
[9] D. Betancourt, J. Castan, Prog. Electromagn. Res. C 38 (2013) 129-140.
[10] S.R. Forrest, Nature 428 (2004) 911-918.
[11] L. Wang, Y. Guo, B. Salam, C.W. Lu, IEEE Asia–Pacific Conf. Antennas and Propagation, (2012) 239-240.
[12] J. Siden, M.K. Fein, A. Koptyug, H. Nilsson, IET Microwaves Antennas Propag. 1 (2007) 401-407.
[13] J. Perelaer, M. Klokkenburg, C.E. Hendriks, U. Schubert, Adv. Mater. 21(2009) 4830-4834.
[14] J. Matyas, R. Olejnik, K. Vlcek, P. Slobodian, P. Urbanek, P. Krcmar, L. Munster, Adv. Mater. Res. 1101 (2015) 245-248.
[15] J. Matyas, L. Munster, R. Olejnik, K. Vlcek, P. Slobodian, P. Krcmar, P. Urbanek, I. Kuritka. Jpn J Appl Phys. 55 (2016) 02BB13-4.

IOP Conf. Series: Journal of Physics: Conf. Series **987** (2018) 012023 doi:10.1088/1742-6596/987/1/012023

Study of spectroscopic properties of nanosized particles of core-shell morphology

T N Bzhalava[1] **and P J Kervalishvili**[2]

[1,2]Engineering Physics Department, Georgian Technical University, 77, Kostava Str., 0175, Tbilisi, Georgia

Email: tamrikobzhalava@yahoo.com

Abstract. Method of studying spectroscopic properties of nanosized particles and estimation of resonance wavelength range for determination of specific and unique "spectral" signatures in purpose of sensing, identification of nanobioparticles, viruses is proposed. Elaboration of relevant models of viruses, estimation of spectral response on interaction of electromagnetic (EM) field and viral nanoparticle is the goal of proposed methodology. Core-shell physical model is used as the first approximation of shape-structure of virion. Theoretical solution of EM wave scattering on single spherical virus-like particle (VLP) is applied for determination of EM fields in the areas of core, shell and surrounding medium of (VLP), as well as scattering and absorption characteristics. Numerical results obtained by computer simulation for estimation of EM "spectra" of bacteriophage T7 demonstrate the strong dependence of spectroscopic characteristics on core-shell related electric and geometric parameters of VLP in resonance wavelengths range. Expected spectral response is observable on far-field characterizations. Obtained analytical EM field expressions, modelling technique in complement with experimental spectroscopic methods should be the way of providing the virus spectral signatures, important in bioparticles characterization.

1. Introduction

Nanoparticles (NPs) are known as zero-dimensional materials-substances of about 100 nm and less sizes. Nanoparticles reveal the properties somewhat intermediate between small molecular clusters of 0.2-1 nm dimensions and bulk. Binary, core-shell structured nanoparticles are finding widespread applications. Optical properties of nanoparticles - gold nanospheres, silica-gold nanoshells, gold nanorods of various nanoparticle sizes or dimensions are highly dependent on the nanoparticle size, shape, and core-shell composition. Gold nanoshells are found to have optical cross-sections comparable to and even higher than the nanospheres and optical resonances fall favorably in the near-infrared region, nanospheres offer resonance wavelengths in the visible region [1].

Target of presented work is nanoparticles of biological origin. Nanobioparticles should reveal the properties common of inorganic analogues and specific of biomaterials. Among nanobioparticles the viruses, the smallest infectious agents are of particular interest of diversified works. The sizes of the most viruses range from 20 to 300 nm, are about 1/100 the size of bacteria. Viruses are also perfectly defined organic nanoparticles which are commonly used as scaffolds or nano-vectors in materials engineering and nanotechnology.

In context of detection and identification priorities, study of viruses, nanosized particles in common physical point of view is proposed. Physical, especially spectroscopic properties of virus-like particles

(VLPs) by examination of optical, electrical and geometrical characteristics are considered. The advanced analysis of physical properties of nanoparticles would be done with electromagnetic (EM) wave and associated microscopy techniques or spectroscopic approaches, including infrared (IR), visible (Vis), ultraviolet (UV), Raman, Sum Frequency Generation (SFG) spectroscopies, and methods of particle analysis, with conventional approaches.

Spectroscopic study of nanobioparticles, estimation of oscillation/vibrational modes, possible resonance wavelength range in purpose of determination of the unique spectra of viruses is the main goal of ongoing research activities. Raman spectra of some viral capsids (STNV, CCMV, M13) were calculated using the atomistic approach based on the idea that each viral capsid would have a set of unique frequencies and mode patterns due to the shape and composition of its capsid [2, 3]. Visible or near-infrared laser via impulsive stimulated Raman scattering (ISRS) techniques applied for excitation of mechanical modes of viral capsids resulted in mechanical resonances, possibility of damage the capsid shell rendering it inactive or even destroyed were considered [4, 5]. Number of works based on the elastic network models (ENMs), the empirical bond polarizability model, the rotation translation block method (RTB) declared the presence of vibrational frequency modes related to viral capsids and possibility to appreciate the activation (resonance or other type) of that by means of proper spectroscopic and laser techniques [2-6].

One of the ways for estimation of resonance frequencies range is the method determining the eigenvalues and corresponding eigenvectors obtained by algebraic system of functional equations which could be written based on the solution of electrodynamic task considering EM wave-VL particle scattering [7-10]. Algebraic system has more complicated form when VL particle is of core-shell structure with different electric parameters. Resonant wavelength range is predicted theoretically in the vicinity of values corresponding the maximums of intensity of particular scattered partial waves [7-8].

Estimation of resonant wavelength range based on determination of far-field characteristics such as scattering or absorption cross sections, the intensity of energy, wholly representing the response on wave-particle interaction is considered as the reasonable and decisive solution preferable for studying the spectroscopic properties and determination of possible spectral signatures of bioparticles, viruses.

2. Method of decision

Method of studying spectroscopic characteristics based on elaborated physical models of viruses represents the new possibility for estimation of spectroscopic properties and resonance wavelength range of viral nanoparticles, virions.

Virion, the extracellular infective form of a virus, is considered as a nanoparticle, consisting of inner core of nucleic-acids (RNA or DNA) and outer protective protein coat called capsid. The structure of a large number of viral capsids are icosahedral or near-spherical with icosahedral symmetry. A regular icosahedron is nature's optimum method of producing a closed shell from identical sub-units. Capsid separates the genome from surrounded area and may be considered as outside surface boundary layer or thick interface with specific geometrical, mechanical and electrical properties. Phenomena occurred on inner and outer surfaces of capsid or within the surface layer seems to be dominant in determining the mechanical, electrical and spectroscopic properties of ultrafine bioparticles as it takes place in metal or semi-conductive materials.

Thus, two main mediums of different structures and properties constitute the virion: capsid of protein capsomeres and nucleic acids into the capsid. Type, number and arrangement of capsomers and length of nucleic acid are essential in defining the size of capsid designed mostly in near-symmetrical geometry, having the unique self-assemble mechanism.

A simple calculation shows, that the ratio of inner (core) and outer spherical volumes of icosahedral capsid of bacteriophage T7 (of diameters inner $d_1 = 42.6$ nm and outer $d_2 = 56.6$ nm) is approximately 0.426 [11], so quite a large portion (0.57) of volume in virion is occupied by the capsid proteins. Capsid's size is large enough than the ''discrete'' nature of protein subunits, therefore the influence of that on capsid whole geometry could probably be less significant. This fact allows virion to be modelled as a spherical core-shell particle of smooth inner or outer surfaces. Core-shell VLP model of spherical geometry could be used as the first approximation of shape-structure of icosahedral un-enveloped virion [9-10]. The structure and geometry of capsids as well as processes happening inside a layer probably dominate in determination of physical, spectroscopic properties of nano-sized particles, virions.

Theoretical approach of studying spectroscopic properties of VLPs is based on classical Maxwell's EM theory, separation of variables method for solving Helmholtz's (wave) equation. Solution of Helmholtz's equation leads to the Bessel's and/or Legendre's equations [12]. EM fields in different areas of VLP are written as the sums of multipole-waves with unknown multipole coefficients. Application of boundary conditions to EM field components on core-shell surfaces and labour-consuming mathematical transformations leads to rigorous theoretical solution of EM scattering problem on single VL particle. Analysis of EM field expressions show that arguments of functions determining the multipole coefficients and scattered fields depend only on relative values of particles diameters over wavelength. Therefore, it makes possible to expand the research area, and the findings of well-established Mie theory [13] considering the light scattering on a homogeneous sphere be applied to nanoparticles of core-shell morphology and particles of biological origin as well.

Spectral response on EM wave-VLP interaction is determined by estimation of far-field ($r >> 2d^2/\lambda$ [14]) characteristics representing the angular distribution of scattered (absorbed) energy. Scattering properties are characterized via expressions such as: the total scattering cross section

$$\sigma_T = 2\pi \frac{1}{k_3^2} \sum_{s=1}^{\infty} \left(\left| \mathrm{A}_s^e \right|^2 + \left| \mathrm{A}_s^m \right|^2 \right) \frac{s^2(s+1)^2}{(2s+1)} \tag{1}$$

the forward and backward cross sections with respect to propagation direction of excitation wave:

$$\theta = 0, \qquad \sigma_F = \frac{1}{k_3^2} \left| \sum_{s=1}^{\infty} i^{-s} \frac{1}{2} s(s+1)(\mathrm{A}_s^e + \mathrm{A}_s^m) \right|^2 \tag{2}$$

$$\theta = \pi, \qquad \sigma_B = \frac{1}{k_3^2} \left| \sum_{s=1}^{\infty} i^{s} \frac{1}{2} s(s+1)(\mathrm{A}_s^e - \mathrm{A}_s^m) \right|^2 \tag{3}$$

Multipole coefficients A_s^e and A_s^m are calculated by expressions: for E -type waves ($H_r^e \equiv 0$)

$$\mathrm{A}_s^e = \frac{\dfrac{i^{s-1}(2s+1)}{s(s+1)} [G_s^e(k_3 b, k_2 b) L_s^e(k_1 a, k_2 a) - L_s^e(k_3 b, k_2 b) G_s^e(k_1 a, k_2 a)]}{\Lambda_s^e(k_3 b, k_2 b) G_s^e(k_1 a, k_2 a) - \Gamma_s^e(k_3 b, k_2 b) L_s^e(k_1 a, k_2 a)} \tag{4}$$

for H -type waves ($E_r^m \equiv 0$)

$$\mathrm{A}_s^m = \frac{i^{s-1} \dfrac{2s+1}{s(s+1)} [G_s^m(k_3 b, k_2 b) L_s^m(k_1 a, k_2 a) - L_s^m(k_3 b, k_2 b) G_s^m(k_1 a, k_2 a)]}{\Lambda_s^m(k_3 b, k_2 b) G_s^m(k_1 a, k_2 a) - \Gamma_s^m(k_3 b, k_2 b) L_s^m(k_1 a, k_2 a)} \tag{5}$$

In expressions (4) - (5), functions $G_s^e(k_1 a, k_2 a)$, $G_s^e(k_3 b, k_2 b)$, $L_s^e(k_1 a, k_2 a)$, $L_s^e(k_3 b, k_2 b)$, $\Lambda_s^e(k_3 b, k_2 b)$, $\Gamma_s^e(k_3 b, k_2 b)$ and $G_s^m(k_1 a, k_2 a)$, $G_s^m(k_3 b, k_2 b)$, $L_s^m(k_1 a, k_2 a)$, $L_s^m(k_3 b, k_2 b)$, $\Lambda_s^m(k_3 b, k_2 b)$, $\Gamma_s^m(k_3 b, k_2 b)$ are defined by Riccati-Bessel Spherical and Associated Legendre functions and derivatives of that [8-10].

Wave number of free space and of (q) medium are correspondingly equal to $k = \omega\sqrt{\varepsilon_0 \mu_0}$, $k_q = k\sqrt{\varepsilon_q \mu_q}$, magnetic permeability $\mu_q \approx 1$, ε_q is a dielectric permittivity, $q = 1$ (core), 2 (shell), 3 (surrounded area of VLP), $\varepsilon_0 = 8.85 \cdot 10^{-12}$ F/m, $\mu_0 = 1.26 \cdot 10^{-6}$ H/m, the frequency of excitation wave is denoted by ω, $a = d_1/2$ and $b = d_2/2$.

Dielectric properties of protein and nucleic-acids are still the subjects of experimental and simulation studies needed high-effective measuring and computing techniques. Dielectric constants are estimated from the molecular dynamics (MD) and normal mode (NM) simulations based on various theories, approximations or experimental measurements. Values of different proteins vary from as low as 1 or 2 to as high as 40 and more [15, 16]. In some studies dielectric constant measurements presumable values are about 6.3 (DNA containing virus) and 3.5 (capsid) for T7 viruses [17].
Proposed theoretical solution, analytical expressions of EM fields make possible to estimate the fields in the areas of core, shell and surrounding areas of VLPs. based on machine learning and modeling techniques have been used for generating a simulated spectrum of nanoparticle of given size and available literature optical constants.

3. Results

Computer simulation based on theoretical findings and expressions (1) - (5) is used for studying the spectroscopic properties of VL particles by estimation of EM field characteristics vs excitation wavelength. Efficiency of model of studying the spectroscopic characteristics is demonstrated for estimation of "spectra" of bacteriophage T7 (figure 1, figure 2). Geometric data of icosahedral virions are gained from data bases [11], dielectric permittivities are experimentally measured [17].

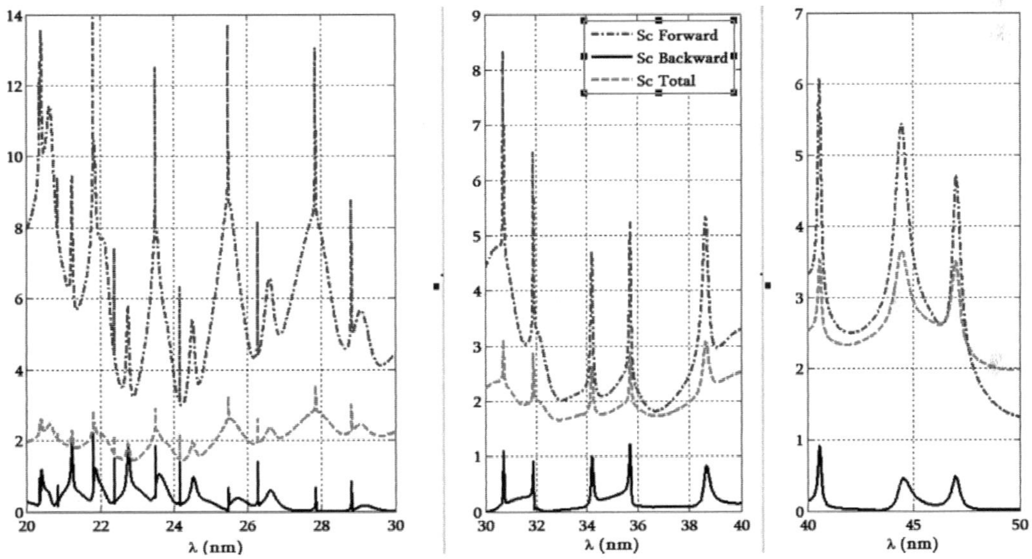

Figure 1. "Spectra" of T7 virus. Scattering cross section (σ/d_2^2) vs excitation wave length (λ). Parameters of core-shell model: diameters of core - d_1= 42.6 nm, shell - d_2=56.6 nm, dielectric permittivity of core - ε_1=6.3, shell - ε_2=3.5, surrounding medium - ε_3=1.

Numerical results are obtained by computer simulation, based on MatLab software (R2011a). Theoretical outcomes and reliability of program units have been tested and compared with results known for spheres [7, 8, 13].
Analysis of result of VLP spectra in different wavelength ranges has revealed the strong dependence of far-field characteristics, scattering cross sections on core-shell related geometric (d_1 and d_2) and electric (ε_1 and ε_2) parameters in resonance wave length ranges (UV for VLP-T7). Expected spectral response is observable on far-field characterizations. Increasing of excitation wavelength leads to diminishing of number of maximums, redistribution and reducing of intensity of scattered wave.

Resonance wavelengths are associated with the values corresponding to the maximums of scattering cross sections and resonance wavelength range could be defined by the formula $\lambda < \pi d_{1,2}\sqrt{\varepsilon_q}$.

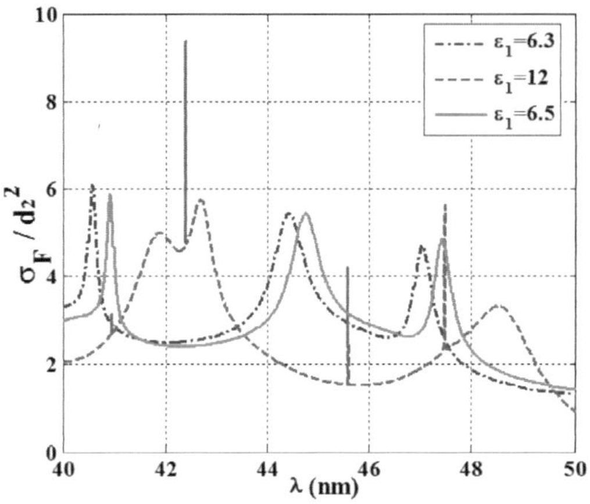

Figure 2. "Spectra" of T7 virus. Forward scattering cross section (σ/d_2^2) vs excitation wave length (λ).
Parameters of core-shell model: diameters of core - d_1 = 42.6 nm, shell - d_2 =56.6 nm, dielectric permittivity of shell - ε_2=3.5, surrounding medium - ε_3=1.

Thus, the proposed model may be used for investigation of spectroscopic properties of nanobioparticles and appreciation of possible resonant wave length ranges correlating with scattering/absorbing efficiency of VLP.
Experimental data obtained by applying exquisite techniques and outcomes of theoretical or simulation models should complement each other and verify factors such as possible anisotropy of core-shell areas of virion, surface roughness and inhomogeneity unforeseen in simplified approaches.

4. Conclusions

Model of studying of spectroscopic properties and estimation of resonance wavelength range of nano-sized particles of core-shell morphology in purpose of sensing and identification of nanobioparticles is considered.
Physical core-shell model of virus-like particle (VLP) is associated with morphology of un-enveloped virion, constructed of protein capsid and nucleic-acids (DNA or RNA).
Rigorous theoretical solution of EM wave scattering on single spherical core-shell particle is applied for determination of EM fields in the areas of core, shell and surrounding medium of VLP, as well as scattering or absorption characteristics.
Numerical results obtained by computer simulation for estimation of EM "spectra" of bacteriophage T7 demonstrate the strong dependence of spectroscopic characteristics on core-shell related geometric and electric parameters in resonance wave length ranges; expected spectral response is observable on far-field characterizations; resonance wavelengths are associated with the values corresponding to the maximums of scattering cross sections.
The shape/structure, inner and outer diameters of capsid, nucleic-acid and protein capsid related dielectric permittivities are considered as the main parameters determining the set of resonance wavelength, as well as the spectral signatures of VLPs, virions. Findings based on well-established theoretical and modelling tools are applicable for studying the spectral properties of nanosized organic, inorganic or hybrid particles of different origin and spherical or core-shell morphology.

Acknowledgements
The work was carried out in Georgian Technical University supported by Shota Rustaveli National Science Foundation (SRNSF) under Grant Agreements FR/430/3-250/13 and MG_36.

References

[1] Jain P K, Lee K S, El-Sayed I H and El-Sayed M A 2006 Calculated absorption and scattering

properties of gold nanoparticles of different size, shape and composition: applications in biological imaging and biomedicine *J. Phys. Chem.* B **110** (14) 7238–48
[2] Dykeman E and Sankey O 2008 Low frequency mechanical modes of viral capsids: an atomistic approach *Phys. Rev. Lett.* PRL **100** 028101
[3] Dykeman E and Sankey O 2010 Atomistic modeling of the low-frequency mechanical modes and Raman spectra of icosahedral virus capsids *Phys. Rev. E. Stat. Nonlin. Soft Matter Phys.* **81** (2) 021918
[4] Tsen K T, Tsen S W and Sankey O 2007 Selective inactivation of micro-organisms with near-infrared femtosecond laser pulses *J. Phys. Condens. Matter.* **19** 472201
[5] Tsen S W *et al* 2014 Studies of inactivation mechanism of non-enveloped icosahedral virus by a visible ultrashort pulsed laser *Virology Journal* **11**:20
[6] Tama F, Gadea F, Marques O and Sanejouand Y 2000 Building-block approach for determining low-frequency normal modes of macromolecules *Proteins Struct. Funct. Genet.* **41** (1) 1-7
[7] Born M and Wolf E 1964 *Principles of optics* (Pergamon Press) p 855
[8] Bohren C and Huffman D 1983, *Absorption and scattering of light by small particles* (John Wiley & Sons) p 530
[9] Kervalishvili P and Bzhalava T 2016 Investigations of spectroscopic characteristics of virus-like nanobioparticles *American Journal of Condensed Matter Physics* **6** (1) 7-16
[10] Kervalishvili P and Bzhalava T 2017 *Spectroscopy of Bioparticles* (Tbilisi: Georgian Technical University ISBN 978-9941-0-9797-3) p 244
[11] VIPERdb database http://viperdb.scripps.edu
[12] *Handbook of mathematical functions with formulas, graphs and mathematical tables* 1964 ed M Abramowitz and I Stegun (National bureau of standards) applied mathematics **55** p 832
[13] Mie G 1908 Beiträge zur Optik trüber Medien, speziell kolloidaler Metallösungen *Ann. d. Physik* **25** 377-445
[14] Ishimaru A 1981 *Wave propagation and scattering in random media* (Moscow: MIR) p 280
[15] Pitera J, Falta M and Gunsteren W 2001 Dielectric properties of proteins from simulation: the effects of solvent, ligands, pH, and temperature *Biophysical Journal* **80** 2546-55
[16] Li L, Li C, Zhang Z and Alexov E 2013 On the dielectric "constant" of proteins: smooth dielectric function for macromolecular modeling and its implementation in DelPhi *J. Chem. Theory Comput.* **9** 2126-2136
[17] Fumagalli L, Esteban-Ferrer D, Cuervo A, Carrascosa J L and Gomila G 2012 Label-free identification of single dielectric nanoparticles and viruses with ultraweak polarization forces *Nature Materials* **11** 808–816

Suppression and enhancement of deep level emission of ZnO on Si^{4+} & V^{5+} substitution

T Srivastava[1], G Bajpai[1] and S Sen[1,2]*

[1]Metallurgy Engineering and Material Science, Indian Institute of Technology Indore, Simrol Campus, Khandwa Road, Indore, India
[2]Electronic Engineering Department and Organic Electronic Research Centre, Ming Chi University of Technology, New Taipei City, Taiwan

Email id: *sens@iiti.ac.in

Abstract. ZnO possess a wide range of tunable properties depending on the type and concentration of dopant. Defects in ZnO due to doped aliovalent ions can generate certain functionalities. Such defects in the lattice do not deteriorate the material properties but actually modifies the material towards infinite number of possibilities. Defects like oxygen vacancies play a significant role in photocatalytic and sensing applications. Depending upon the functionality, defect state of ZnO can be modified by suitable doping. Amount and nature of different dopant has different effect on defect state of ZnO. It depends upon the ionic radii, valence state, chemical stability etc. of the ion doped. Two samples with two different dopants i.e., silicon and vanadium, $Zn_{1-x}Si_xO$ and $Zn_{1-x}V_xO$, for x=0 & 0.020, were synthesized using sol-gel method (a citric acid-glycerol route) followed by solid state sintering. A comparison of their optical properties, photoluminescence and UV-Vis spectroscopy, with pure ZnO was studied at room temperature. Silicon doping drastically reduces whereas vanadium doping enhances the green emission as compared with pure ZnO. Suppression and enhancement of defect levels (DLE) is rationalized by the effects of extra charge present on Si^{4+} & V^{5+} (in comparison to Zn^{2+}) and formation of new hybrid state (V3d O2p) within bandgap. Reduction of defects in $Zn_{1-x}Si_xO$ makes it suitable material for opto-electronics application whereas enhancement in defects in $Zn_{1-x}V_xO$ makes it suitable material for photocatalytic as well as gas sensing application.

1. Introduction

Increasing demands for short-wavelength photonic devices and high power in past decades, leads to the development of several wide band gap materials. Wide-bandgap semiconductors, such as 6H-SiC (3.0eV, at 2 K) and GaN (3.5 eV), have come to the forefront in the past decade but due to their toxicity and high cost, there was a need to replace it [1]. Currently, ZnO is attracting attention for its multiple applications like UV light-emitters, transparent electronics, varistors, surface acoustic wave devices, piezoelectric transducers, gas-sensing and as a window material for display and solar cells [2]. It has advantages over SiC and GaN because of its availability in bulk, single-crystal form, its larger exciton binding energy and its tunable physical properties.

It is known that different types of defects are responsible for green, yellow and orange-red emissions [3]. Defects in ZnO due to doped aliovalent ions can generate certain functionalities [4]. Such defects in the lattice do not deteriorate the material properties but actually modifies the material towards infinite number of possibilities. Impurities and deliberate doping with aliovalent (differently charged) ions have marked effects on the defect equilibria which modifies colour emission and opto-electronic properties of ZnO. Reports found red emission in Co doped ZnO[5], yellow-orange-red emission in Mn doped ZnO[6], blue emission in Cu doped ZnO[7] etc.

Density functional theory (DFT) calculations show that the band structure of ZnO is determined by O 2p and Zn 3d orbitals, and the upper levels of the valence band are constituted by O 2p levels. The defect levels in ZnO are usually categorized into four types, namely, oxygen vacancies (VO) and zinc vacancies (VZn) which are generally surface defects, and interstitials (Zn and O) and antisites which exist in the bulk of the material [8,9]. The majority of defect levels belong to oxygen and zinc vacancies. The nature of defects in ZnO depends upon various factors like synthesis technique, dopant, morphology etc [10].

In the present study, we have reported the sol-gel processed pure ZnO and 2% Si^{4+} & 2% V^{5+} doped ZnO. The aim of this study is to analyse the effect of dopant on defect state, colour emission and optical properties of ZnO.

2. Experimental

$Zn_{(1-x)}M_xO$ nanopowders, for M= Si^{4+} and V^{5+}, for x=0 & 0.020 have been synthesized by standard Pechini sol-gel process followed by solid state sintering. Pure ZnO, Si^{4+} doped ZnO and V^{5+} doped ZnO has been named as ZnO, ZS2 and ZV2 respectively in the study. ZnO powder (Alfa Aesar, purity 99.9%) was firstly dissolved in HNO_3. For silicon doping, Tetraethyl Orthosilicate [$(C_2H_5O)_4Si$] (Alfa Aesar (99%)) solutions were dissolved in deionized water. For vanadium doping, V_2O_5 was dissolved in NH_4OH. These silicon and vanadium solution is added to Zn solution sepertely to achieve ZnO, ZS2 and ZV2. These three solutions were vigorously stirred for several hours. Polymeric solution was formed by heating 1 g of citric acid (Alfa Aesar, 98%) and 1 mL of 20% concentrated glycerol (Alfa Aesar, 99.9%) in de-ionized water at 70 °C for 4 h. These pure Zn, Zn/Si and Zn/V solutions were then sepertaely added to the polymeric solution. The resultant solutions were stirred and heated on hot plates at ~60 °C. The Zn/Si ions attached homogenously with the polymeric chains. The solution gradually dehydrated to form gel in ~4hrs. The gels were burnt on hotplates in ambient conditions. The resultant powders were decarbonized and denitrified by heating in air at 450°C for 6 h. The obtained powder was pressed into pellets and heated to 600°C for 2 hours.

X-ray diffraction studies were carried out using a Bruker D2-Phaser diffractometer. Agilent UV–vis spectrometer (model Carry 60) was used to analyze electronic bandgap. Room temperature photoluminescence measurement was investigated using Dongwoo Optron DM 500i.

3. Result and Discussion

XRD spectra revealed a dominant hexagonal wurtzite ZnO structure with some minor reflections similar to zinc blend as shown in figure 1. However, there were no secondary phases found related to simple or complex oxides of Zn, Si & V. There is a clear shifting in XRD peaks of ZS2 and ZV2 in comparison to ZnO which shows change in lattice parameter with Si^{4+} and V^{5+} doping. Table1 shows change in lattice parameter with doping.

Table 1: Lattice parameters change with doping.

Sample name	a=b (Å)	c(Å)
ZnO	3.25	5.206
ZS2	3.253	5.16
ZV2	3.249	5.204

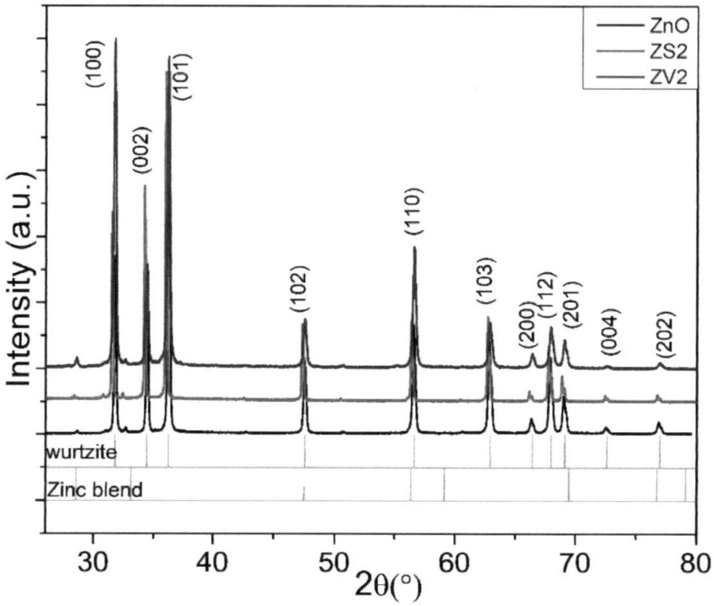

Figure 1: XRD spectra of ZnO, ZS2 and ZV2.

Optical absorption studies indicate that band gap increases with Si^{4+} and V^{5+} substitution. The bandgaps were calculated using Tauc plot as shown in figure 2. Pure ZnO has bandgap of 3.16 eV which increases to 3.19 nm for ZS2 and ZV2. Urbach energy, E_u, is a measure of structural disorderness. E_u was estimated to confirm reduction in lattice disorderness or strain.

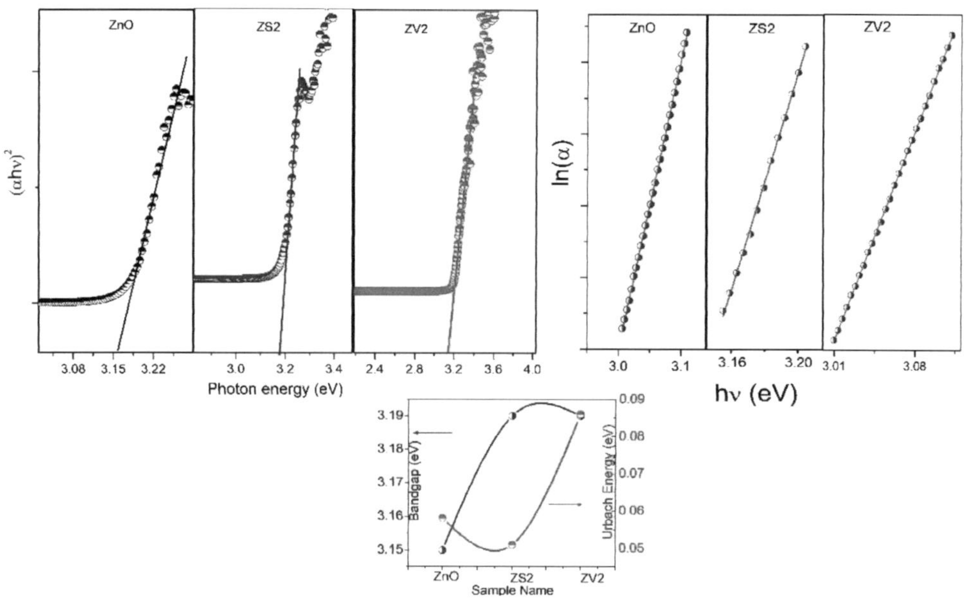

Figure 2: Linear fitting for bandgap calculation, Linear fitting for Urbach energy calculation and Bandgap & Urbach energy variation with x.

The absorption coefficient, α(ħν), below the optical gap (tail absorption) depends exponentially on the photon energy as:

$$\alpha(\hbar\nu) = \alpha_0 \exp(\hbar\nu/Eu);\ \text{where, } \alpha_0 \text{ is constant.}$$

For pure ZnO, urbach energy was 58 meV which decreases to 53 meV for ZS2 and increases for ZS2 to 83 meV. In ZS2, decrease in Urbach energy effectively increases the bandgap. In case of vanadium doping (ZV2), Urbach energy increases which do not affect the bandgap but it signifies enhancement in lattice disorder which results into strained lattice. The bandgap enhancement in ZV2 might be due to interaction between d level of vanadium and bands (conduction and valence band).

Room temperature Photoluminescence (figure 3) has been studied to understand the changes in defect state and their related colour emission with doping in ZnO. Defect state plays a very significant role in determining opto-electronic properties of the material. Figure 3 shows the photoluminescence spectra of the pure ZnO and doped ZnO (ZS2 and ZV2). ZnO photoluminescence spectra consist of a near band UV emission (NBE) at 380 nm and a deep level visible (DLE) emission extending laterally from 450 nm to 750 nm. The DLE of pure ZnO is centred at 518 nm whereas for ZS2 and ZV2, it is centred at 620 nm and 558 nm respectively. Deep level emission in ZS2 has been suppressed significantly whereas in case of ZV2, there is a drastic enhancement of DLE. As per conventional understanding, intensity of NBE and DLE has been calculated and plotted. A marked increase in I(NBE)/I(DLE) ratio for ZS2 and decrease for ZV2 confirms the suppression of defect with Si^{4+} and enhancement of defect with V^{5+}.

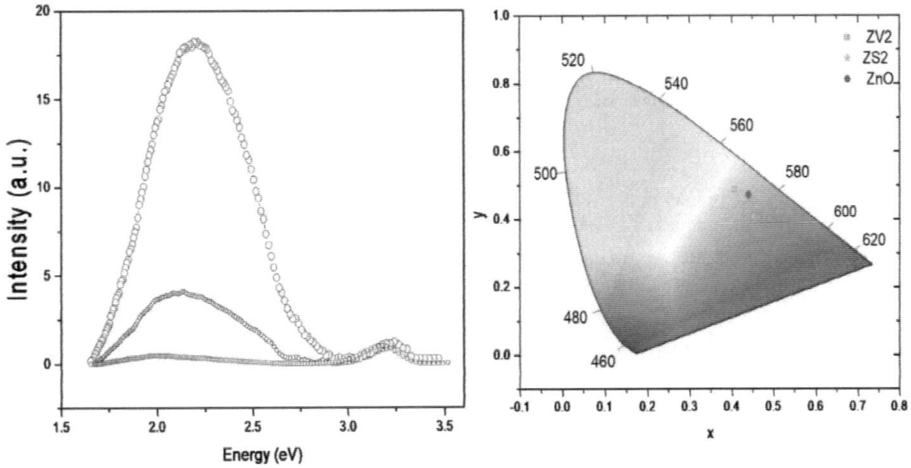

Figure 3: Photoluminescence spectra and Chromaticity coordinates of ZnO, ZS2 and ZV2.

To further confirm this, chromaticity coordinates have been calculated and plotted on 1931 standard chromaticity diagram. The coordinates are (0.44, 0.47), (0.45, 0.41) and (0.40, 0.48) for ZnO, ZS2 and ZV2 respectively. It confirms that ZS2 has orange red emission whereas ZV2 has enhanced green emission.

In case of ZS2, Si^{4+} has higher charge than the Zn^{2+} host atom. It has been reported that oxygen vacancies in ZnO are responsible for green emission. Due to higher charge of Si^{4+}, it attracts more oxygen in the lattice and therefore suppress oxygen vacancies of ZnO lattice. Due to this, DLE has reduced drastically with silicon doping. In case of V^{5+}, inspite of extra charge (in comparison to Zn^{2+}) there is drastic enhancement in green emission. As discussed above, higher charge of dopant will lead to reduction in oxygen vacancies and quenching of green emission. This signifies, oxygen vacancies are not the source of green emission in ZV2. Density functional theory confirmed that vanadium doping in ZnO will leads to new defect state formation due to hybridization of 3d of vanadium and p orbital of oxygen (V3d-O2p). Transition from conduction band to new *pd* hybrid state might be the possible reason for enhancement in green emission.

4. Conclusion

ZnO, 2% Si doped ZnO (ZS2) and 2% V doped ZnO (ZV2) has been prepared using Sol-gel method. 2% silicon and vanadium are completely soluble in ZnO. No extra phase is seen in XRD pattern. Bandgap enhancement in case of ZS2 is due to reduced Urbach energy and lattice disorder whereas in case of ZV2, bandgap enhancement is due to interaction between d-level of vanadium and bands (conduction and valence band of ZnO). Deep level emission (DLE) in case of ZS2 reduced drastically in comparison to pure ZnO which is might be due to higher charge of Si^{4+} (in comparison to Zn^{2+}). In ZV2, DLE enhanced significantly due to the transition from conduction band to new hybrid state created due to interaction of V3d-O2p. Silicon doping makes ZnO, a potential material for photonics and opto-electronics application due to its tunable colour emission and enhanced transparency to visible light. Vanadium doping makes ZnO, a potential material for photocatalytic and gas sensing application by creating more trap states.

5. Acknowledgement

The authors thank Dr. Vipul Singh for providing PL spectroscopy and Dr. Pankaj Sagdeo for providing UV–Vis facility.

References:

[1] Look D C 2001 Recent advances in ZnO materials and devices *Mater. Sci. Eng. B* **80** 383–7

[2] Pearton S J, Norton D P, Ip K, Heo Y W and Steiner T 2003 Recent progress in processing and properties of ZnO *Superlattices Microstruct.* **34** 3–32

[3] Djurišić A B, Leung Y H, Tam K H, Hsu Y F, Ding L, Ge W K, Zhong Y C, Wong K S, Chan W K, Tam H L, Cheah K W, Kwok W M and Phillips D L 2007 Defect emissions in ZnO nanostructures *Nanotechnology* **18** 095702

[4] Norton D P, Heo Y W, Ivill M P, Ip K, Pearton S J, Chisholm M F and Steiner T 2004 ZnO: growth, doping & processing *Mater. Today* **7** 34–40

[5] Loan T T, Long N N and Ha L H 2009 Photoluminescence properties of Co-doped ZnO nanorods synthesized by hydrothermal method *J. Phys. Appl. Phys.* **42** 065412

[6] Romeiro F C, Marinho J Z, A. Silva A C, Cano N F, Dantas N O and Lima R C 2013 Photoluminescence and Magnetism in Mn2+-Doped ZnO Nanostructures Grown Rapidly by the Microwave Hydrothermal Method *J. Phys. Chem. C* **117** 26222–7

[7] Zhu H, Iqbal J, Xu H and Yu D 2008 Raman and photoluminescence properties of highly Cu doped ZnO nanowires fabricated by vapor-liquid-solid process *J. Chem. Phys.* **129** 124713

[8] K. Kavitha M, B. Jinesh K, Philip R, Gopinath P and John H 2014 Defect engineering in ZnO nanocones for visible photoconductivity and nonlinear absorption *Phys. Chem. Chem. Phys.* **16** 25093–100

[9] Sheetz R M, Ponomareva I, Richter E, Andriotis A N and Menon M 2009 Defect-induced optical absorption in the visible range in ZnO nanowires *Phys. Rev. B* **80** 195314

[10] McCluskey M D and Jokela S J 2009 Defects in ZnO *J. Appl. Phys.* **106** 071101

IOP Conf. Series: Journal of Physics: Conf. Series **987** (2018) 012025 doi:10.1088/1742-6596/987/1/012025

Hydrogenization of fullerene as a method of storage of hydrogen

Zh.E. Otarbay[1], M.T. Gabdullin[1], Kh.A. Abdullin[1], D.V. Shur[2], D.V. Ismailov[1], Ye. Yerlanuly[1], D.S. Kerimbekov[1]

[1]National Nanotechnology Laboratory of the Open Type, al-Farabi 71/23, Almaty, Kazakhstan, 050040
[2]Institute of Material Science Problems named after N.N. Francevich, NAS of Ukraine, Kruzhanovskogo 3, Kiev, Ukraine, 03680

otarbay@physics.kz

Abstract. In the present work the method of catalytic hydrogenation of fullerenes is considered. To determine the optimal regime for the hydrogenation of fullerenes at a hydrogen pressure of 12 MPa and a temperature in the range of 200-600°C, the mass of fullerite C60 was equal to 0.501 g. During the conducting of this experiment, it was found that the system reached its maximum interaction speed at 2150 minutes, which corresponded to temperature of 433°C. As a result, the temperature, at which the rate of interaction of hydrogen with the fullerene molecule is maximized, was defined and the given value of temperature is in the range of 425-455°C, where the total hydrogen content is about 8.2% by weight.

1. Introduction

The search for an ecologically clean and highly efficient cheap energy carrier, an easily accessible and almost inexhaustible source of energy has long ago and firmly taken the leading place in the list of problems affecting the future of not only further development but also the existence of all mankind. One of the promising ways of obtaining thermal energy is the use of hydrogen produced from water using environmentally friendly primary energy sources. Hydrogen is an ideal fuel with a high calorific value and a non-hazardous combustion product - water vapor [1].

2. Methods

Storage of hydrogen is one of the intermediate links in the life cycle of hydrogen from its production to consumption. The development of the most economical and efficient ways of storing hydrogen is one of the main technological problems of hydrogen energy.

To store hydrogen various types of hydrogen batteries are suggested - hydrogen sources, in which at room temperature hydrogen atoms are in a bound state with a sorbent material, either chemically or in an encapsulated form. As the temperature rises, such bonds of hydrogen with the sorbent are destroyed, thereby releasing hydrogen atoms. Today due to their low specific density and large specific surface area carbon nanomaterials are used to store hydrogen. For example, activated carbon [2], graphene [3-4], multi-walled and single-walled carbon nanotubes [5-6], fullerenes [7-8], etc. Thus, the study of hydrogen-sorption characteristics of fullerene is very relevant.

It is known that fullerene can exist as a gas (at temperatures of 450ºC and above) or a solid. Under normal conditions fullerene is in the form of a solid, which is called fullerite. Hydrogenation of solid fullerene can be carried out by various methods such as direct non-catalytic hydrogenation [9], catalytic hydrogenation in toluene solution [5], radical hydrogenation with the C_2H_5I promoter [10] and many other methods presented in [11-14].

To carry out work on the hydrogenation of fullerene samples a method was developed based on the determination of the hydrogen-sorption characteristics of the samples at different pressures and temperatures. Typically, the hydrogen sorption capacity is measured by a volumetric method in installations of the Sievert type. This known method is sufficiently accurate, if we work with samples of hydrogen sorbing materials having a strictly defined density. Otherwise, especially if the specific density is low, the value of uncertainty in the density can give a very large contribution to the error in determining the hydrogen sorption capacity.

The structure of the experimental setup consists of a stainless tube (reaction chamber), stove, systems of evacuation and puffing working gas of hydrogen through a special purification filter [15]. This installation is intended primarily for studying the hydrogen-sorption characteristics of carbon materials and composites on their basis with low specific density using volumetric method in the pressure range from 0.01 to 30 MPa H_2 and at the boiling point of liquid nitrogen and when heated to a temperature of about 1000°C.

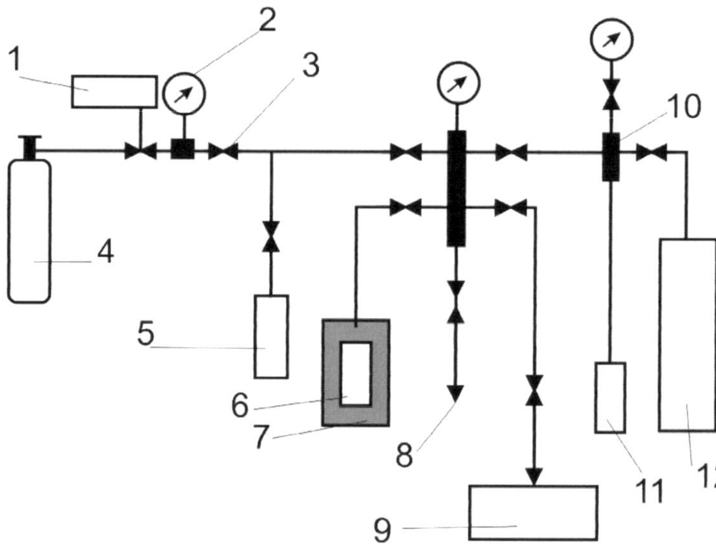

Figure 1. Schematic diagram of the experimental setup for the measurement of hydrogen sorption [15]: 1 - refueling port; 2-manometers; 3 - a hydrogen valve; 4 - cylinder with compressed hydrogen of technical purity; 5 - metal hydride accumulator compressor with a capacity of 220 liters of hydrogen; 6- experimental reactor; 7 - thermostat; 8 - the main line of discharge into the atmosphere; 9 - vacuum system; 10 - low pressure collector; 11 - buffer capacity of 50 cm^3; 12 - buffer capacity of 300 cm^3.

3. Results and discussion

In this paper hydrogen-sorption characteristics of fullerene are studied by a non-catalytic method at the Sievert type installation. The essence of the experimental procedure is as follows: if the hydrogen capacity of the sample is known in advance, for example, the sample is not hydrogenated and it is located in a closed volume filled with hydrogen with certain parameters, then the percentage of hydrogen in the sample can be determined from the changes in the parameters of the gas phase and comparison with the initial values.

The sorption capacity for hydrogen was determined by the following formula [15]:

$$a = \frac{\Delta n}{m} + \frac{P}{ZRT}\frac{1}{\rho},$$

where Z = Z (P, T) is the compressibility factor for gaseous hydrogen, R is the gas constant, $\Delta n = n_0 - n$ is the difference, and ρ is the specific density.

In the process of studying hydrogen-sorption characteristics of fullerenes experimental work on hydrogenation was carried out. The experiment was carried out in two stages: the first - preliminary annealing. The second is the hydrogenation of fullerite.

Experimental work on hydrogenation was carried out to obtain hydrofullerite. It is known that in direct non-catalytic hydrogenation hydrofuelerites $C_{60}H_2$-18 are obtained at known hydrogen pressure parameters of 50-85 MPa and temperatures in the range of 300-350^{0}C. Such high pressures of the working gas (hydrogen) lead to the polymerization of fullerene molecules, the destruction of their frames and the hydrogenation of the formed fragments (it was shown by analyzing mass spectrometry and dehydrogenation of products obtained under these conditions). Thus, in order to avoid the polymerization and destruction of fullerene molecules, a necessary condition is a reduction in the pressure of the working gas, and with it, therefore, an increase in the synthesis temperature to create hydrogenation conditions. For these purposes, an experiment was performed at the catalytic hydrogenation of fullerene molecules in a toluene solution at hydrogen pressures of 12 MPa. As a result, it was found that the samples of hydrofullerene are not destroyed. Based on these considerations, the working pressure in subsequent experiments of non-catalytic method was the hydrogen pressure of 12 MPa.

To determine the optimal mode for carrying out the hydrogenation reaction of a fullerene molecule at a hydrogen pressure of 12 MPa, a 75-hour experiment was carried out. In the course of this experiment, the temperature, at which the rate of interaction of hydrogen with the fullerene molecule was maximized, was determined. To carry out the experiment we used a sample of fullerite C_{60} with a mass of 0.501 g. The reactor was evacuated for one hour to remove impurities from the fullerite. After that, hydrogen was fed from the metal hydride storage to the reactor up to the pressure of 12.5 MPa and held for 4 hours. Then, the temperature in the reactor was raised to 200°C (at a speed of 1 degree per minute). The pressure in the reactor increased by 0.3 MPa (reaching 12.8 MPa). This can be explained by the decomposition of solvates formed during the crystallization of fullerite. To stabilize the baro-temperature regime the sample was held for another 4 hours (T=200°C, P=12.8 MPa). At 480th minute the temperature rise in the reactor was continued from 200 to 600°C at a rate of 0.125 degrees per minute.

In the course of this experiment it was found that the system reached its maximum interaction speed at 2150th minute, which corresponded to a temperature of 433°C (Figure 2).

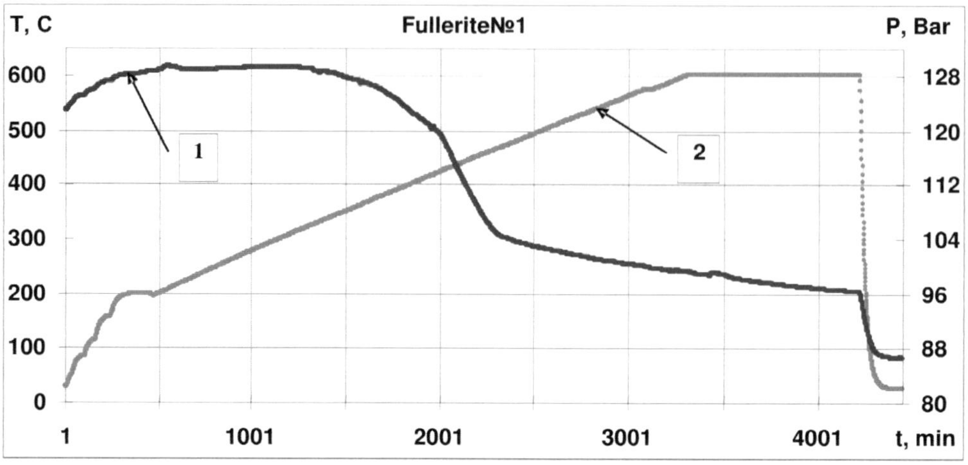

Figure 2. Dependence of the pressure change in the reactor with a temperature change at a rate of 0.125 degrees per min. (1 - curve of pressure change in the reactor, 2 - curve of temperature change in the reactor).

This corresponds to a change in temperature from 425 to 455 ° C. With a further increase in temperature the rate of interaction began to decrease.

In this experiment holding of the sample for 10 hours at P=12 MPa and a temperature of 200°C resulted in a change in the sample mass by 0.6% by weight. The total amount of hydrogen absorbed by the sample during the experiment was fixed at 8.2 wt.% of H (Figure 3)

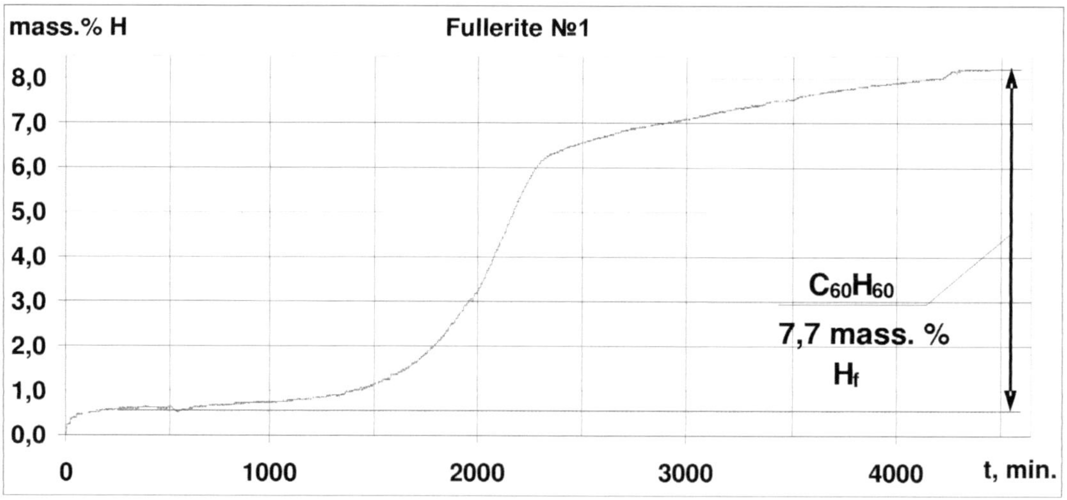

Figure 3. Experimental curve of the interaction of gaseous hydrogen with fullerite. The arrow indicates the fraction of fullerized hydrogen (H_f).

Another experiment showed that preliminary annealing of fullerite in a vacuum at 350°C leads to a more intense interaction of it with hydrogen gas. A rapid (within 3 hours) temperature rise to 550°C at a hydrogen pressure of 12.5 MPa leads to saturation of the sample with hydrogen to 7.3% by weight of H during this period.

Experiments have shown that hydrogen absorbed by fullerite after depressurization gradually spontaneously desorbs under normal conditions. Such hydrogen mobility can mean that the process of chemisorption of hydrogen by all fullerene molecules of C_{60}will proceed uniformly with the formation of energetically equivalent molecules, or at least fullerene molecules having an equal number of hydrogen atoms throughout the lattice volume. In this case the topochemistry of the hydride formation reaction will correspond to the volume formation of the hydride.

Based on the conducted experiments, it can be concluded that the hydrogenation of a fullerene molecule is an energy-controlled process, in which the temperature of the system determines the number of hydrogen atoms on the shell of the fullerene molecule. This conclusion is of great importance for the further study of the fullerene-hydrogen system and, especially, for its application.

As already mentioned above, carbon nanomaterials such as activated carbon, carbon nanofibres, graphene, multi-walled and single-walled carbon nanotubes, fullerenes, etc. are used to create the capacity of the hydrogen storage due to its low specific density and large specific surface area.

Activated carbon (ACs) is synthesized by carbonization of organic precursors followed by an activation step using, for example, annealing in an oxidizing atmosphere or chemical treatment. Such results of the synthesis procedure in materials with a very high specific surface area up to 3300 m^2/g, which have excellent properties of hydrogen adsorption at 77 K. ACs is capable of adsorbing about 6.0 wt% of H_2 at a pressure of 40 bar at 77 K [16]. As for carbon nanotubes, were first discovered in 1991 by Professor Iijima [17] and after attracted much attention as a promising material for hydrogen storage applications. Before the adsorption of hydrogen, the measurements of nanotubes are usually subjected to some processing to make the inner part accessible for penetration of the gas and to increase the available surface area, in CNT there is basically a specific surface area in the range of 600-2600 m^2 / g [18]. At the temperature of liquid nitrogen CNT capable of adsorbing up to 8.0 wt.% of H_2 [19]. Nevertheless, at room temperature hydrogen the capacity of CNTs for storage of H_2 is less than 1mass% at pressures up to 100bar [20-21]. Considering various modifications of carbon nanostructures, it is necessary to focus on fullerenes and their hydrides, since they have unique adsorption properties and during our experiment it was recorded that the amount of hydrogen absorbed by the sample reached 8.2% by weight of H_2, which exceeds the ability to adsorb in the above works. Thus, during the conducting of the experimental work on the hydrogenation of fullerite, hydrofullerenes with various hydrogen contents were experimentally obtained. The optimal hydrogenation regime of C_{60}, leading to complete hydrogenation of the fullerene C_{60} molecule,was determined. It was also found that the optimum conditions for the hydrogenation of fullerite C_{60} are P=12 MPa and T=425÷ 455°C.

References

[1] Gel'fand B Ye, Popov O Ye and Chayvanov B B 2008 *Vodorod: parametry goreniya i vzryva, (Fizmatlit, in rus)* p 288

[2] Schlapbach L. And Zuttel A 2001 *Hydrogen-storage materials for mobile applications. Nature* vol 6861,p 353-358

[3] Novoselov K S 2005 *Two-dimensional gas of massless Dirac fermions in graphene*, Nature 438, p 197 DOI:10.1038

[4] Srinivas G; Zhu Y W; Piner R; Skipper N; Ellerby M and Ruoff R 2010 *Synthesis of graphene-like nanosheets and their hydrogen adsorption capacity (Carbon)* p 630-635

[5] Ye Y; Ahn C C.; Witham C, Fultz B; Liu J.; Rinzler A G.; Colbert D; Smith K A; Smalley R. E.1999 *Hydrogen adsorption and cohesive energy of single-walled carbon nanotubes (Applied Physics Letters)* vol 74(16), p 2307-2309

[6] Zhang X R, Cao D P and Chen J F 2003 *Hydrogen adsorption storage on single-walled carbon nanotube arrays by a combination of classical potential and density functional theory. (Journal of Physical Chemistry)* vol 107(21),p 4942-4950

[7] Dresselhaus S, Williams K and Eklund P C 1999 *Hydrogen absorption in carbon materials (MRS Belliti)* vol.24, p 45-50.

[8] Shigematsu K, Abe K, Mitani M and Tanaka K *Catalytic hydrogenation of fullerene C60*

(Chem. Express) vol.7,p 37-40

[9] Jin C, Hettich R, Compton R, Joyce D, Blencoe J and Burch T 1994 *Direct solid-phase hydrogenation of fullerenes (J. Phes. Chem)* vol 98, p 4215.

[10] Attalla M I, Vassallo A M, Tattam B N and Hanna J V 1993 *Preparation of hydrofullerenes by hydrogen radical induced hydrogenation (J.Phys. Chem)* vol 97, p 6329-6331.

[11] Henderson C C and Cahill P *A C60h2 - Synthesis of the Simplest C60 Hydrocarbon Derivative. Science* vol 259, p 1885-1887

[12] Ballenweg S, Gleiter R and Kratschmer W *Hydrogenation of Buckminsterfullerene C-60 Via Hydrozirconation - a New Way to Organofullerenes (Tetrahedron Letters)* vol 34, p 3737-3740

[13] Ruchardt C, Gerst M, Ebenhoch J; Beckhaus H D, Campbell E E and Pitter S *Bimolecular Radical Formation through H-Transfer .3. Transfer Hydrogenation and Deuteration of Buckminsterfullerene C-60 by 9,10-Dihydroanthracene and 9,9',10,10'[D(4)]Dihydroanthracene. Angewandte Chemie-International Edition in English –* vol 32,p 584-586

[14] Darwish A D, Abdulsada A K, Langley G J, Kroto H W, Taylor R, Walton D R, *Polyhydrogenation of [60]-Fullerenes and [70]-Fullerenes// Journal of the Chemical Society-Perkin Transactions* vol 12, p 2359-2365

[15] Schur D V, Gabdullin M T, Zaginaichenko S Yu, Veziroglu T N, Lototsky M V,. Bogolepov V A and Savenko A F 2016 *Experimental set-up for investigations of hydrogen-sorption characteristics of carbon nanomaterials (International Journal of Hydrogen Energy* vol 41, p 401-406

[16] Fierro V, Szczurek A, Zlotea C, Mareche J F, Izquierdo M T, Albiniak A, Latroche M, Furdin G and Celzard A 2010 *Experimental evidence of an upper limit for hydrogen storage at 77 K on activated carbons (Carbon)* vol 48, p 1902-1911

[17] Iijima S *Helical Microtubules of Graphitic Carbon. Nature* vol 354, p 56-58

[18] Schur D V, Dubovoy A G, Zaginaichenko S Yu, Adejv V M, Kotko A V., Bogolepov V A, Savenko A F and Zolotarenko A D *Production of carbon nanostructures by arc synthesis in the liquid phase (Carbon)* vol 45(6), p 1322-1329

[19] Ye Y, Ahn C C, Witham C, Fultz B, Liu J, Rinzler A G, Colbert D, Smith K A and Smalley R E *Hydrogen adsorption and cohesive energy of single-walled carbon nanotubes(Applied Physics Letters)* vol 74(16), p 2307-2309

[20] Xu W C, Takahashi K, Matsuo Y, Hattori Y, Kumagai M, Ishiyama S, Kaneko K and Iijima S,2007 *Investigation of hydrogen storage capacity of various carbon materials. (International Journal of Hydrogen Energy)* vol 32(13), p 2504-2512

[21] Panella B, Hirscher M and Roth S 2005 *Hydrogen adsorption in different carbon nanostructures(Carbon)* vol 43(10), p 2209-2214

IOP Conf. Series: Journal of Physics: Conf. Series **987** (2018) 012026 doi:10.1088/1742-6596/987/1/012026

Comparison between layering $NbSe_2$ and rod characteristic of MgB_2 by investigation of elastic constants

Asiye Shokri, Ahmad Yazdani* and Behrad Barakati

Department of Physics, Tarbiat Modares University, P.O. Box 14115-175, Tehran, Iran

*yazdania@modares.ac.ir

Abstract. The delicate balancing of strong anisotropy on strength of hybridisation resulted to CDW- order "T_{CDW}=33K" and finally emerging superconductivity at "T_c= 7.2K" are the most intriguing question in characteristic behaviour of $NbSe_2$. On other hand, the original mechanism of MgB_2 old superconductor, which has unlike the cuprates a lower anisotropy on strength hybridisation is still unknown. We believe this could result to bond exchange and larger coherence length of the grain boundary to current. Since the cause and the mechanism of band strengths of two original layering and rod structures are consequence of bond- rupturing-atomic displacement, here the stability of crystalline structure of inter atomic potential through the elasticity-compressibility is investigated. Consequently, in order to clear out the strong difference between the layering $NbSe_2$ and domination of rod-character of MgB_2 the stability of both crystal structures through the cohesive energy c/a, c_{zz} and c_{33} are investigated. The proposed investigations are more evident on different characteristic behaviour of calculated parameters.

1. Introduction

As a member of transition metal dichalcogenids, it has been more than fifty years that 2H-$NbSe_2$ has attracted many attentions due to its electronic properties and phase transition [1]. 2H-$NbSe_2$, a hexagonal structure with space group $P6_3/mmc$ and structural parameters of a=3.6Å, and c=12.6Å contains two $NbSe_2$ units as ABAB stacking and it is a type II superconductor under 7.1K [2,3]. On other hand electronic instabilities driven by Fermi surface nesting, are because of the two dimensional nature of materials like 2H-$NbSe_2$. The interplay between charge density wave order (T_{CDW} =33K [4]) and superconductivity (T_c=7.1K) and also coexistence of these properties has attracted so many attentions [5]. In 2015, a group showed the decrease of critical temperature of superconductivity in this material by decreasing the layers to T_c=3.1K in achieving a monolayer. It is while they found that by decreasing the layers, T_{CDW} is increasing to 145K for the monolayer [6]. H. Suderow et al. in 2005 measured critical temperature up to 19GPa and found a maximum at 10.5GPa [7]. It is while different works on MgB_2 show that the critical temperature decrease with increasing of the pressure [8]. MgB_2 with the highest critical temperature among conventional BCS and intermetallic compounds (T_c=39K) [9], has hexagonal structure with space group P6/mmm (a=3.08 and c=3.52Å) [10]. Wires and thin films of this material have been constructed for decades [11]. Although the Be and Mg belong to a same column in periodic table, but the different stiffens [12] made MgB_2 rod and wire shape unlike BeB_2 which is to be a layered material like $NbSe_2$. Relative to layering of low T_c=0.7 of BeB_2, as it is presented by others [12], and because of the different behavior of critical temperature in presence of the pressure, the difference between layering character of $NbSe_2$ and rod-formation character of MgB_2 superconductors could be the intriguing question. The directional bond formation or a strong anisotropy of different "B-B" length do to smaller- intro- layer to the inter-layer distance of "B-B" could be the cause where we are following characteristic behavior of MgB_2 for the cause of rod/wire character. Insignificancy of T_c=0.7 in BeB_2 is surprisingly in a challenge which could be due to its layering- character as its common feature with $NbSe_2$.

In this work, we have compared the structural properties of these two materials. For the best of our knowledge there is no study, comparing the structural properties of $NbSe_2$ and MgB_2 from the elastic constants point of view

2. Calculation methods

Calculations of the electronic structure were carried out using the linearized augmented plane wave(LAPW) method as performed in the WIEN2k package [13]. Calculating the elastic constants of these hexagonal structures were performed using Hex-elastic package [14] based on the second-order derivative (E''(δ)) of the polynomial t of energy (E=E(δ)) vs strains, at zero strain (δ=0). The generalized gradient approximation exchange- correlation function of Perdew et al [15] was used in the present work.

3. Result and discussion

As a starting point of discussion, the calculated equilibrium lattice parameters of MgB_2($NbSe_2$) were calculated to be a=3.06Å(3.38Å) and c=3.54Å(12.96Å). Diagram of density of states for both structures are shown in figure 1. The c/a optimization calculations were done and c/a for constant volume was optimized to find the optimized a and c lattice parameters (figure 2). Elastic constants of crystals provide a link between the mechanical and dynamical behaviors of solids, give important information about the elastic response of a crystal to an external pressure and also can predict the anisotropic character of binding. Five independent elastic constants are listed in table 1. To investigate the stiffness, hardness and structural anisotropy of these structures, bulk, shear and young modulus based on voigt, Reuss and Hill approximations are presented in table 2. Young's modulus, the ratio of stress and strain, indicates the stiffness. It means the material with larger Young's modulus is stiffer. As can be seen from table 2, MgB_2 is stiffer than $NbSe_2$. If we want to know how much a material is resistant against shape changes, we should discuss about the hardness which is related to bulk modulus (resistance against volume changes) and shear modulus (resistance against reversible deformations). Then, it can be seen that MgB_2 is more resistant. Strain vs energy in figure 3 shows the C_{zz} elastic constant diagram which says by this strain, volume and symmetry will not change. The structural anisotropy can have different definitions. The amount of deviation from 1 can be a criterion for anisotropy according to equation (1). It shows that MgB_2 has anisotropic structure due to f=1.86.

$$f = \frac{C_{11} + C_{12} - 2C_{13}}{C_{33} - C_{13}} \qquad (1)$$

Table1: The calculated elastic constants C_{ij} (in GPa) for MgB_2 and $NbSe_2$.

	C11	C12	C13	C33	C55
MgB_2	430.8062	49.9489	38.6537	255.1515	73.0260
$NbSe_2$	167.4299	82.9819	79.8618	209.5128	35.2905

IOP Conf. Series: Journal of Physics: Conf. Series **987** (2018) 012026 doi:10.1088/1742-6596/987/1/012026

Table2: Calculated bulk(B), shear modulus(S) and Young's modulus E (all in GPa), in Voigt (V)–Reuss (R)–Hill (H) approximations

	B(V)	B(R)	B(H)	S(V)	S(R)	S(H)	E(V)	E(R)	E(H)
MgB_2	152.36	143.07	147.72	133.26	109.82	121.54	309.54	262.34	286.145
$NbSe_2$	114.42	113.45	113.93	42.672	41.2	41.941	113.86	110.27	112.071

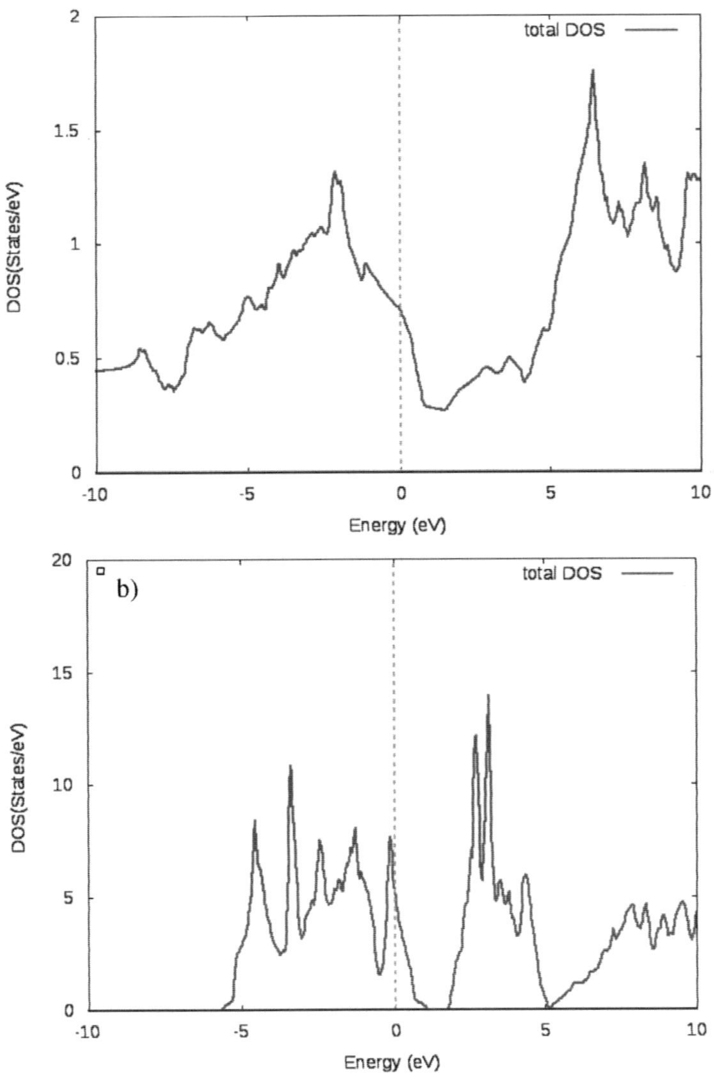

Figure 1: density of states for a) MgB_2 and b) $NbSe_2$

IOP Conf. Series: Journal of Physics: Conf. Series **987** (2018) 012026 doi:10.1088/1742-6596/987/1/012026

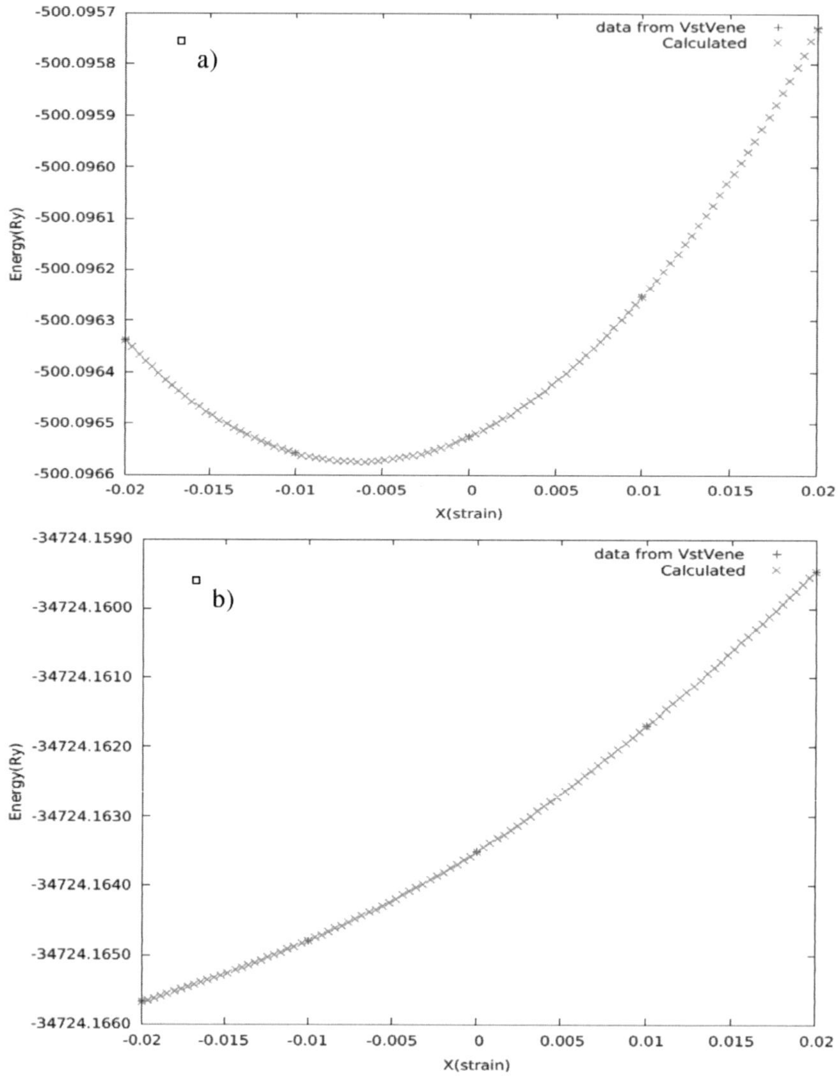

Figure 2: C_{11} diagram for a) MgB_2 and b) $NbSe_2$ structures

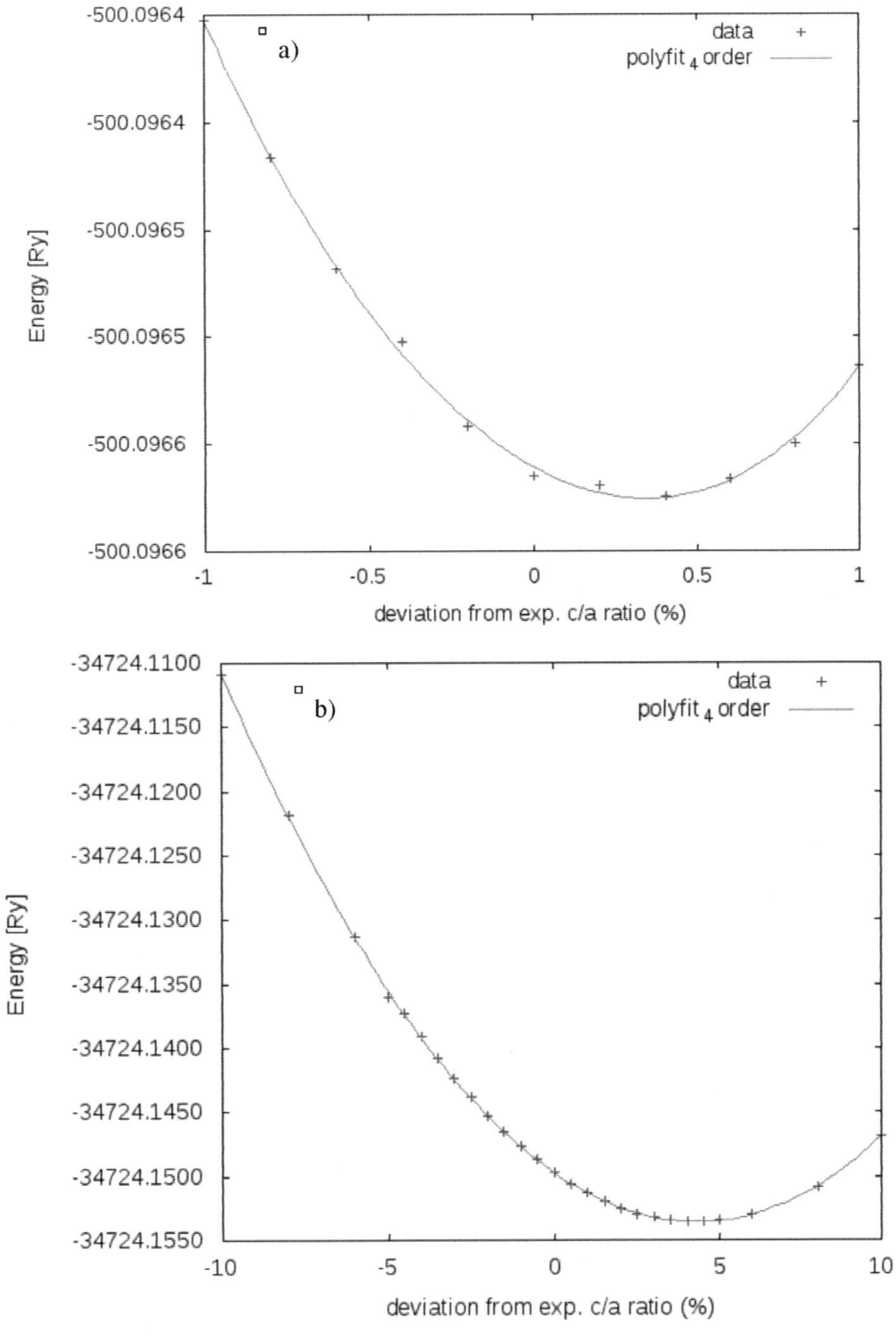

Figure 3: c/a vi energy for a) MgB_2 and b) $NbSe_2$

4. Conclusion

As a consequence, due to Young, Bulk and shear modulus our results show that in compare with $NbSe_2$, because of having larger E, B and S, MgB_2 is stiffer, more resistant against volume changes and reversible deformations. Accordingly, more especially due to Bulk modulus, it is clear that MgB_2 has a continuous and anisotropic structure, while someone could access to a monolayer of $NbSe_2$ easier because of more easily shearing. Finally, it could be a predictable reason for different (rod and layering) structures of these two materials.

References

[1] Withers RL, Wilson JA. An examination of the formation and characteristics of charge-density waves in inorganic materials with special reference to the two-and one-dimensional transition-metal chalcogenides. 1986 Sep 10 *Journal of Physics C: Solid State Physics*. **19**(25):4809.

[2] Mattheiss LF. Band structures of transition-metal-dichalcogenide layer compounds. 1973 Oct 15. *Physical Review B*. **8**(8):3719.

[3] Lee HN, McKinzie H, Tannhauser DS, Wold A. The Low Temperature Transport Properties of $NbSe_2$. 1969 Feb. *Journal of Applied Physics*. **40**(2):602-4.

[4] Xi X, Zhao L, Wang Z, Berger H, Forró L, Shan J, Mak KF. Strongly enhanced charge-density-wave order in monolayer $NbSe_2$. 2015 Sep 1. *Nature nanotechnology*. **10**(9):765-9.

[5] Le Tacon M, Bosak A, Souliou SM, Dellea G, Loew T, Heid R, Bohnen KP, Ghiringhelli G, Krisch M, Keimer B. Inelastic X-ray scattering in YBa2Cu3O6. 6 reveals giant phonon anomalies and elastic central peak due to charge-density-wave formation. 2014 Jan 1. *Nature Physics*. **10**(1):52-8.

[6] Xi X, Zhao L, Wang Z, Berger H, Forró L, Shan J, Mak KF. Strongly enhanced charge-density-wave order in monolayer $NbSe_2$. 2015 Sep. *Nature nanotechnology*. 1;**10**(9):765-9.

[7] Suderow H, Tissen VG, Brison JP, Martínez JL, Vieira S. Pressure Induced Effects on the Fermi Surface of Superconducting 2 H− NbSe $_2$. 2005 Sep. *Physical review letters*. 9;**95**(11):117006.

[8] Buzea C, Yamashita T. Review of the superconducting properties of MgB_2. 2001 Nov. *Superconductor Science and Technology*. 5;**14**(11):R115.

[9] Nagamatsu J, Nakagawa N, Muranaka T, Zenitani Y, Akimitsu J. Superconductivity at 39 K in magnesium diboride. 2001 Mar. *Nature*. 1;410(6824):63-4.

[10] Mijatovic D, Brinkman A, Oomen I, Veldhuis D, Rijnders G, Hilgenkamp H, Rogalla H, Blank DH. MgB_2 thin films and Josephson devices. 2003 Jun. *IEEE transactions on applied superconductivity*. **13**(2):3245-8.

[11] Grasso G, Penco R, Braccini V, Putti M, Fanciulli C, Tumino A, Brisigotti S, Berta S, Demencik E, Pietranera D, Vignolo M. Current MgB_2 wire performance and their industrial development.Tromans D. Elastic anisotropy of HCP metal crystals and polycrystals. 2011 Mar. *Int. J. Res. Rev. Appl. Sci*. **6**(4):462-83.

[12] Tromans D. Elastic anisotropy of HCP metal crystals and polycrystals. 2011 Mar. *Int. J. Res. Rev. Appl. Sci*. **6**(4):462-83.

[13] Blaha P, Schwarz K, Madsen GK, Kvasnicka D, Luitz J. 2001. *wien2k*. An augmented plane wave+ local orbitals program for calculating crystal properties.

[14] M. Jamal, 2012, Hex-elastic, http://www.wien2k.at/reg_user/unsupported.

[15] Perdew JP. Kieron Burke and Matthias Ernzerhof. 1996. *Phys. Rev. Lett*.**77**:3865.

IOP Conf. Series: Journal of Physics: Conf. Series **987** (2018) 012027 doi:10.1088/1742-6596/987/1/012027

Dynamic Consolidation and Investigation of Nanostructural W-Cu / W-Y Cylindrical Billets

B.Godibadze[1], A.Dgebuadze[1], E. Chagelishvili[1], G. Mamniashvili[2], A. Peikrishvili[3]

[1] G.Tsulukidze Mining Institute, 7, Mindeli St., 0186, Tbilisi, Georgia.
[2] Tbilisi State University, Andronikashvili Institute of Physics. 6, Tamarashvili St., 0177, Tbilisi, Georgia
[3] F.Tavadze Institute of Metallurgy and Materials Science, 10 Mindeli St., 0186, Tbilisi, Georgia.

E-mail: bgodibadze@gmail.com

ABSTRACT: The main purpose of presented work is to obtain W-Cu & W-Y cylindrical bulk nanostructured billets by explosive consolidation technology (ECT) in hot condition, with low porosity near to theoretical densities and improved physical / mechanical properties.
Nanocomposites were subjected to densification into cylindrical steel tube containers using hot explosive consolidation (HEC) technology to fabricate high dense cylindrical billets.
The first stage: Preliminary explosive densification of the precursor powder blend is carried out at room temperature with a loading intensity up to 10GPa to increase the initial density and to activate the particle surfaces in the blend.
The second stage investigation were carried out for the same already predensified billets, but consolidation were conducted in hot conditions, after heating of samples in between 940-1100^{0}C, the intensity of loading was equal to 10GPa.
Consolidated different type of W-Cu composition containing 10-40% of nanoscale W, during investigation showed that the combination of high temperatures (above 940°C) and two-stage shock wave compression was beneficial to the consolidation of the incompatible pair W-Cu composites, resulting in high densities, good integrity and good electronic properties. The structure and property of the samples obtained, depended on the sizes of tungsten particles. It was established that in comparison with W-Cu composites with coarse tungsten the application of nanoscale W precursors and depending of content of W gives different result.
Tungsten is a prime material candidate for the first wall of a future fusion reactor. In this study, the microstructure and microhardness of tungsten-yttrium (W-Y) composites were investigated as a function of Y doping content (0.5÷2 wt. %). It was found that the crystallite sizes and the powder particle sizes were increased as a result of the increase of Y content. Nearly fully dense materials were obtained for W-Y alloys when the Y content was higher than 0.5 wt. %. Investigation revealed that the Y rich phases were complex (W-Y) oxides formed during the sintering process. Also very interesting to use doping chromium with yttrium-containing alloys. e.g. (W - 10÷12 Cr -0.5÷2 Y) wt. %. The extent up to which yttrium acts as an active element improving the adherence and stability of the protective Cr 2 O 3 layer formed during oxidation is assessed.
The structure and characteristics of the obtained samples depends on the phase content, distribution of phases and processing parameters during explosive synthesis and consolidation.
Cu – (10-30%) W powder mixtures were formed into cylindrical rods using a hot shock wave consolidation (HSWC) process. Different type of Cu - W precursor composition containing 10, 20 and 30% of nanoscale W were consolidated near theoretical density under 900°C; The loading intensity was under 10 GPa. The investigation showed that the combination of high temperatures (above 800°C) and two stage shock wave compression was beneficial to the consolidation of the W-Cu & W-Y composites, resulting in high densities, good integrity and good electronic properties.

1. INTRODUCTION

In recent years, tungsten (W)-based heavy alloys have received increased use in both commercial and industrial applications. Most heavy alloys consist of W particles embedded in matrix of other metals or their alloys such as iron, nickel, or copper [1]. In particular, W–copper (Cu) composites may have potential uses as heat dissipation materials in the microelectronics field [2], diverter plates in fusion reactors [3], or in special industrial (i.e., aerospace) applications.

Experimental studies relating the mechanical properties of samples formed from nanocrystalline precursor powders show that these ultra-fine-grained materials are fundamentally different from their normal, coarse-grained counterparts. These materials often have very unusual properties: they are ultra-hard and wear resistant, have an ideal compatibility of strength and elasticity, and are characterized by super-plasticity. When the average grain size is less than or equal to the wavelength of visible light, the material will also have unique optical, thermal, electrical and magnetic properties as well. Therefore, a decrease of the grain size and concomitant control of the defect substructure of the grains seems to be a promising way to improve properties of these materials. At present, there are various methods (e.g., cold or hot isostatic compaction in vacuum or other media) for the manufacturing of monolithic specimens using precursor powders ranging from micro- to sub-micrometer- to nanometer-sized powders (i.e., covering the visible-light spectrum from infrared to ultraviolet wavelengths).

All existing conventional technologies, alongside with imparting positive properties, introduce certain negative features. Nanometer-scale grains are very sensitive to heating; with increasing temperatures, these powders begin to grow rapidly. Typically, this grain growth is non-uniform and its overall impact causes imperfections and nonuniformity in the nanostructure and, as a result, monolithic materials formed under such conditions do not have the unique physical and mechanical properties that are otherwise would be intrinsic to nanostructured materials.

Usually, decreasing the compaction or sinter temperature during low-temperature manufacturing processing does not lead to a desirable outcome. In this case, the relatively large free surface area of the powder precludes the attainment of high-density samples. Additionally, at low temperatures, the required interfacial grain-to-grain boundaries do not form; this is especially true during the compression and consolidation of refractory and ceramic powders. Thus, the as-pressed samples are characterized with high level of porosity and, therefore, inadequate physical or mechanical properties.

Nevertheless, sufficient experience has been accumulated to provide solutions to some of the aforementioned problems. The idea is to apply high temperatures, up to 1,500 K, to the samples and carry out rapid consolidation at the hot, elevated temperature, conditions. Heating of the powders or alloys before loading assists in increasing the sample's plasticity. As a result, common boundaries, interfacial solid solutions, intermediate layers (for the case of joining bulk alloys), and another beneficial feature form. The short heating period, ≤ 50 s, prevents or suppresses grain growth processes. A further novelty of the proposed, non-conventional approach relies on the fact that the consolidation of solid samples in a cylindrical geometry from sub-micrometer and nanometer-sized W–Cu & W-Y powders is performed in two stages:

a) First stage: preliminary explosive compression of the precursor powder blend is carried out at room temperature with a loading intensity of 5-10 GPa to increase the initial density and to activate the particle surfaces in the blend;
b) Second stage: the same, already pre-densified cylindrical sample is reloaded by an primary explosive shock wave with a loading intensity of 10 GPa, but at a temperature between 800-1100°C.

It is expected that the effect of the first consolidation stage is to primarily compact the precursor powder without causing a change in its microstructure. However, unlike that in conventional compaction carried out under quasi-static conditions where the powders are subjected to elevated temperatures for extended times, the second, hot consolidation stage is performed on the pre-compacted samples by an electric resistance heating method. The heating takes place at a high rate (about 10 to 20 K/s), reaching the process temperature in approximately 0.3 to 1 min. The shortened timescale significantly decreases the probability of the thermally activated grain growth process (as it was demonstrated in the preliminary experiments), while the high intensity shock wave loading imparts fluidity to the grain surfaces thereby increasing their plasticity and, thus generate the particle-particle bonds that would otherwise not form under the quasi-static conditions.

In this study we undertook the consolidation of copper–20 wt.% tungsten (Cu–20W) powder mixtures into cylindrical rods using both hot shock wave consolidation (HSWC) and hot vacuum compaction (HVC) processes. Two types of Cu–W precursor compositions, one type with a nanometer-scale W and another with coarser grain sizes of > 1 μm W were consolidated to near theoretical density at 800 and 1000 °C. The shock wave loading intensity was about 10 GPa; the loading intensity during static compression was 33.9 MPa (346 kg/cm2).

The intent of the investigations was to determine if the use high temperatures and the use of two-stage shock wave processing method would be beneficial, resulting in high densities, good integrity, and good electrical properties. The effects of the distribution and precursor size of the W on the structure and property of the samples were of interest. Of further interest was a determination of the mechanical properties (elastic modulus and internal friction losses) as a function of precursor W size, processing method, and processing temperature. Lastly, the electro-magnetic properties (electrical resistivity and diamagnetic susceptibility) were measured. These results are described.

2. EXPERIMENTAL PROCEDURES

The W-Cu & W-Y precursor powder blends were prepared from commercially available elemental powders. In turn these powders were consolidated into cylindrical rods using hot shock wave consolidation (HSWC) or hot vacuum consolidation (HVC) methods. Two types of Cu–W precursor compositions with two powders having a nanometer size and two powders with grain sizes of > 1 μm were consolidated to near theoretical density. The shock wave loading intensity was about 10 GPa, whereas the loading during static HVC was 33.9 MPa (346 kg/cm^2). The temperatures during compaction were 800 or 1000 °C.

The explosive compaction process is carried out using a cylindrical (axi-symmetric) scheme of dynamic loading. Typically, the shock waves are initiated with the use of industrial explosives and their mixtures with varying amounts of ammonium nitrate.

After consolidation, the HSWC samples were subjected to microstructural characterization and microhardness measurements. The electro-magnetic properties (electrical resistivity and magnetic susceptibility) along with, the mechanical properties (elastic modulus and internal friction) were also measured.

3. RESULTS AND DISCUSSION

Using scanning electron microscopy (SEM) it was found that while the nanometer powders did indeed contain nanometer-scale particles, they were heavily agglomerated into multi-micrometer agglomerates. Additionally, it was found that both powders have similar agglomerate and particle size. Figures 2(a) and 2(b) illustrate the HSWC sample microstructures of the Cu–20%W billets fabricated with nanometer-sized W shown in (a) (left column), and 5-6-μm-sized W (b) (right column), respectively, at increasing magnifications after the completion of the second-stage of HSWC at 800 °C. Based on a comparative examination of the samples, it is practically impossible to find any major difference in their microstructures. While the W appears mostly agglomerated, it seems that, in general, both HSWC samples have a uniform distribution of the two phases; that is, the W phase is uniformly distributed in the Cu matrix. Note that the larger 5-6 μm-sized W grains have undergone considerable fragmentation.

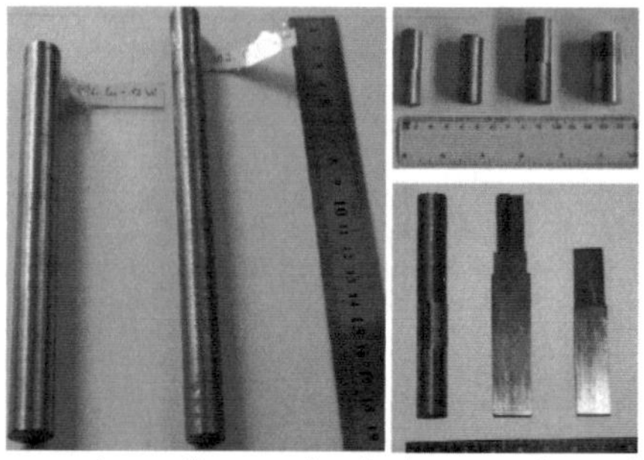

Figure 1. Compacted Cu-W composites. after mechanical processing, with content different percent of tungsten.

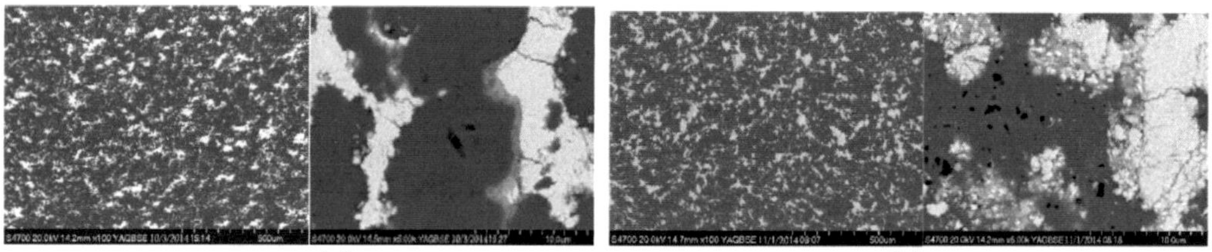

(a) **(b)**

Figure 2. Microstructures of the HSWC Cu–20%W samples at increasing magnifications; (a) - Cu–20%W (Nano) and (b) Cu–20%W (5-6 μm). Pre-consolidation temperature was 800 °C and the shock loading intensity was about 10GPa.

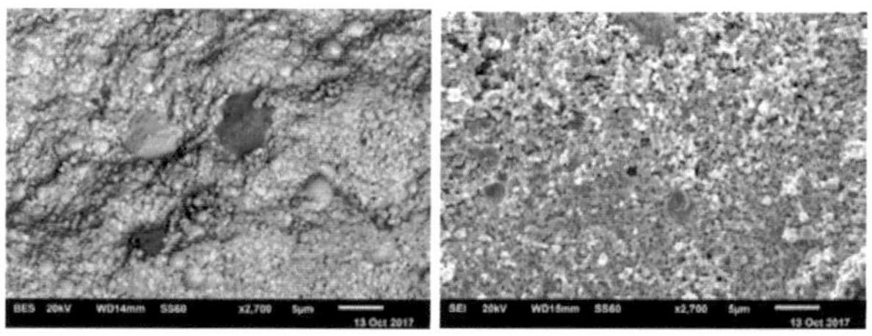

Figure 3. Microstructures of Sample W-1.85%Y-13.15%Cr.

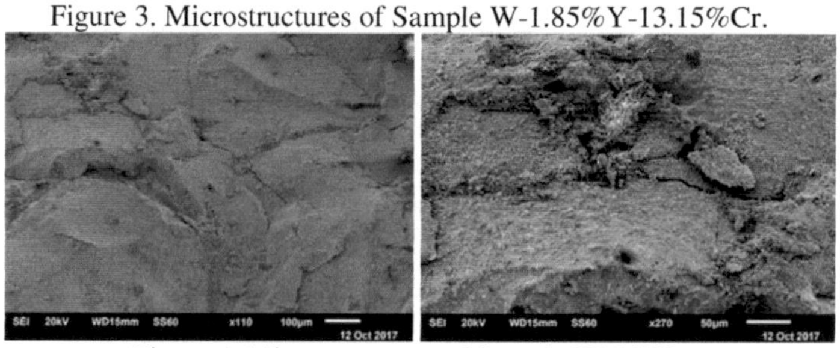

Figure 4. Microstructures of Sample W-2%Y.

3.1. Electronic and Magnetic Properties

The electrical resistivity measurements for the samples consolidated by both HSWC and HVC methods were carried out by the standard four-point method and are presented in Figure 5. As it is seen from the figure, the composites fabricated by HSWC are characterized by higher resistivity values than those obtained by the HVC method. The resistivity further decreases with increasing temperature for the HVC samples.

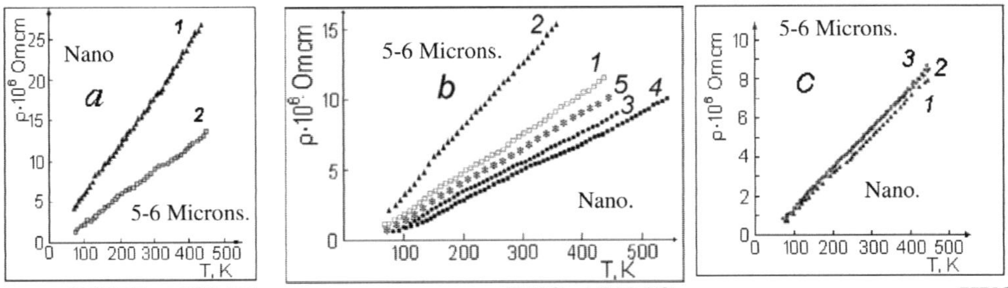

HSWC at 800 °C **HVC at 800 °C** **HVC at 1000 °C**

Figure 5. The quasi-static consolidation pressure during HVC was equal to 33.9 MPa (346 kg/cm2). Shown in (a) are the 800-°C HSWC samples; curve-1 Nano-W and curve-2 5-6Microns W, respectively. Shown in (b) are the 800-°C HVC samples; curve-1 Osram W, curve-2 Frantsevich W, and curves-3 through 5 are Treibacher, Zhang-Yuan W, and Treibacher/Zhang-Yuan W mixture, respectively. Shown in (c) are the 1000-°C HVC samples; curve-1 Treibacher W, curve-2 Zhang-Yuan W, curve-3 Osram W, respectively. (Treibacher – Nano powder & Frantsevoch 5-6- Microns).

Based on the unusual resistivity properties of the nanometer-sized powder samples, their mechanical behavior, including the relative stiffness (elastic modulus) and loss mechanisms (internal friction) were also measured. The measurements were conducted using an acoustic spectrometer on specimens of a rectangular shape (thin plates) in which the quarter-wave bending, and vibration mode could be excited [4]. The natural frequency of vibrations was in the range of 1-5 kHz. The vibrations were excited by the electrostatic method. Measurements of the temperature dependence of the internal friction (Q^{-1}) and elasticity modulus were carried out from 100 to 500 K as the samples were heated at a rate of 1 K/min. The maximum amplitude of the deformation of the vibrating specimens did not exceed ~ 10^{-6}.
Both HSWC samples show a monotonic increase of the modulus with decreasing temperature; the nanostructured sample has a much higher elastic modulus (***I***) than the micro-meter sized sample (at least by five times). This is in good agreement with the relaxation data for Q^{-1}. The micrometer sample also shows a loss peak at a lower temperature that the nanometer sample.

4. CONCLUSIONS

An HSWC process was used to consolidate nanostructured W precursor powders into Cu–20%W composites to near full density. The composites have better and more uniform mechanical properties and electronic characteristics than those measured for conventional (micrometer grain size) composites. Electrical resistivity measurements of the Cu–20%W composites indicate the formation of nanostructured features at the grain boundaries. This effect, manifested as a higher electrical resistivity, is more pronounced for the HSWC samples than those fabricated with the HVC method.
It was further established that the Cu–W composites containing nanometer-scale W have a stronger diamagnetic susceptibility response and are characterized with a lower dependence of the susceptibility on the applied magnetic field than composites containing micrometer grain size W or pure copper.
The proposed studies are focused on synthesis, property measurements and performance evaluation of these alloys and to provide a roadmap to their commercial utilization in industry. From the commercial standpoint it must be mentioned that Nano-structural cylindrical billets of W-Y-Cr & W-Cu composites (tubes, rods. Pallets, rings) with dimensions up to ϕ= 40mm and L=250mm will be fabricated.
Fractures analysis of the tested samples (W-Y & W-Y-Cr) showed a primarily intergranular breakage between the grains boundaries. This result supports the brittle behavior observed during the tests. A

grain morphology was observed. Coarse W grains and rounded smaller grains containing dispersed yttrium nanoparticles generally located at the grain boundaries, in habited the grain corrosion and growth. So, in this direction we are continue study of structure and consolidation process too.

Acknowledgement:
The authors gratefully acknowledge the financial support of the Shota Rustaveli National Science Foundation (SRNSF Grant Agreement № YS15_2.2.10_101). http://www.rustaveli.org.ge/en/

References

[1] A. Bose and R. M. German. 1996. Developments in the Sintering of Tungsten Heavy Alloys. Sintering Theory and Practice. *John Wiley & Sons, Inc.,* New York, NY. Pp.256-58
[2] R. M. German, K. F. Hens, J. L. Johnson, and Y. Bin. , 1993, Powder Metallurgy Processing of Heat Dissipation Components for Microelectronic Applications. Advances in Powder Metallurgy and Particulate Materials. Specialty Materials and Composites. *Proceedings of the 1993 International Conference on Powder Metallurgy and Particulate Materials, MPIF Press,* Princeton, NJ. pp. 189-202
[3] J. Boscary, S. Suzuki, K. Nakamura, T. Suzuki, and M. Akiba. 1999. "Thermal Fatigue Tests on CVD– W/Cu Divertor Plates, Fusion Engineering and Design," *Proceedings of the 1997 4th International Symposium on Fusion Nuclear Technology*, Vol. **39-40**, *Pt. A, Elsevier,* Lausanne, Switzerland, pp. 537-542
[4] F. K. Akopov, N. A. Arabadzhian, N. D. Kvataya, G. S. Oniashvili, I. V. Chkhartishvili. 2004. "Magnetic and Acoustic Spectroscopies of Sintered Composite Materials," *Phys. Met. Metallogr.*, Vol. **97**, No. **3**, pp.266-268
[5] A. I. Gusev. "The Effects of Nanocrystallic State in Solids," *Usp. Fiz. Nauk*, Vol. **168**, 1998, pp.55-82

IOP Conf. Series: Journal of Physics: Conf. Series **987** (2018) 012028 doi:10.1088/1742-6596/987/1/012028

Decoration of wide bandgap semiconducting materials for enhancing photoelectrochemical efficiency of PEC systems.

N Bakranov[1], A Zhabaikhanov[1], S Kudaibergenov[1] and N Ibraev[2]

[1] Kazakh National Research Technical University after K I Satpayev, Almaty, Kazakhstan

[2] The Karaganda State University of the name of academician E A Buketov, Karaganda, Kazakhstan

E-mail: bakranov@gmail.com

Abstract. The production of photoanodes based on wide-band gap materials such as TiO_2 is economically viable because of the low cost of synthesis methods. Contrary to economic aspects, wide-band gap semiconductor materials have a significant disadvantage due to low sensitivity to photons of visible light. To increase the photoactive parameters of the material of the electrodes in the visible range, the methods for decorating nanomasses of titanium dioxide by narrow-gap semiconductors are used. One of the most suitable narrow-gap semiconductor materials are CdS and Fe_2O_3. Controlled deposition of such materials on wide-gap semiconductors allows to regulate both the diffusion time of charge carriers and the band structure of TiO_2/Fe_2O_3 and TiO_2/CdS composites. The dimensions of the structure of the photoelectrode material of the cell have a large influence on the characteristics of the photocatalyst created. Thus, in the hematite structures of nanometre dimension, the rate of recombination of charge carriers fades away in comparison with bulk structures. Reducing the size of CdS structures also positively affects the nature of the photocatalytic reaction.

1. Introduction

Sunlight and water are the most common sources of clean renewable energy on Earth [1]. The Sun as a source of electromagnetic waves refers to inexhaustible natural resources, so the use of solar radiation is a key direction of alternative energy today. The power of solar energy incident on the earth's surface at any time is 130 million 500 MW [2]. Water splatted by the action of sunlight, is an inexhaustible source of hydrogen, the production of which belongs to one of the most important criteria of hydrogen energy. The main factor in the production of hydrogen from water under the action of photons is the maximum absorption of photocatalytic materials falling on the Earth's surface (4% ultraviolet (UV), 46% visible, 50% infrared spectrum). Among the developed methods of extracting hydrogen fuel by means of photon energy, the method of photocatalytic water splitting is one of the most promising. The use of widely distributed elements in the earth's crust to create photoactive systems absorbing a wide spectrum of solar radiation will solve the main problem of hydrogen evolution from water. As a result of photocatalytic decomposition of water, light energy is converted into chemical energy, breaking the energy barrier, with a large positive change in free energy Gibbs (ΔG = 238 kJ/mole) [3], similar to the process of energy transformations in the photosynthesis of green plants [4]. Photocatalysts based on widely distributed elements have semiconductor properties. Such semiconductors have a large band gap, the photoabsorption capacity of which lies in the UV range. Thus, there are a number of challenges to researchers developing systems consisting of low-cost materials capable of exhibiting photocatalytic properties in a wide light range.

The semiconductor materials used in photocatalytic reactions with hydrogen evolution make important demands on the potentials of the levels of the conduction and valence bands. Thus, to realize the half-reaction of the decomposition of water, which consists in the reduction of H^+ to H_2, the potential of the conduction band of the semiconductor should be more negative than the level of hydrogen evolution ($E_{(H_2/H_2O)}$). While the half-reaction of the oxidation of the water molecule occurs under the condition of the position of the valence band potential in a more positive value of the oxygen release level ($E_{(O_2/H_2O)}$) [5]. Therefore, complete photoelectrochemical splitting of water by semiconductor photocatalysts is realized by using materials of potential regions of conductivity that are more negative than the level of water recovery and potentials of valence bands of more positive levels of water oxidation. The photocatalytic release of hydrogen from water occurs as a result of an oxidation-reduction reaction that occurs both on the surfaces of photocatalytic particles (powder) dispersed in water (electrolyte) and on the surfaces of photoelectrodes. Consequently, two types of photocatalytic systems of water splitting are distinguished: photoelectrochemical and photochemical cells [1].

The photocatalysis process on semiconductor materials includes the following steps:

1) absorption of photons of light whose energy is equal to or greater than the width of the forbidden band of the semiconductor, which leads to the generation of an electron-hole pair (excitons) in the semiconductor material;

2) rupture of the exciton by migration of photoinduced charge carriers to the surface of the material;

3) carrying out the oxidation-reduction reaction by charge carriers of the water molecule [6].

Since the first successful photoelectrochemical splitting of water, performed by Honda and Fujishima in the early 1970s [1], a wide range of materials exhibiting photocatalytic activity has been developed. According to the properties of electronic configurations, such materials form four groups: 1) metal oxides with the configuration d0, such as Zr^{4+}, Nb^{5+}, Ta^{5+}, W^{6+}, Mo^{6+} and Ti^{4+}; 2) metal oxides with a configuration d10 (Zn^{2+}, Ga^{3+}, Ge^{4+}, Sn^{4+}, Sb^{5+} and In^{3+}); 3) metal oxides with the configuration f0, for example, Ce^{4+}; 4) non-oxide photocatalysts (Ta_3N_5, CdS) [2, 3]. Metal oxides, such as TiO_2, ZnO, Fe_2O_3, etc., are widely used as the main material of photoanodes of catalytic systems [4].

Titanium dioxide is a wide-band semiconductor material, the photosensitivity of which lies in the UV region. Based on the fact that an insignificant part of the UV spectrum of solar radiation falls to the Earth's surface (about 4%), titanium dioxide has a low level of absorption of light passing through the layers of the atmosphere. Despite the low photosensitivity, TiO_2 is a promising photocatalytic material due to its physicochemical properties - high chemical resistance, low photocorrosion. The process of photocatalytic decomposition of water by dioxide is initiated by the absorption of a photon with an energy equal to or greater than the width of the forbidden band of the semiconductor. After electron irradiation, electron-hole pairs are formed. The photogenerated electrons migrated to the conduction band perform the function of the reducing agent H^+ to H_2, and the holes migrating to the surface of the semiconductor decompose H_2O to O_2 and H^+.

By using the evaluation of the operation of a photoelectrochemical or photochemical cell by direct measurement of the quantum yield, it is necessary to take into account both the exact volume of the released gas and the exact number of photons incident on the photocatalyst, which under real conditions is hard to reproduce. It is convenient to evaluate the photoelectrochemical or photochemical system by measuring the so-called photocatalytic efficiency η. The photocatalytic efficiency η of the electrochemical system is calculated by the formula:

$$\eta = {((1.229V - V_b)J)}/{P} \quad (1),$$

1.229V is the standard potential needed to separate molecular hydrogen and oxygen from water, V_b is the bias voltage supplied by the external power source, J is the current density through the photoelectrochemical cell (A/m^2), and P is the power of the electromagnetic radiation emitted by the light source (W/m^2) [5].

Experimental data obtained during the decomposition of water by low-dimensional titanium dioxide showed the efficiency of the process of a component of the order of 0.4% when the material was irradiated with a light source with a filter of 1.5 AM-100mW/cm^2 [6]. It was found that the size and shape of TiO_2 particles play an important role in the formation of the overall efficiency of the photoelectrochemical cell of hydrogen evolution [7]. For example, nanoparticles of titanium dioxide (a mixture of the anatase and rutile phases) precipitated as a thin film with a thickness of 15μm onto the transparent conducting surface have a photocatalytic activity 10 times lower than in the case of TiO_2 nanotubes having an average length of 3μm [8]. An important factor affecting the photocatalytic activity of a photoelectrode made on the basis of nanotubes or nanorods of titanium dioxide is the length of the structures. This feature is due to the total area of light absorption and the area of contact of the photoelectrode with the electrolyte. The downside of increasing the length of nanotubes is the increase in the path travelled by the charge, hence, the time of electronic transport. In view of the fact that the recombination time of most charge carriers (of the order of 60-80%) in the material of a photocatalyst based on titanium dioxide lies in the nanosecond range [9], an increase in carrier transport time increases the probability of recombination of excitons, whereas with an increase in the path traversed by an electron the probability of particles entering the charge-capture zone increases, which has a negative effect on the photocurrent. The suppression of the number of charge carrier capture zones is achieved by improving the crystallinity of the structures by annealing. The density of the photocurrent passing through the photoelectrochemical cell has a direct effect on the efficiency of the cell.

The creation of composite low-dimensional materials to expand the light sensitivity of wide-gap semiconductors requires the observance of the rules of band transitions between the main and decorating semiconductors, the sequence of layers of the material relative to the incident light, etc. Thus, when choosing a narrow-band semiconductor material, the positions of the Fermi levels and the levels of the valence bands and conduction bands, as well as the type of conductivity. The production of low-dimensional TiO_2/Fe_2O_3 composite materials with a given hierarchy was achieved by decorating the titania nanostructures with hematite particles. The tandem architecture of composite samples was performed taking into account the distributions of the spectral sensitivities of wide-band/narrow-band semiconductors. Thus, the TiO_2 layer absorbing only the UV spectrum was located first relative to the front of the incident light. Thus, a visible part of the light that was not absorbed by the wide-gap semiconductor penetrated the hematite layer to generate additional charge carriers in it.

In this paper, the efficiency of the photocatalytic activity of the bare structures of wide-band titanium dioxide and nanocomposites such as TiO_2/Fe_2O_3 and TiO_2/CdS have investigated.

2. Experimental section

Synthesis of low-dimensional TiO_2 rods was performed by a hydrothermal method at a temperature of 180°C for 12 hours. A solution for depositing TiO_2 nanorods onto transparent conductive substrates (FTO glass) was prepared from 30 ml of hydrochloric acid, 30 ml of deionized water and 1 ml of titanium butoxide. After carrying out hydrothermal synthesis, the substrates with the structures deposited on them were washed with distilled water. Further, the obtained samples were annealed at 500°C for 2 hours.

Low-dimensional titanium dioxide films were formed by applying a paste consisting of 0.96g of nanoparticles (25 nm in size) TiO_2, 2 ml of polyethylene glycol to transparent conductive surfaces. After applying the TiO_2 paste layers, the samples were annealed at a temperature of 500°C for 2 hours.

2.1. Decoration of TiO_2 nanostructures by hematite

To obtain a nanosized film of Fe_2O_3, an ethanol solution of 0.02 M $FeCl_3$ was prepared. The deposition of low-dimensional layers of iron oxide was carried out for 1 minute by precipitation of Fe ions contained solution a rotating substrate at a speed of 1000 rpm. a spin coating method, followed by calcination of the structures at a temperature of 350°C. Calcination of coatings was carried out to remove chlorine and form iron oxide. Thus, up to 5 layers of the film Fe_2O_3 were deposited. An annealing at 450°C, following the deposition of the fifth (last) layer of a narrow-band semiconductor, was carried out to create TiO_2/Fe_2O_3 heterostructures.

After deposition of Fe-containing layers on the surface of nanorods or thin-film TiO_2 layers and subsequent calcination in an oxygen atmosphere, the objects obtained formed a TiO_2/Fe_2O_3 heterostructure.

2.2. Synthesis of TiO_2/CdS heterostructures

Thin films based on TiO_2/CdS composites were obtained by two-step hierarchical deposition of titanium dioxide nanoparticles and subsequent deposition of a CdS layer. The hierarchy of TiO_2/CdS composites is formed both taking into account the absorption by the nanostructures of titanium dioxide of the UV spectrum, and the creation of a narrow-gap semiconductor-electrolyte contact.

Deposition of CdS structures on the surface of TiO_2 nanomass was carried out by immersion of the wedge in a vessel with dissolved 0.2M cadmium nitrate and 0.2M thiourea in ethyl alcohol, followed by calcination of the sample at a temperature of 90°C. The calcination time of the components of the CdS structure was 1 hour. After the formation of the CdS film, the samples were washed in distilled water.

3. Result and discussion

The morphology of the sample obtained by hydrothermal synthesis of titanium dioxide was studied by scanning electron microscopy (figure 1). A uniform array of TiO_2 structures consists of rods, with a diameter of 150-200 nm, mainly directed perpendicular to the substrate.

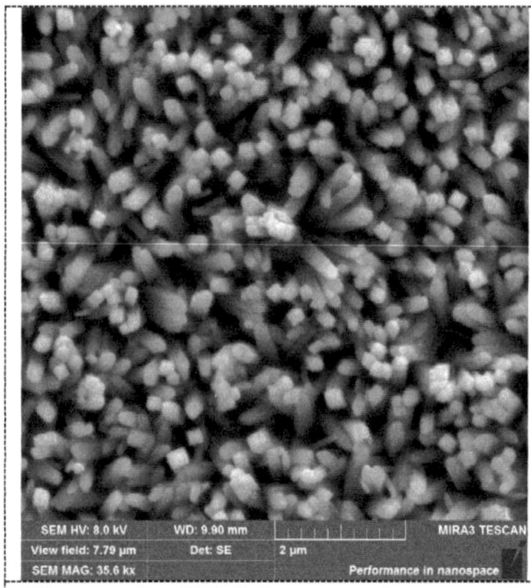

Figure 1. SEM image of nanorods TiO_2 obtained by the hydrothermal method.

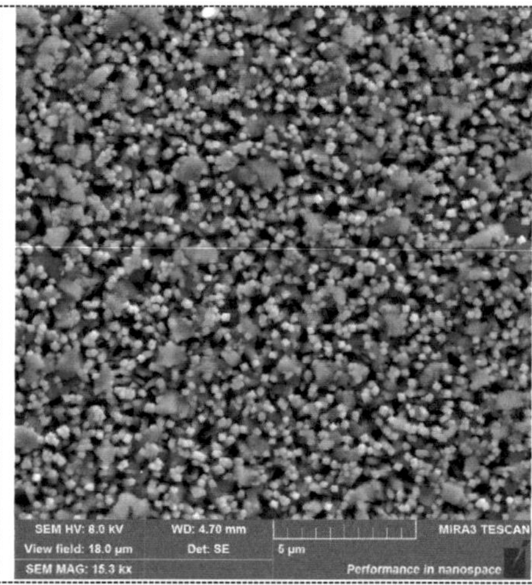

Figure 2. SEM image of a low-dimensional TiO_2/Fe_2O_3 composite

After deposition of Fe-containing layers on the surface of nanorods or thin-film TiO_2 layers and subsequent calcination in an oxygen atmosphere, the objects obtained formed a TiO_2/Fe_2O_3 heterostructure. The morphology of the surface of the resulting nanocomposites is shown in figure 2. The crystals of Fe_2O_3, having an average length of one side of the surface equal to 300 ± 50 nm, are combined on the average with six TiO_2 nanorods, which can facilitate the rapid separation of charge carriers. When carrying out photocatalytic experiments using a photoanode based on TiO_2/Fe_2O_3.

The photocatalytic activity of the obtained materials was studied by a potentiometric method. The photoelectrochemical cell consisted of a platinum cathode, an Ag/AgCl reference electrode, and an anode obtained by depositing semiconductor materials on a conductive glass. The distance between the platinum cathode and the photoanode was 2 cm. The size of the active zone of the photoanode was chosen equal to 1 cm^2 for a convenient calculation of the current density. For this purpose, an epoxy resin was applied to the entire area, except for a square with a side of 1 cm, a photoelectrode. The contact between the output of the potentiostat and the photoelectrode was produced by graphite glue (figure 3).

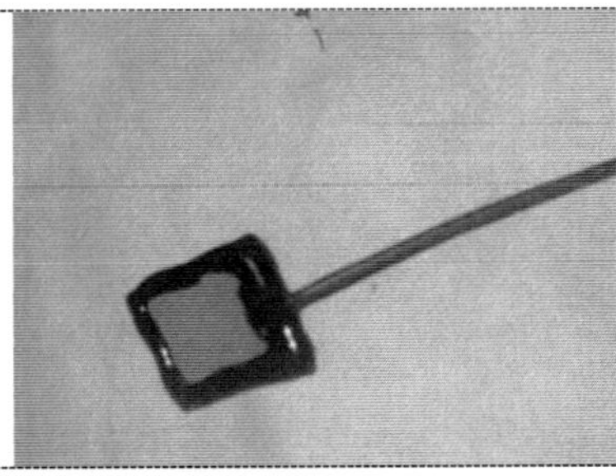

Figure 3. Photographic image of the photoanode

When studying the photocatalytic activity of TiO_2/Fe_2O_3 composite materials, the increase in the current density of the sample in comparison with the current densities of one-component TiO_2 semiconductors (figure 4) was not detected. The reason for this may be a more positive position of the conductivity level of the narrow-band semiconductor material Fe_2O_3.

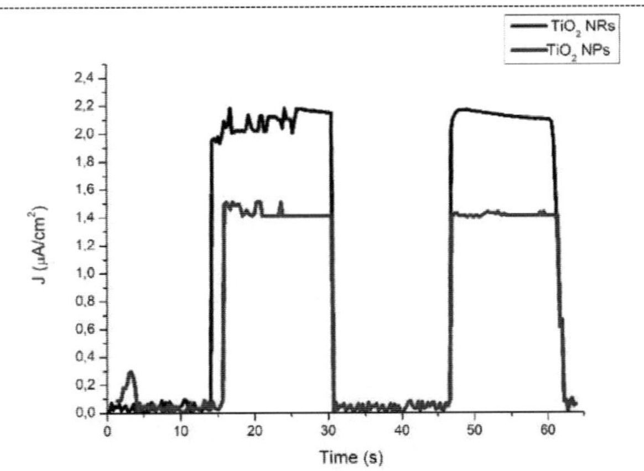

Figure 4. JT diagram of photocurrent curves through low-dimensional TiO_2 systems (NR_s black curve, NP_s blue curve).

The morphology of the coating surface of nanostructures of a wide-gap semiconductor TiO2 material by a uniform layer of a narrow-gap semiconductor CdS is shown in figure 5. It is understood that the transverse dimension of the formed CdS film exceeds the transverse dimensions of the TiO_2 arrays since The rods and titanium dioxide particles are completely coated with a CdS layer. The dimensions of the particles forming the CdS film averagely 500 nm (figure 5).

Figure 5. SEM image of the morphology of the surface layers of composite materials TiO_2/CdS.

Energy-dispersive analysis of the samples showed the presence of titanium, oxygen, cadmium and sulphur elements on the composite surface (figure 6). The elements Ca, Si, P, V, identified as a result of EMF analysis, refer to the substrate material.

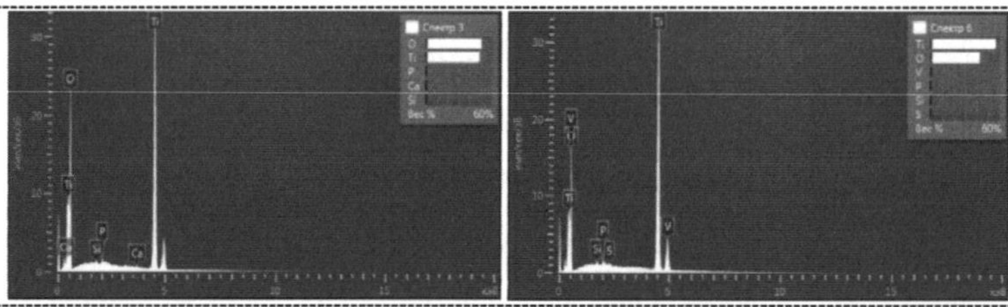

Figure 6. EDX analysis of the elemental distribution of the TiO_2/CdS composite..

Photoelectrochemical measurements of TiO_2 CdS composite materials carried out in potentiostatic mode using distilled water as electrolyte showed a sharp increase in current densities after performing wide-gap semiconductor decorations with low-dimensional CdS layers. At the same time, by depositing CdS particles on the surface of nanorods TiO_2, the current density of the system increased by an average of 4 times (from 2 μA when one-component array was irradiated to 8 μA when TiO_2/CdS composites were irradiated). Decoration of nanoparticles by a TiO_2 thin film of CdS made it possible to increase the photocurrent of a photoelectrochemical cell from 1.4 μA to 5 μA (figure 7). This effect confirms the increase in the photocatalytic activity of the composite photoanode TiO_2/CdS in comparison with the anode made on TiO_2 structures.

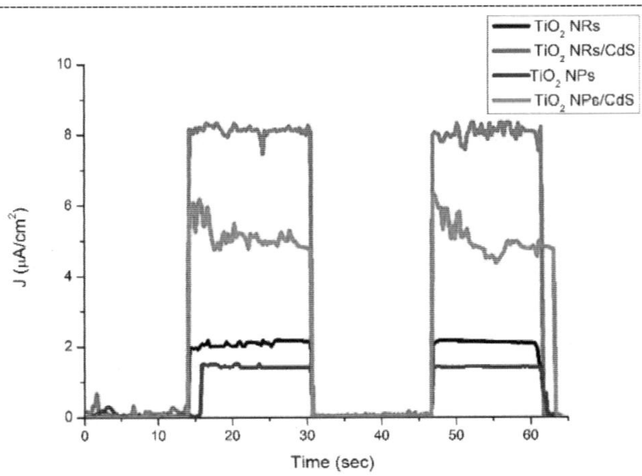

Figure 7. JT diagram of photocurrent curves through low-dimensional TiO_2 systems (NRs, NPs), TiO_2/CdS composites. A blue curve was obtained upon irradiation of a photoanode made of TiO_2 nanoparticles. The red curve is obtained as a result of irradiation of TiO_2/CdS nanolayers.

Calculation of the efficiency of obtained photocathodes

The efficiency of the photocatalytic cell is calculated by the formula (1).

Thus, substituting the values of the current densities J, the experimental efficiency η of the obtained composites is:

η $TiO_2/Fe_2O_3 = 0.0378$;

η $TiO_2/CdS = 0.0945\text{-}0.1512$

4. Conclusion

The production of a photoanode based on TiO_2 nanostructures, TiO_2/CdS, TiO_2/Fe_2O_3 composites was successfully carried out by a number of simple techniques, such as hydrothermal deposition and electrochemical synthesis. The study of photocatalytic properties of arrays of nanorods, nanoparticles and nanoplasts showed a comparable current density when irradiated with a light source. After surface modification of the structures of titanium dioxide with nanoparticles of narrow-band semiconductor materials, the similarity of the photocurrent values disappears. Thus, the general picture of the increase in the photocurrent density confirms the positive effect of the decoration of structures by narrow-band semiconductor materials on the photoelectrochemical processes of the cell. Decoration of wide-gap semiconductor materials by layers of hematite also favourably contributed to an increase in the efficiency of the cell as a whole, which confirms the theory of band transitions in semiconductors and creates prerequisites for further work in the band engineering of semiconductor composites.

References

[1] Fujishima A Electrochemical evidence for the mechanism of the primary stage of photosynthesis 1971 Book (Editor Bulletin of the chemical society of Japan) pp 1148-1150

[2] Maeda K, Domen K New non-oxide photocatalysts designed for overall water splitting under visible light 2007 Journal of Physical Chemistry C 111 22 pp 7851-7861

[3] Ismail A A, Bahnemann D W Photochemical splitting of water for hydrogen production by photocatalysis: A review 2014 Solar Energy Materials and Solar Cells 128 pp 85-101

[4] Wolcott A Smith W A, Kuykendall T R, Zhao Y P, Zhang J Z Photoelectrochemical study of nanostructured ZnO thin films for hydrogen generation from water splitting 2009 Advanced Functional Materials 19, 12 pp 1849-1856

[5] Shankar K, Basham J I, Allam N K, Varghese O K, Mor G K, Feng X J, Paulose M, Seabold J A, Choi K S, Grimes C A recent advances in the use of TiO_2 nanotube and nanowire arrays for oxidative photoelectrochemistry 2009 Journal of Physical Chemistry C 113, 16 pp 6327-6359

[6] Hartmann P, Lee D K, Smarsly B M, Janek J Mesoporous TiO_2: comparison of classical Sol-Gel and nanoparticle based photoelectrodes for the water splitting reaction 2010 Acs Nano 4, 6 pp

3147-3154
[7] Szymanski P, El-Sayed M A Some recent developments in photoelectrochemical water splitting using nanostructured TiO_2: a short review 2012 Theoretical Chemistry Accounts 131, 6
[8] Park J. H, Kim S, Bard A J Novel carbon-doped TiO_2 nanotube arrays with high aspect ratios for efficient solar water splitting 2006 Nano Letters 6 1 pp 24-28
[9] Tang J W, Durrant J R, Klug D R Mechanism of photocatalytic water splitting in TiO_2 reaction of water with photoholes, importance of charge carrier dynamics, and evidence for four-hole chemistry 2008 Journal of the American Chemical Society 130 42 pp 13885-13891

IOP Conf. Series: Journal of Physics: Conf. Series **987** (2018) 012029 doi:10.1088/1742-6596/987/1/012029

Structural and optical properties of SiC-SiO_2 nanocomposite thin films

I.Bozetine*, A. Keffous, S. Kaci, H. Menari and A. Manseri

Centre de Recherche en Technologie des Semi-conducteurs pour l'Energétique (CRTSE), Division Couches Minces, Surfaces et Interfaces (CMSI) 02 Bd Frantz Fanon, BP.140 Alger 7 merveilles, Alger, Algérie

*Corresponding author: BelkhodjaAisma@crtse.dz

Abstract. This study deals with the deposition of thin films of a SiC-SiO_2nanocomposite deposited on silicon substrates. The deposition is carried out by a co-sputtering RF magnetron 13.56 MHz, using two targets a polycristallin 6H-SiC and sprigs of SiO_2. In order to study the influence of the deposition time on the morphology, the structural and optical properties of the thin films produced, two series of samples were prepared, namely a series A with a 30 min deposition time and a series B of one hour duration. The samples were investigated using different characterization techniques such as Scanning Electron Microscope (SEM), X-ray Diffraction (DRX), Fourier Transform Infrared Spectroscopy (FTIR), Secondary Ion Mass Spectrometry (SIMS) and photoluminescence. The results obtained, reveal an optical gap varies between 1.4 and 2.4 eV depending on the thickness of the film; thus depending on the deposition time. The SIMS profile recorded the presence of oxygen (^{16}O) on the surface, which the signal beneath the silicon signal (^{28}Si) and carbon (^{12}C) signals, which confirms that the oxide (SiO_2) is the first material deposited at the interface film - substrate with an a-OSiC structure. The photoluminescence (PL) measurement exhibits two peaks, centred at 390 nm due to the oxide and at 416 nm due probably to the nanocrystals of SiC crystals, note that when the deposition time increases, the intensity of the PL drops drastically, result in agreement with dense and smooth film.

1. Introduction

Silicon carbide (a-SiC) layers have attracted much attention in recent years because of their potential applications in various kinds of microelectronic and optoelectronic devices **[1]**, such as image sensors, photodiodes and solar cells. For photovoltaic applications, Si nanocrystals (SiNCs) embedded in a dielectric SiC matrix can be used as absorber material for an upper cell of crystalline silicon based tandem solar cell **[2]**. Si-rich hydrogenated silicon carbide (a-SiC) thin films prepared by co-sputtering of Si and 6H-SiC targets by DC magnetron has attracted great interest in optoelectronics and next generation photovoltaic **[1]**. It is possible to generate nanocrystals (NCs) through annealing of a-SiC thin films. Using the multilayer approach proposed by Zacharias **[2]**, size and nanocrystal density can be controlled by a stoichiometric diffusion barrier layer and the composition of the Si-rich layer. Studies on the effect of Si anisotropic effective masses in different corresponding matrix material system showed the greatest potential of a SiC matrix, compared to SiO_2 and Si_3N_4, as a top high band gap cell in an all-Si tandem solar cell. This is due to a lower barrier height of SiC and thus a broad extension of the mini bands into the matrix material, and likely a higher carrier mobility **[4]**. Recently, investigations of magnetron sputtered a-SiC:H single layers and superlattices consisting of a-SiC annealed at temperatures between 800 °C and 1100 °C showed a crystallization of both Si and SiC**[1]** and **[5]**. However, in the multilayers a dependency of the crystallization temperature of ion the amount of Si in the a-SiC interlayers was reported. The most widely used method for the deposition of a-SiC is PECVD by the decomposition of silane (SiH_4) and a hydrocarbon gas, mostly methane (CH_4). Numerous microstructural investigations of the as-deposited films were done, showing that the composition and microstructure of the resulting thin films are strongly dependent on the deposition conditions **[6]**. Consequently, the optical properties which result from the microstructure vary greatly **[7]**. Two different deposition regimes for the deposition of a-SiC thin films are commonly distinguished, namely a high power and a low power regime. In the high power regime, hydrocarbon bonds are cracked in the plasma, thus $C–H_n$ bonds are directly deposited on the substrate. Films with arbitrary Si to C ratios can be achieved with high power densities. On the other hand, the low power regime refers to a plasma power density that is sufficient to crack SiH_4, but that is not high enough to crack the C–H bonds in CH_4. No primary decomposition of hydrocarbon in the plasma occurs at low power densities, therefore incorporation of C in the film can only result from chemical reactions between $Si–H_n$ species created in the plasma and the hydrocarbon molecules **[8]**. In this work it was also shown the role of incorporated SiO_2 into SiC thin films by co-sputtering SiO_2 and SiC using RF (13.56 MHz) magnetron sputtering method and its structural and optical properties.

2. Experimental procedure

Thin films of SiC-SiO_2 composite were elaborated by RF magnetron sputtering technique by using 6H-SiC as target obtained from ACM-France and 2 sprigs of SiO_2. The SiC-SiO_2 films with thicknesses of 0.1 – 0.15 µm have been deposited on p-type Si(100) for structural measurements and glass substrates for optical measurements. All substrates have been cleaned with ethanol and 10% hydrofluoric acid before deposition. The deposition pressure in sputtering chamber was 10-2 mbar. A mixture of hydrogen (H_2) and argon (Ar) gases with flow rates of 10 sccm, respectively, and a plasma power of 120 W have been used. The deposited films SiC-SiO_2 were investigated by Scanning Electron Microscope (SEM), X-ray Diffraction (DRX), Fourier Transform Infrared Spectroscopy (FTIR) and Secondary Ion Mass Spectrometry (SIMS) and photoluminescence measurements. Figure 1 illustrates the deposition chamber of SiC-SiO_2.

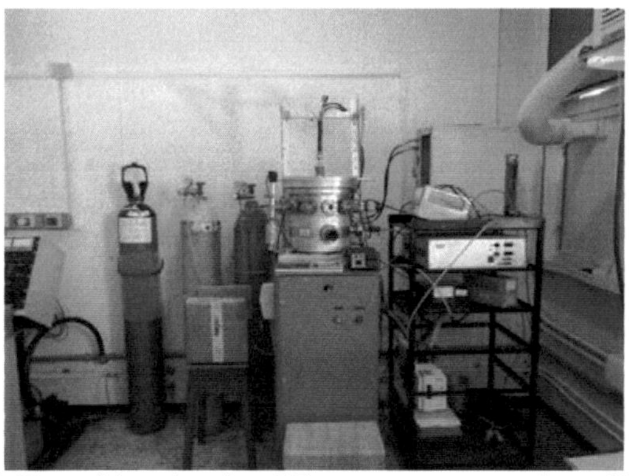

Figure 1. Deposition chamber RF magnetron sputtering of SiC-SiO_2

3. Results and discussion.

3.1. Structural properties

*3.1.1. SIMS Analysis.*There is the formation of a Si-O-C alloy. Oxygen (O) being on the surface, below it Si and C. The oxide deposits at first sce the interface of substrate Si. The layer deposited for 1 hour is more uniform compared to that deposited for 30 min. The ratio: Si / C vary between 1.4 and 2.4, which is necessary because we have to make an alloy a-SixOyCz. Thickness of the layer is between 0.1 and 0.15 µm depending on the duration of the deposit. We note that 12C is the first atomic surface plane (SiC-SiO_2) that is different when SiC is deposited on its own (Figure 2, 3 and 4).

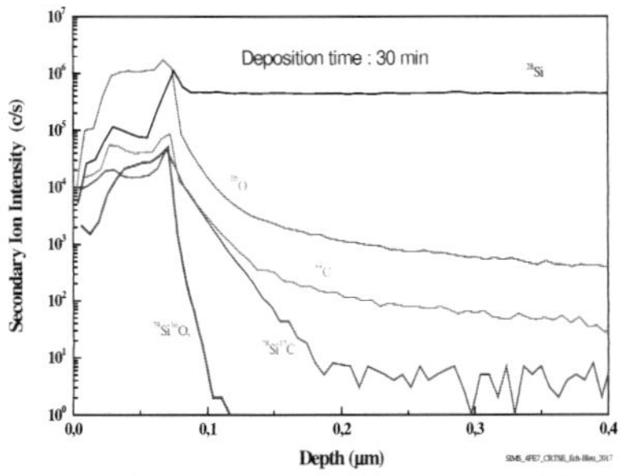

Figure 2. SIMS Profile of SiC+SiO_2 deposited onto Si for 30 min deposition time

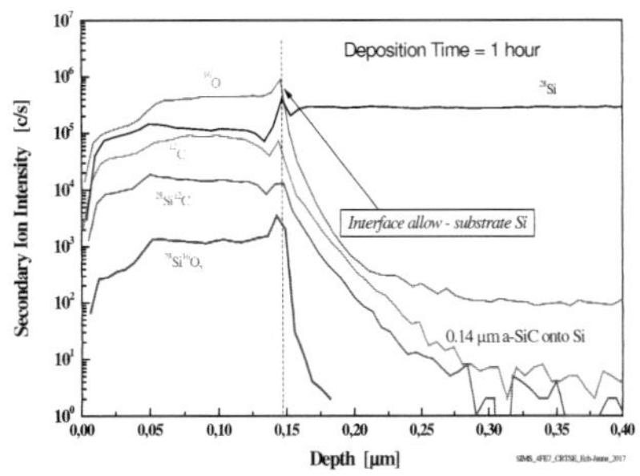

Figure 3. SIMS Profile of SiC+SiO_2 deposited onto Si for 1 hour deposition time

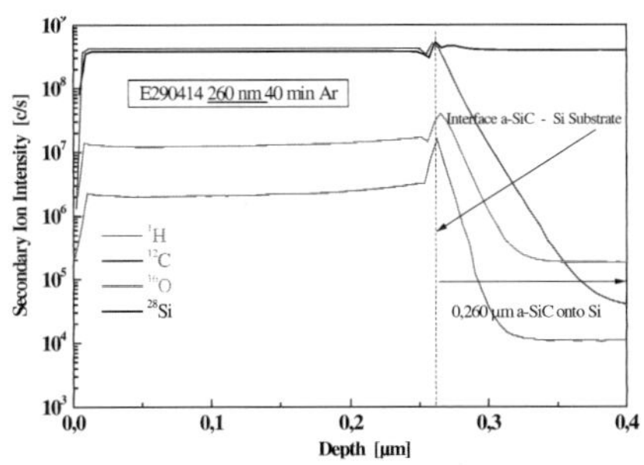

Figure 4. SIMS Profile of SiC deposited onto Si without SiO_2

3.1.2. FTIR Analysis. Figure 5 shows infrared spectra of SiC-SiO_2 films. Peak centred at 609 cm^{-1} represents the Si-Si bond. Peak centred at 738 cm^{-1} is the amorphous Si-C bond. Peak centred at 894 cm^{-1} is the SiOH bond. Peak centred at 1108 cm-1 corresponding to Si-O-Si bending vibrations in silicon oxide. Peak centred at 1467 cm^{-1} is the double bond C = C (Disorder). Peak centred at 1682 cm^{-1} is the double bond C = C (Graphite) and at 1108.30 cm^{-1}, peak is ascribed bending vibrations of Si-O-Si.

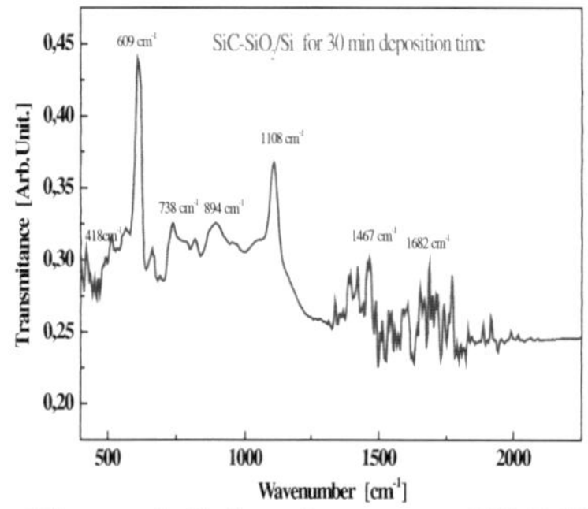

Figure 5. Infrared spectra of SiC-SiO_2 for 30min deposition time

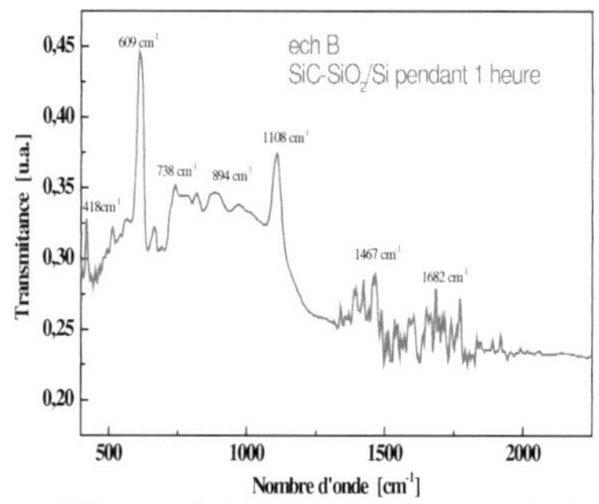

Figure 6. Infrared spectra of SiC-SiO_2 for 1 hour deposition time

3.1.3. DRX Analysis. The XRD analysis allows identifying the nature single-phase of the sample, crystalline quality of the material analyzed and to measure the crystalline parameters. Figures 7 and 8 show XRD pattern of SiC-SiO_2 thin films deposited onto Si. The structure of SiC-SiO_2 composite is identified using the XRD diagram of ASTM 72-0018 as hexagonal (6H) with a preferential orientation (202). The comparison between the XRD patterns of polycrystalline 6H-SiC and that of thin SiC-SiO_2 layer clearly indicates the crystalline structure of this later and a same structure 6H [9].

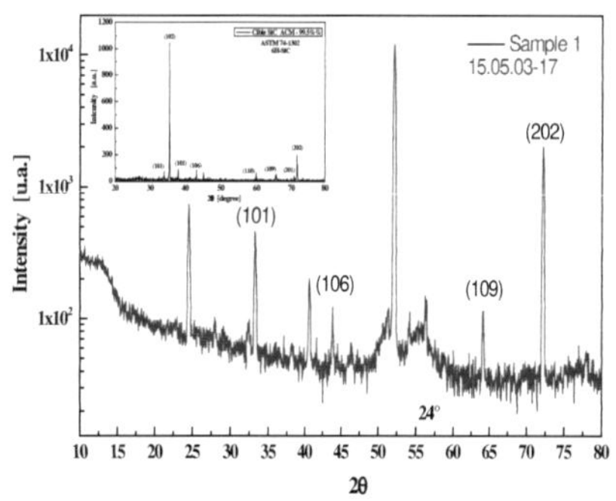

Figure 7. XRD pattern of SiC-SiO_2 for 30 min deposition time

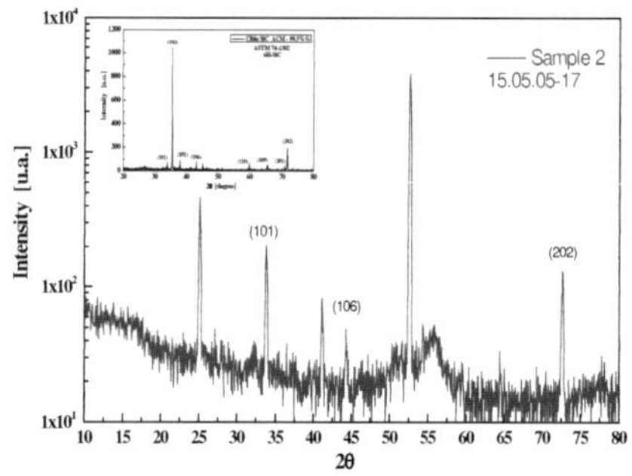

Figure 8. XRD pattern of SiC-SiO_2 for 1 hour deposition time

*3.1.4. SEM Analysis.*Figure 9 depicts a SEM plan view and cross sectional of the thin film SiC-SiO_2 deposited on silicon substrate for 30 min deposition time where it shows that the material exhibits a granular form with an average grain size diameter assessed to be round 100 nm and a regular thin film. Figure 10 shows a plan view and cross-sectional SEM image of a thin SiC- SiO_2 layer deposited on the Si(100) substrate for 1 hour time deposition which revealed a planar surface with a grains wich have a diameter round 400 nm.

IOP Conf. Series: Journal of Physics: Conf. Series **987** (2018) 012029 doi:10.1088/1742-6596/987/1/012029

Figure 9. Plan view SEM (a) and Cross sectional (b) image of SiC-SiO_2 for 30 min deposition time

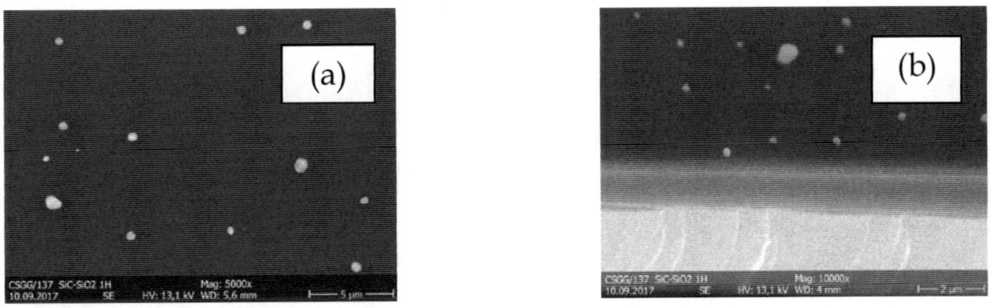

Figure 10. Plan view SEM (a) and Cross sectional (b) image of SiC-SiO_2 for 1hour deposition time

3.2. Optical properties: Photoluminescence

Figure 11 shows PL spectra of two sample SiC-SiO_2 deposited at 30 min (**curve a**) and 1 hour (**curve b**), respectively. The PL spectra exhibit a blue luminescence with two peaks centred at 390 nm (3.18 eV) and 416 nm (2.98 eV). It shows that PL intensity decreases with increasing thickness of the films SiC-SiO_2. Lio et al. suggested that this band could be the result of quantum confinement effects in SiC-SiO_2 film **[10]**. Xu et al. attributed this emission of SiC with incorporation SiO_2 particles to the radiative recombination from some direct transitions such as self-trapped excitons in the SiC surface **[11, 12]**. When the excitation takes place in SiC-SiO_2 film, the excited conduction band electrons diffuse into Si-Ox bonding layers, so as they interact with defects resulting in radiative recombination **[13]**.

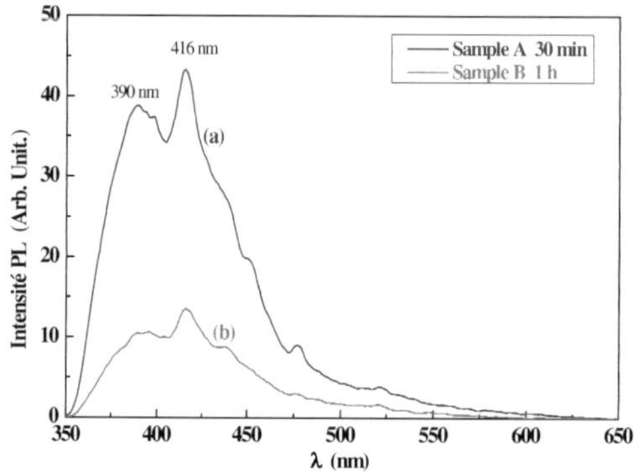

Figure 11. Photoluminescence measurement at different deposition time SiC-SiO_2 deposited onto Si

4. Conclusion

The thin films of a SiC-SiO_2 nanocomposite have been deposited on silicon substrates. byrf magnetron sputtering with systematic variation of the time of depostion. The influence of the deposition time on the morphology, the structural and optical properties of the thin films produced has been studied.an optical gap varies between 1.4 and 2.4 eV depending on the thickness of the film was obtained; thus depending on the duration of deposition. The SIMS profile recorded the presence of oxygen on the surface, which the signal beneath the silicon signal and carbonsignals, whichconfirmsthat the oxide (SiO_2) is the first materialdepositedat the interface film - substratewith an a-OSiC structure. The photoluminescence (PL) measurementexhibitstwopeaks, centredat 390 nm due to the oxide and at 416 nm due probably to the nanocrystals of SiCcrystals, note thatwhen the deposition time increases, the intensity of the PL drops drastically, result in agreement with dense and smooth film.

References

[1] Song E.C, Cho Y.H, Conibeer G, Huang Y 2008 Thin Solid Films 516 3824.
[2] Zacharias M, Heitmann J,. Scholz R,. Kahler U,. Schmidt M and Bläsing 2002 *J, Appl. Phys. Lett.***80** 661.
[3] Jiang C.W and Green M.A 2006 J. Appl. Phys.99 114902.
[4] Song D, Cho E.C, Conibeer G, Huang Y, Flynn C and Green M.A 2008*J. Appl. Phys.* 103 083544.
[5] El Khakani M A, Guay D, Chaker M and. Feng X H, 1995 *Phys. Rev. B***51** 4903.
[6] Chernyshov S.V, Terukov E.I.,. Vassilyev V.A and. Volkov A.S 1991*J. Non-Cryst. Solids* **134** 218.
[7] Solomon I, Schmidt M.P and Tran-Quoc H, 1988*Phys. Rev. B***38** 9895.
[8] Künle M, Janz S, Eibl O, Berthold C, Presser V and Nickel K G, 2009*Mater. Sci. Eng. B* 159–60 355.
[9] Keffous A, Gabouze N, Cheriet A, Belkacem Y and Boukezzata A, 2010*Applied Surface Science***256** 5629.
[10] Lio L, Bao X, Yang Z and Min N, 1995*Appl. Phys. Lett.* **66** 2382.
[11] Xu J, Mei J, Rui Y, ChenD, Cen Z and. Li W 2006*J. Non-Cryst. Solids* **352** 1398.
[12] Mnivannan A,. Fujishima A and Subba Rao G.V 1988*Phys. Chem.* **92** 1522.
[13] Xu J, Yang L, Rui Y, Mei J and Zhang X 2005 *Solid State Commun.* 133 565.

Acknowledgements

This work was funded under the Algerian General Directorate of Scientific Research and Technological Development (DGRSDT).

IOP Conf. Series: Journal of Physics: Conf. Series **987** (2018) 012030 doi:10.1088/1742-6596/987/1/012030

Scattering Resonances in bilayer graphene

Michele Pisarra, Antonio Sindona and Pierfrancesco Riccardi

Dipartimento di Fisica, Università della Calabria and INFN – Gruppo Collegato di Cosenza – via P. Bucci cubo 33C, 87036 Arcavacata di Rende, Cosenza (CS), Italy

E-mail: pierfrancesco.riccardi@fis.unical.it

Abstract. We present density functional calculation of the empty states of graphene single- and bi-layer. Particular attention is given to the recently discovered scattering resonances of graphene. When two layers are assembled, these states hybridize, giving rise to more complex states that are strongly localized in the interlayer regions. These states are at the origin of bands of graphite that show a pronounced dispersion in the direction perpendicular to the planes and have been observed over decades in experiments.

1. Introduction

The electronic structure of free-standing graphene and of graphene based materials has been intensively investigated. Most of these investigations have been concerned with the occupied energy levels, or have covered a limited energy range above the Fermi level (E_F), say, of the order of a few eV [1-6]. Therefore, the knowledge of the unoccupied energy bands is still largely lacking and appears important for the understanding of the interfacial properties of graphene on substrates [7,8]. In particular, the electronic structure of free-standing graphene has been intensively inspected by countless approaches, ranging from tight-binding methods to density functional theory (DFT) codes, and it has been recently classified using group theory algebra [5,6]. At energies above the vacuum level some studies [6] have predicted the existence of two-dimensional (2D) discrete states embedded in the three-dimensional (3D) continuum, showing that electrons in these confined 2D states have a non-negligible probability to escape into the vacuum, thus turning the states into resonances.

To study how the peculiar 2D properties of graphene are reflected in 3D derived systems, recently we considered the local density DFT band structure of graphene and graphene grown on Ni(111) [8]. We focused on the unoccupied energy levels up to 50 eV above E_F, and compared their energy-momentum dispersions with available Angle Resolved Secondary Electron Emission (ARSEE) observations [7,8]. Indeed, experiments of secondary electron emission have been proven to be very well suited for the characterization of surfaces and nanomaterials [8-11]. We showed how the dispersions obtained in ARSEE experiments on graphene/Ni(111) can be related to hybridized energy states found in DFT calculations, discussing the capability of SEE to probe the interfacial coupling of 2D materials with a substrate. Our results provided a first evidence of the scattering resonance predicted in ref. [6], showing that its dispersion is only slightly modified by the interaction with the substrate.

As a part of our research effort in understanding the properties of the unoccupied electronic structure of graphene based materials, we performed also state of the art density functional theory (DFT) calculations of the electronic band structure of graphene few-layers and graphite, focusing on the unoccupied energy levels above the Fermi level. Here we present the calculated band structure for graphene bilayer. Similarly to the case of graphene on nickel [8], the calculations show the sensitivity of the scattering resonance to the environment surrounding the plane. When two layers are assembled, these states hybridize, giving rise to more complex states. The dispersion of these states is similar to that observed in graphene and also to that of states of 3D graphite that have been observed in LEED and ARSEE experiments.

2. Calculations

We performed ground state (GS) calculations on 1-3 layer graphene slabs and graphite, using a DFT approach based on the local density approximation (LDA) [12] for the exchange and correlation part of the total Hamiltonian, as implemented in the Abinit code [13,14]. We used the plane-wave basis set (with a cut-off energy of 25 Hartree) to expand the valence electron density, while the core electrons of the carbon atoms were replaced by Troullier-Martins norm-conserving pseudo-potentials [15]. The few-layer graphene structures were obtained by superimposing graphene sheets with the ABAB stacking of graphite at the graphite interlayer spacing, namely 3.354 Å. The separation distance between the repeated systems in the perpendicular direction was large enough to exclude interaction between the replicas, i.e., we kept the same total volume in all slab calculations, and adopted a vacuum distance d larger than 30 Å. Brillouin Zone (BZ) integrations were performed over unshifted $60 \times 60 \times 1$ and $30 \times 30 \times 17$ Monkhorst-Pack grids [16], for 2D-slab and 3D calculations, respectively. The converged electron density was finally used to compute the one electron Kohn-Sham (KS) eigenstates and eigenvectors on selected k-points along high-symmetry BZ-paths.

3. Results

The calculated band structure of Graphene in Fig.1a is in agreement with other calculations found in literature (see for example refs.[1,6]).

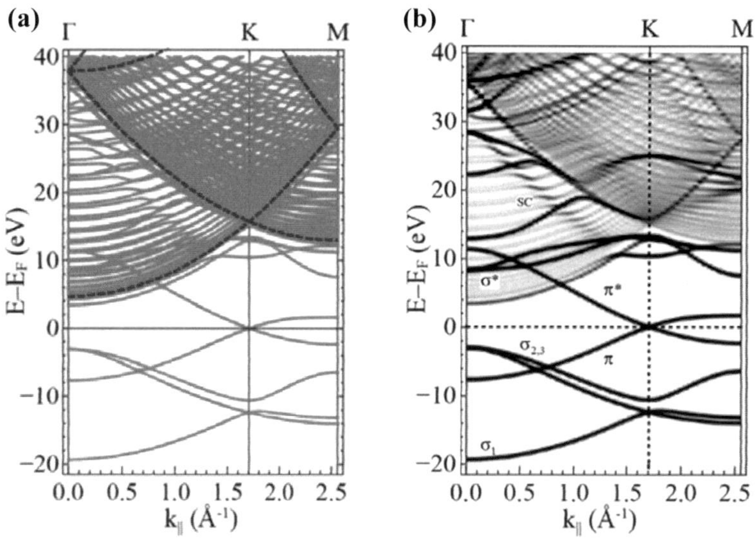

Figure 1. (a) Energy bands of graphene (red solid lines), and 2D free-electron bands (blue dashed lines) along the ΓKM path. **(b)** Projected k-DOS of graphene: the darker regions mark states with a large localization onto the graphene plane.

We can identify the Dirac cone at the K point, formed by the π and π^* bands that touch at $E_F = 0$. The empty part is composed by a quasi-continuum of states (which extrapolates to a continuum in the $d \to \infty$-limit) starting at about 4 eV with a parabolic-like dispersion vs $k_\parallel$. The quasi-continuum produces 5 high density regions (dashed blue lines), dispersing as: $\frac{\hbar^2|\boldsymbol{k}_\parallel+\boldsymbol{G}_\parallel|^2}{2m} + \varphi$. These are 2D free electron bands typical of the hexagonal honeycomb lattice, where $\boldsymbol{G}_{\|}$ is a surface reciprocal lattice vector and $\varphi = 4.6$ eV is the energy of the vacuum level relative to E_F. Similar parabolic dispersions have been observed experimentally as intensity patterns in SEE from graphene layers grown on SiC [17]. Superimposed to the continuum there are discrete states typical of the graphene sheet [5,6]. The lowest lying of these states enter the continuum from below and are clearly seen to retain their individuality. At higher energies the discrete states are fully immersed into the continuum and some of them are no longer visible. To detect them with sufficient accuracy, we focused on the localization of the KS wave-function $\psi_{n,\boldsymbol{k}}(x, y, z)$ on the graphene plane, for the sampled wave-vectors **k** and band-energies $E_n(\boldsymbol{k})$. Accordingly, we considered the probability densities $\rho_{\mathbf{k},n}(z) = \int_{\mathrm{UC_{2D}}} dx\, dy \left|\psi_{\mathbf{k},n}(x, y, z)\right|^2$ associated to the KS-states along the out of plane direction $\hat{z}$, where the in-plane integration was performed over the 2D unit cell of the graphene layer ($\mathrm{UC_{2D}}$). In the attempt to improve the visualization as a whole of the band structure of graphene, we further integrated the $\rho_{\mathbf{k},n}(z)$ over an interval $\delta z = 4\AA$ centered at the graphene plane. The resulting energy distributions were convoluted with a Gaussian function G_σ of standard deviation $\sigma = 0.5$ eV. In this way, we could introduce a **k**-resolved density of states projected onto the C atoms:

$$\mathbf{k}\text{-DOS(E)} = \sum_n G_\sigma(E - E_n(\boldsymbol{k})) \int_{-\delta z/2}^{\delta z/2} \rho_{\mathbf{k},n}(z) dz. \qquad (1)$$

In Fig.1b we report a false color plot of the projected **k**-DOS obtained by sampling the *ΓKM* path in the BZ. This representation improves the visualization of discrete states according to their localization on the graphene plane. The darker regions correspond to states with larger localization integrals. With the aim of studying the transition from a pure 2D material as graphene to graphite we performed similar **k**-DOS calculations on few-layers graphene. In the following we will discuss the **k**-DOS for the bilayer, reported in Fig.2 , where the integration of the probability density has been carried out over an interval containing the whole slab.

4. Discussion

We start our analysis with the graphene **k**-DOS in Fig.1b; some wave functions are strongly concentrated near the graphene plane, whereas others show a more extended tailing into the vacuum. Particularly visible in Fig.1 are, for instance, the states labeled by σ^* and π^* which at the Γ point are found at 8.4 eV and 11.4 eV, above the Fermi level. Looking at the **k**-DOS for 2-layer in Fig. 2, we recognize features similar to those present in graphene. In particular, Fig.2 shows that the dispersion of the states that are more strongly localized on the graphene plane, like for instance the σ^* and π^* states, are not changed by the interlayer interaction. Due to the strong localization of their wave-functions on each graphene plane , these states are preserved in shape with increasing number of layers (even though there is a correspondence between the states in single-layer graphene and in fewlayer graphene each band in graphene gives rise to multiples bands in a fewlayer slab, the number being equal to the number of layers in the slab). These states can be connected to similar bands in graphite that show negligible dispersion in the direction perpendicular to the carbon planes [18]. On the other hand, the interaction between the graphene planes leads to substantial changes in those states that are less localized on the planes, and determines the onset of new states that are not observed in single-layer graphene.

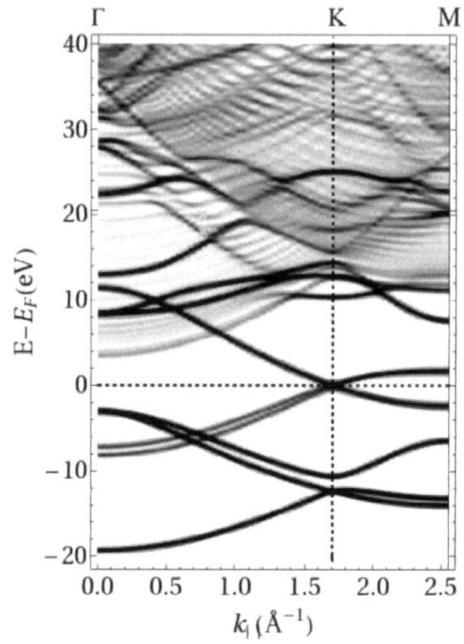

Figure2:Projected k-DOS of 2-layer graphene. the yellow crosses mark the energy-momentum positions of the states used for the $\rho_{n,\mathbf{k}}(z)$ plots in Fig.3

A strongly dispersive band labeled SC is observed in Fig.1b, extending from ~ 28 eV at the Γ point to ~ 13 eV at the K point. The dispersion of the SC-band corresponds to the narrow line of maximum probability for electron transmission, associated to the lowest lying dispersing branch of the scattering resonance of graphene (see Fig.4 of Ref.[6]). The SC-probability densities are presented in Fig.3a for two representative **k**-points along the ΓK path of the BZ, namely the Γ point ($E - E_F \approx$ 28.5 eV) and the point lying at $k = 1.02$ Å^{-1} ($E - E_F \approx 15.7$ eV). They become zero at the graphene plane, and have two maxima at some z-values in the middle of the interlayer spacing in graphite. Their shapes then recall those of the π^* band states, but their widths are almost double. Thus, the SC-states are expected to show strong modifications when the graphene layers are assembled to form few-layer slabs or graphite. Indeed a closer look at Fig.2 shows that, unlike the mono-layer graphene case, two high density bands having a nearly parallel dispersion, starting at $25-28$ eV at Γ and dispersing towards $13-15$ eV at K, are observed in 2-layer graphene. A sample of their probability densities is given in Fig. 3b for 2-layer graphene, where a wave-vector $k = 0.77$ Å^{-1} along the Γ K line is chosen and the two bands are clearly resolved in energy (to 17.9 eV and 19.8 eV above the Fermi level, respectively).

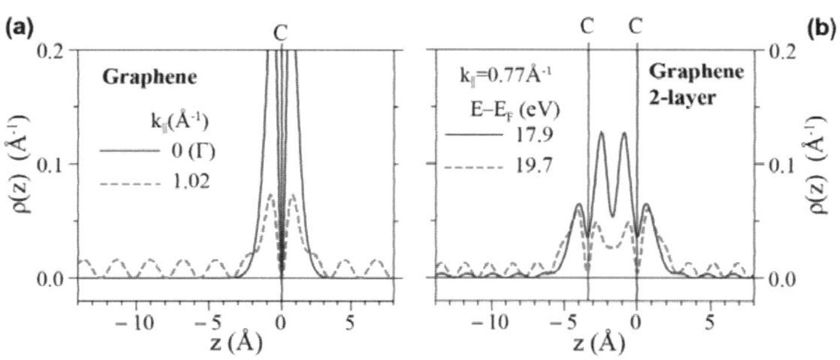

Figure 3: (**a**) Probability density profiles for the SC state of graphene at two representative **k**-points along the ΓK path. (**b**) the two states generated from the SC in 2-layer graphene are given at $k_{||} = 0.77$Å^{-1}

These high energy features find correspondence in graphite, as bands with a pronounced $k_\perp$ dispersion and probability density strongly concentrated in the interlayer region [18]. These states of graphite have been observed in several ARPES and ARSEE experiment, as well as other techniques [19,20], and clearly resolved as two features in VLEED [18]. In all these experiments strongly dispersive bands from about 28 eV at Γ to 13 eV at K have been spotted, that are reminiscent of the dispersion of the scattering resonance in the mono-layer graphene given in Fig.1. Furthermore, similar spectral features have been observed as prominent ones also on graphene layers grown on SiC [17] and Ni(111) [21], as well as on other metal substrates [22].

References

[1] Bertoni G, Calmels L, Altibelli A, Serin V 2005 *Phys. Rev. B* **71** 075402-8.
[2] Bose S, Silkin V M, Ohmann R, Brihuega I, Vitali L, Michaelis C H et al. 2010 *New J. Phys.* **12** 023028 -10.
[3] Armbrust N, Gudde J, Jakob P, Hofer U 2012 *Phys. Rev. Lett* **108** 056801-5.
[4] Craes F, Runte S, Klinkhammer J, Kralj M, Michely T, Busse C 2013 *Phys. Rev. Lett* **111** 056804-5.
[5] Kogan E and Nazarov V U 2012 *Phys. Rev. B* **85** 115418-5.
[6] Nazarov V U, Krasovskii E E, Silkin V M 2013 *Phys. Rev. B* **87** 041405(R)-5.
[7] Pisarra M, Riccardi P, Cupolillo A, Sindona A, Caputi L S 2012 *Nanoscience and Nanotechnology Letters* vol. **4** 1100
[8] Pisarra M, Riccardi P, Sindona A, Cupolillo A, Ligato N, Giallombardo C, Caputi L 2014 *Carbon* **77** 796
[9] Sindona A, Rudi S A, Maletta S, Baragiola R A, Falcone G, Riccardi P *Surf. Sci.* 2007 **601**, 1205 (2007)
[10] Commisso M, Bonanno A, Oliva A, Camarca M, Xu F, Riccardi P, Baragiola R A 2005 *Nucl. Instr. & Meth. B* **230** 438
[11] Cupolillo A, Pisarra M, Sindona A, Commisso M, Riccardi P 2010 *Vacuum* **84** 1029
[12] Goedecker S, Teter M and Huetter J 1996 *Phys. Rev. B* **54** 1703
[13] Gonze X et al. 2009 *Computer Phys. Commun.* **180** 2582
[14] Gonze X et al. 2005 *Z. Kristallogr.* **220** 558
[15] Troullier N and Martins J L 1991 *Phys. Rev. B* **43** 1993
[16] Monkhorst H J and Pack J D 1976 *Phys. Rev. B* **13** 5188
[17] Hibino H, Kageshima H, Guo F Z, Maeda F, Kotsugi M and Watanabe Y 2008 *Appl. Surf. Sci.* **254** 7596
[18] Strocov V N, Blaha P, Stranberg H I, Rohlfing M, Claessen R, Debever J M, Themlin J M 2000 *Phys. Rev.B* **61** 4994
[19] Maeda F, Takahashi T, Ohsawa H, Suzuki S, Suematsu H 1988 *Phys. Rev. B* **37** 4482
[20] Mahatha S K, Menon K R S, Balasubramanian T 2011 *Phys. Rev. B* **84** 113106
[21] Riccardi P, Pisarra M, Sindona A, Cupolillo A, Ligato N, Caputi L 2013 *Nanoscience and Nanotechnology Letters* **5** 1191
[22] Nagashima A, Tejima N, Oshima C 1994 *Phys. Rev. B* **50** 17487

IOP Conf. Series: Journal of Physics: Conf. Series **987** (2018) 012031 doi:10.1088/1742-6596/987/1/012031

Characterization of local atomic structure in Co/Zn based ZIFs by XAFS

Yulia Podkovyrina[1], Vera Butova[1], Elena Bulanova[1], Andriy Budnyk[1], Maria Kremennaya[1], Alexander Soldatov[1], Carlo Lamberti[1,2,3]

[1] The Smart Materials Research Center, Southern Federal University, Sladkova 178/24, 344090, Rostov-on-Don, Russia
[2] Department of Chemistry, INSTM Reference Center and CrisDi interdepartmental Center for crystallography University of Turin, via Giuria, I-10125, Torino, Italy
[3] Department of Physics and NIS interdepartmental Center, University of Torino, via Giuria 1, I-10125, Torino, Italy

E-mail: podkovyrina@sfedu.ru

Abstract. The local atomic structure in bimetallic Co/Zn zeolitic imidazolate frameworks (ZIFs) was studied using X-ray Absorption Fine Structure (XAFS) spectroscopy and theoretical calculations. The experimental Co K-edge and Zn K-edge XANES (X-ray Absorption Near Edge Structure) spectra of $Zn_{1-x}Co_xC_8H_{10}N_4$ samples (x = 0.05, 0.25, 0.75) synthesized by microwave synthesis were compared with the data for the ZIF-67 (x=1) and ZIF-8 (x=0). Theoretical XANES spectra for the bimetallic ZIFs were calculated. It was shown that in bimetallic ZIFs the Co and Zn atoms have the similar local environment.

1. Introduction

Unique structural properties, such as high porosity, surface area, and unexpected thermal and chemical stability of ZIFs are successfully applied in catalysis, gas separation and storage [1, 2]. ZIF-8 is expected to be effectively applied for adsorption of iodine isotopes which are a by-product of nuclear fuel cycle [3]. Most ZIF structures having been studied over the last years are based on a single metal. The fabrication of bimetallic ZIFs, however, is very likely to open up an opportunity for introducing new unique properties in ZIFs which could be efficient for a number of applications. This implies the necessity to thoroughly investigate the local environment of metal centers in heteroatomic ZIFs.

This work focusses on research into the local atomic structure of $Zn_{1-x}Co_xC_8H_{10}N_4$ materials by applying the XANES technique, due to its sensitivity to changes in the local environment of absorbing atoms [4-6].

2. Synthesis and Characterization

In this study we synthesized $Zn_{1-x}Co_xC_8H_{10}N_4$ (x = 0.05, 0.25, 0.75) samples from zinc nitrate hexahydrate, cobalt nitrate hexahydrate and 2-methylimidazole by using microwave radiation as described in [7]. ZIF-8 (x = 0) and ZIF-67 (x = 1) were synthesized analogously using respective metal salts. The resulting solutions were heated in a microwave oven (MARS 6) for 120 min at 140° C. Then the obtained powders were collected by centrifugation, washed several times with methanol and dried at 60°. The crystal structure of $Zn_{1-x}Co_xC_8H_{10}N_4$ samples were characterized by means of powder X-ray diffraction (XRD) using Bruker D2 Phaser diffractometer (Cu k_α, λ=1.5406 Å). The experimental Co K-edge (7709 eV) and Zn K-edge (9659 eV) XANES spectra were measured using in-house X-ray spectrometer Rigaku R-XAS Looper.

3. Results and discussion

The XRD patterns of the synthesized $Zn_{1-x}Co_xC_8H_{10}N_4$ samples display sharp and prominent peaks analogous to those of ZIF-8 and ZIF-67 presented in Figure 1. This suggests that the bimetallic ZIFs have the same crystalline structure as that of ZIF-8. Elemental mapping through scanning electron microscopy coupled with energy dispersive X-ray spectroscopy (SEM-EDS) presented in Figure 2 shows that Co and Zn atoms are uniformly distributed in the $Zn_{0.5}Co_{0.5}C_8H_{10}N_4$ sample. The experimental XANES spectra of Co K-edge and Zn K-edge were measured for $Zn_{1-x}Co_xC_8H_{10}N_4$ samples (Figure 3). They are identical to the absorption spectra for the ZIF-8 and ZIF-67, which allows us to conclude that the local environment around Zn and Co atoms in the structure of bimetallic ZIFs is identical to the structure of monometallic ZIFs. Consequently, cobalt ions occupy partially zinc positions with tetrahedral coordination. This allows us to assume that oxidation state of cobalt ions is +2 due to their tetrahedral coordination, which is not the case with Co^{3+} ions.

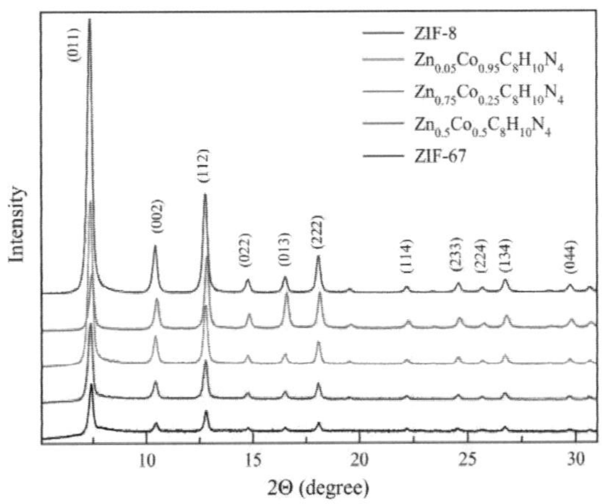

Figure 1. XRD data for bimetallic Zn_1-$Co_xC_8H_{10}N_4$ samples (x = 0.05, 0.25, 0.5), ZIF-8 and ZIF-67

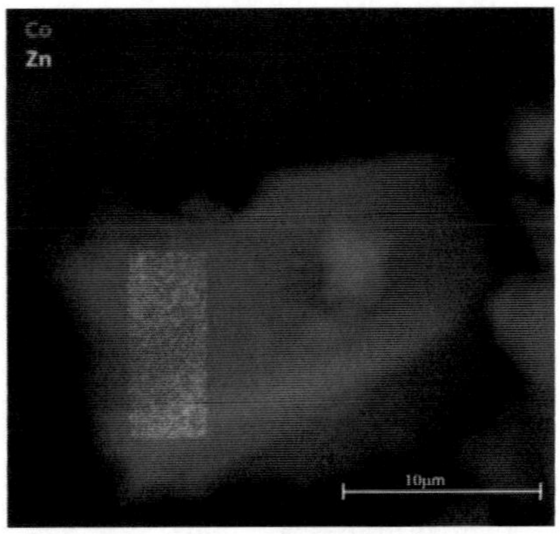

Figure2. SEM-EDS mapping for $Zn_{0.5}Co_{0.5}C_8H_{10}N_4$

A well-resolved pre-edge feature in Co K-edge XANES spectra (~7,700 eV) is characterized by an 1s→3d electronic transition. In accordance with the selection rule (Δl = ± 1), such an electronic transition is forbidden in centrosymmetric compounds having octahedral environment around absorbing atoms.

Thus, the pre-edge feature on the XANES spectra of the Co-ZIF-67 and $Zn_{1-x}Co_xC_8H_{10}N_4$ materials clearly indicates that the local environment of cobalt atoms is not centrally symmetric and confirms the tetrahedral symmetry. The position of the absorption edge and the white line (~ 7708 eV) on the XANES spectra for the ZIF-67 (x=1) sample and the bimetallic $Zn_{1-x}Co_xC_8H_{10}N_4$ samples coincide, which means that the bimetallic ZIFs as well as the original Co-based ZIF-67 material contain divalent cobalt ions.

To prove that the local environment of the metallic atoms in ZIF-8 and ZIF-67 does not change after adding another metal to the structure, the theoretical spectra (Figure 4) were calculated using the finite difference method implemented in FDMNES [8] software. The calculations were performed for atomic clusters of 6 Å around the metal atom (Co or Zn). To simulate the XANES spectra we used crystallographic data for ZIF-8 and ZIF-67 from the Crystallography Open Database [9, 10]. It should be noted that theoretical spectra have some spectral features that are not seen in the experimental spectra, being the result of experimental resolution (1.9 eV for Co K-edge and 4.1 eV for Zn K-edge).

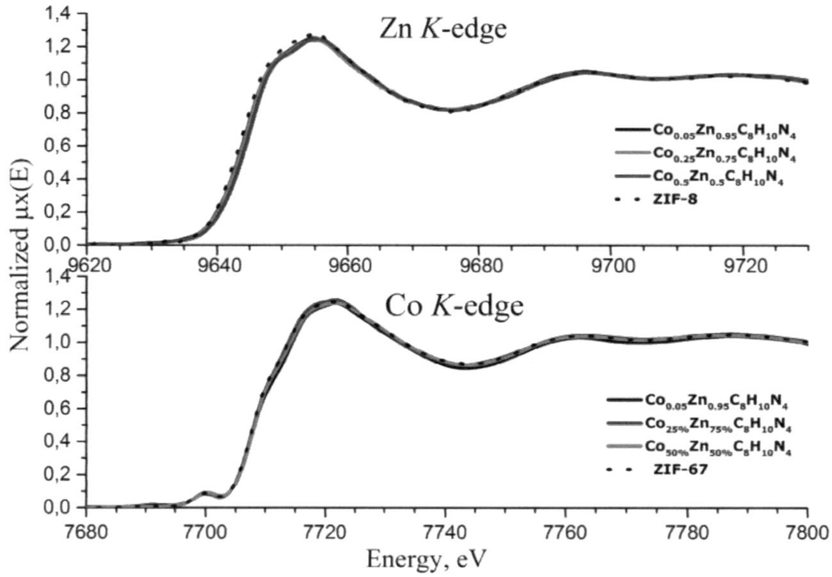

Figure 3. The experimental Co *K*-edge and Zn *K*-edge XANES spectra of $Zn_{1-x}Co_xC_8H_{10}N_4$ samples compared with ZIF-8 and ZIF-67

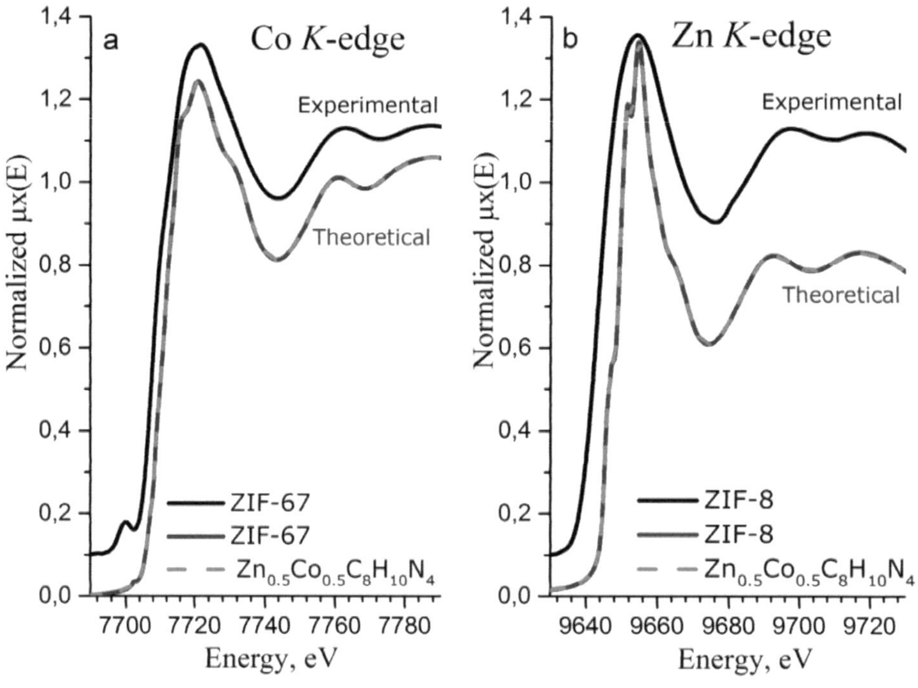

Figure 4. Comparison of the experimental and theoretical Co *K*-edge (a) and Zn *K*-edge (b) XANES spectra for ZIF-67 and ZIF-8 with theoretical spectra for bimetallic $Zn_{0.5}Co_{0.5}C_8H_{10}N_4$

To calculate the XANES spectra for bimetallic $Zn_{0.5}Co_{0.5}C_8H_{10}N_4$, we used the crystallographic data available for ZIF-67 and ZIF-8, and replaced the Co or Zn atoms in the nearest environment of the absorbing atom in the ratio 50/50. We found that the theoretical spectra for the monometallic ZIF-8 and ZIF-67 structures and bimetallic $Zn_{0.5}Co_{0.5}C_8H_{10}N_4$ were virtually identical.

IOP Conf. Series: Journal of Physics: Conf. Series **987** (2018) 012031 doi:10.1088/1742-6596/987/1/012031

4. Conclusions

XRD patterns of synthesized $Zn_{1-x}Co_xC_8H_{10}N_4$ samples (x = 0.05, 0.25, 0.75) displayed sharp and prominent peaks analogous to those of ZIF-8 suggesting that the bimetallic ZIFs have the same crystalline structure. The XANES absorption spectra of the Zn and Co K-edges for bimetallic Zn1-xCoxC8H10N4 samples were found to be identical to the absorption spectra for the pure ZIF-8 and ZIF-67, which allows us to conclude that the local atomic structure around Zn and Co atoms in the structure of bimetallic ZIFs is identical to the structure of monometallic ZIFs.

Acknowledgements

This work was supported by the Government of the Russian Federation (Mega-grant no. 14.Y26.31.0001). P.Y.S and K.M.A. are a grateful for the grant from the Russian Foundation for Basic Research RFBR 16-32-00568. We also acknowledge Alexander Guda and Victor Shapovalov (SFedU) for the assisting during XANES measurements.

References

[1] Butova V V, Soldatov M A, Guda A A, Lomachenko K A and Lamberti C 2016 *Russ. Chem. Rev.* **85** 280

[2] Yao J and Wang H 2014 *Chem. Soc. Rev.* **43** 4470-93

[3] Sava D F, Rodriguez M A, Chapman K W, Chupas P J, Greathouse J A, Crozier P S and Nenoff T M 2011 *J. Am. Chem. Soc.* **133** 12398-401

[4] Bordiga S, Groppo E, Agostini G, van Bokhoven J A and Lamberti C 2013 *Chem. Rev.* **113** 1736-850

[5] Bordiga S, Bonino F, Lillerud K P and Lamberti C 2010 *Chem. Soc. Rev.* **39** 4885-927

[6] Tulchinsky Y, Hendon C H, Lomachenko K A, Borfecchia E, Melot B C, Hudson M R, Tarver J D, Korzyński M D, Stubbs A W, Kagan J J, Lamberti C, Brown C M and Dincă M 2017 J. Am. *Chem. Soc.* **139** 5992-7

[7] Butova V V, Budnik A P, Bulanova E A and Soldatov A V 2016 *Mendeleev Commun.* **26** 43-4

[8] Guda S A, Guda A A, Soldatov M A, Lomachenko K A, Bugaev A L, Lamberti C, Gawelda W, Bressler C, Smolentsev G, Soldatov A V and Joly Y 2015 *J. Chem. Theory Comput.* 11 4512-21

[9] Karagiaridi O, Lalonde M B, Bury W, Sarjeant A A, Farha O K and Hupp J T 2012 J. *Am. Chem. Soc.* **134** 18790-6

[10] Banerjee R, Phan A, Wang B, Knobler C, Furukawa H, O'Keeffe M and Yaghi O M 2008 *Science* **319** 939-43

IOP Conf. Series: Journal of Physics: Conf. Series **987** (2018) 012032 doi:10.1088/1742-6596/987/1/012032

Wireless digital pressure gauge based on nanomaterials

Dilyara Abay, Zhuldyz Otarbay, Madengul Token, Nazim Guseinov, Mukhit Muratov, Maratbek Gabdullin, Daniyar Ismailov
National Nanotechnology Laboratory of the Open Type, Al-Farabi Kazakh National University, Almaty 050040, Kazakhstan

dilyara.abai@mail.ru
Abstract. In the article studies the efficiency of using nanostructured nickel copper films as thin films for bending sensors. Thin films of nickel-copper alloy were deposited using magnetron sputtering technology followed by the appropriate masks. Scanning electron microscopy (SEM) and energy- dispersive X-ray spectroscopy (EDS) techniques were used to examine structure and surface of the Ni Cu coatings. The results of the bending sensors result indicated that the Ni Cu thin film strain gauge showed an excellent sensitive.

Keywords: resistive strain gauge, nanostructured films, nickel copper alloy, digital pressure gauge,

1. Introduction

Pressure measurement is a basic operation in technological processes. Disadvantages of measuring equipment in most cases can lead to errors in the measurements. Only reliable information about current parameters of the technological process determines the output from production lines of quality products. This is achieved by using modern methods of obtaining measurement elements. Recently, in the field of electronics, special interest was given to work aimed at creating nanosized arrays of porous structures on the surface of nickel- copper samples for further application of the resulting formations for practical purposes. Advanced thin film sensors, which are sputter deposited directly onto the surface of heating components, have thickness on the order of a few micrometers; therefore, they leave intact the surface structural integrity and add negligible mass to the surface and create minimal disturbance of the gas flow over the surface [1]. Thin film strain gauge can provide more effective method of accurate strain and other mechanical parameter measurement at high temperature environment. Nickel has a significant effect on the physical and mechanical properties of Cu-Ni alloys [2,3,4]. While tensile strength, 0.2% proof strength, hot strength, solidus and liquidus temperature and corrosion resistance increase with nickel content, thermal and electrical conductivity decrease. Tensile strength increases with nickel content, elongation remains almost constant after a slight decrease (up to 5% Ni) [5]. The desired pattern of the strain gauge is deposited by masking the non-conductive areas. The advantages of strain gauge technology: fast response time, ease of compensation of temperature effects, the relative freedom from the effects of acceleration [6]. One of the urgent problems in Kazakhstan is the transfer of all utilities to digital electronics and the most relevant of these is the measurement of pressure in the pipes of the city heating. Since high accuracy is required to control the pressure in the pipes, it is necessary to have appropriate instruments that have a high sensitivity. The task was to create a sensitive element based on nanostructured nickel-copper composite.

2. The experimental part

The coatings were applied on the experimental test bench to obtain functional metallic and oxide coatings by using a magnetron sputtering system based on a vacuum installation VUP-5M [7]. Vacuum universal post is designed to apply metal oxide thin-film coatings in vacuum to the surface of products made from various materials. In magnetron sputtering system, the ionization efficiency of the electrons is increased by increasing their path length by applying a transverse magnetic field normal to the electric field. Magnetron sputtering makes it possible to utilize the cathode discharge power very efficiently to generate high current densities at relatively low voltages and lower operating pressure to yield deposition rates which are higher than those in the nonmagnetron sputtering systems. Magnetron sputtering system operated in the following modes: voltage on the cathode was 500 V, the discharge current was 100 mA with a duration of 15 minutes, the pressure in the vacuum chamber was maintained at about 0.5 PA (Fig. 5.), the working gas was argon, the magnetron target was nickel-copper. The discharge power was maintained at 50 W.

The structure of the stand includes: the vacuum camera, the diffusion pump, a vacuum pump, the volume cylinder, the switching valves of the vacuum system, the thermocouple sensor PMT-2, the ionization sensor PMI-2 as shown in the diagram (Fig. 1).

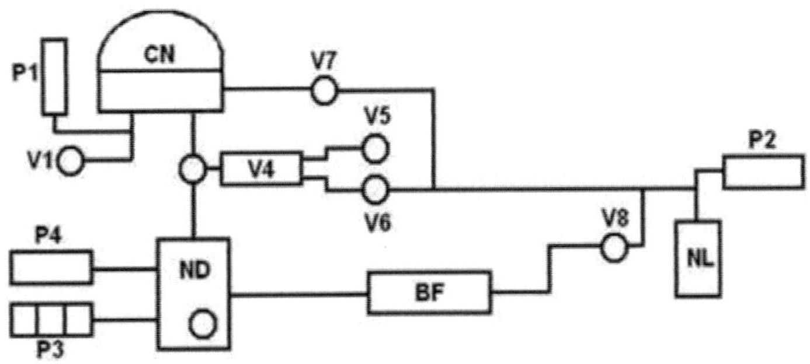

Figure 1. The scheme of the vacuum system VUP-5

The thin-film strain gauge was made on samples, as the bending element was the kapton material as shown in Figure 2.

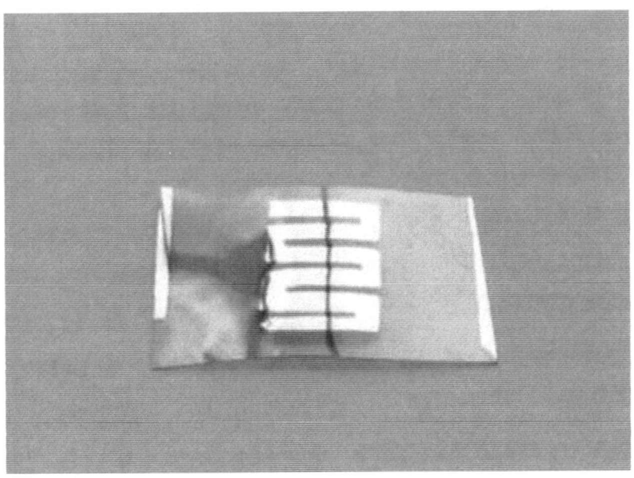

Figure 2. A sample of Ni-Cu thin film resistive strain gauge

At the beginning of the spraying process an inert gas argon is supplied. When an electromagnetic field is created, a plasma is generated. The particles begin to break away from the target and movement is chaotic. The concentrated metal particles begin to break away and are deposited on the surface of the substrate. For the experiment we used 3 different materials as a substrate. At first, we sprayed on the glass to find out the suitable mode, then on silicon to see the thickness of the film with a scanning electron microscope, finally we used the kapton to look at the bend.

3. Discussion of result

The strain gage which we obtained by the magnetron sputtering method is attached to a beam then connected to an amplifier. The beam is under the action of a uniformly distributed load (Figure 3). When the beam is subjected to deformation from the neutral layer, one part of the cross section of the beam is subjected to stretching, another part – to compression.

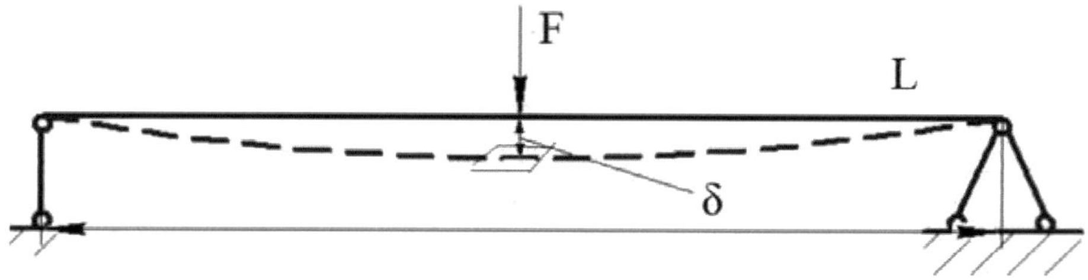

Figure 3. Calculate the beam of bending.

From two ends, evenly increasing, we set the moment of forces. Until the middle of the beam reaches maximum bending (for the experiment the maximum bend is f max = 3 mm), we simultaneously record the voltmeter's reading. We compare the value of the diagram with the theoretical bend. The maximum value of the bend will be in the middle, since the internal force of the beam will be a pure bend. Below are the formulas for measuring sensitivity and elongation. As shown in the formulas 1,2 we calculate the sensitivity and the relative extension.

$$K_F = \frac{\Delta R/R0}{Ex}; \quad (1)$$

$$E_x = \frac{\Delta L}{L_0}; \quad (2)$$

The dependence of relative resistance change from the relative extension shown in chart 1. It is possible to evaluate the sensitivity of the lode cell. As the graph shows we have reached the maximum for metal films.

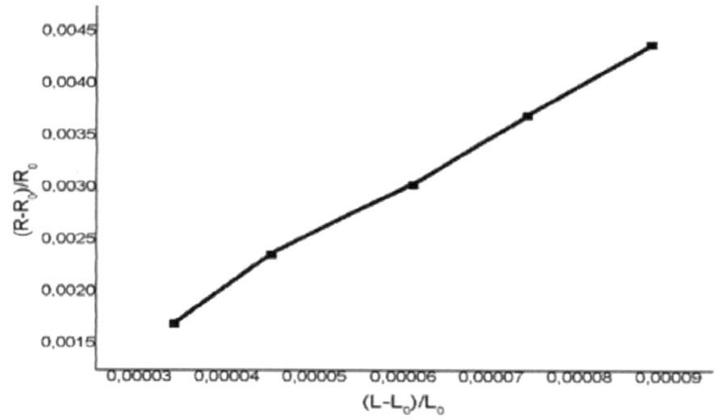

Chart 1.The dependence of relative resistance change from the relative extension

Scanning electron microscopy (SEM) was used to examine surface of the Ni Cu coatings. Thus, the films were deposited on the substrate at a deposition rate of 1.2 A / s, according to the SEMimages as shown in Fig. 4, one film with a thickness of 110.7 nm, a second film with a thickness of 205.4 nm. Samples have a uniform surface as shown in Figure 5.

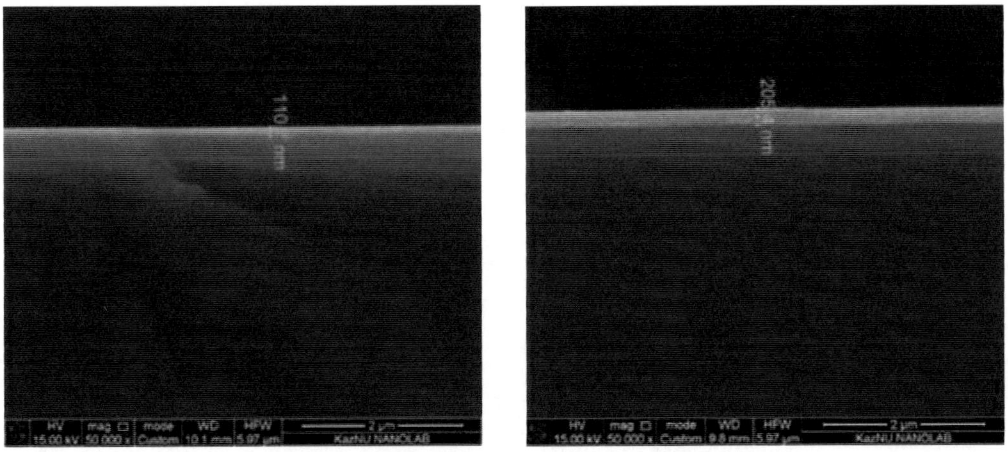

Figure 4. SEM images of the cleaved Cr on which the Ni+Cu alloy is deposited to measure the thickness

Figure 5. SEM images of the cleaved Cr surface of sample

In the experiment, the mask was made of aluminum foil. Energy-dispersive X-ray spectroscopy (EDS) techniques was used to examine elemental analysis. As seen from the elemental analysis(Fig 6), it amounted to Nickel 54% and Cu 45%.

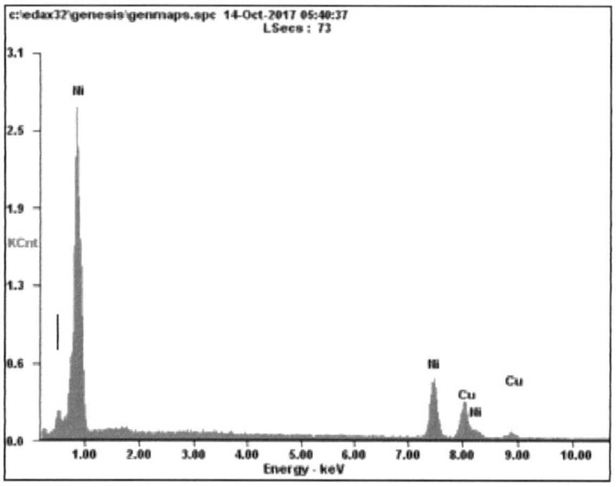

Element	Wt%	At%
NiK	54.55	56.50
CuK	45.45	43.50
Matrix	Correction	ZAF

Figure 6. Elemental analysis

In this work, results indicated that measuring devices based on nanostructured materials have a good sensitivity. The sensitive strain gauge on the basis of nanostructured nickel-copper alloy with sensitivity Kf=5 at a voltage of 3V was made. The coatings were applied by using a magnetron sputtering method. Scanning electron microscopy was used to study the film surfaces. Energy-dispersive X-ray spectroscopy (EDS) techniques was used to examine elemental analysis. The calculation the beam of bending was made.

References

[1]. Miyazaki K., Islam N. Nanotechnology systems of innovation. An analysis of industry and academia researchactivities //Technovation. - 2007. - № 27. - P.661-675
[2]. Tkachev A. G. Apparatus and methods for synthesis of solid-state nanostructures / A. G. Tkachev and I. V. Zolotukhin, Moscow: Mashinostroenie-1, 2007.
[3]. Plasma technology / N..Sosnin, S. A. Ermakov, P. A., Topolyansky. – SPb : publishing house of St. Petersburg Polytechnical Institute. University press, 2008. – 406 p.
[4]. Gulyaev, I. P. Plasma processing of dispersed materials : monograph : in 2 h / I. P. Gulyaev // Part I. Physical basis. – Khanty-Mansiysk : SOUTH, 2013. – 115 p.
[5]. Noise Control in Strain Gage Measurements. Tech.Note TN-501-2. – Vishay Micro- Measurements, 2007. – 8 p.
[6]. Ohring. M. (2002) Materials Science of Thin Films: Deposition and Structure. 2nd Edition, Academic, SanDiego.
[7]. Alexandre A.B. A new method to calibrate pressure gauges for pneumatic applications / A.B. Alexandre,AutelianoA. dos S. Junior, G.R.Taina//ABCM Symposium Series in Mechatronics.Vol.5 2012pp.1367-137

IOP Conf. Series: Journal of Physics: Conf. Series **987** (2018) 012033 doi:10.1088/1742-6596/987/1/012033

All optical programmable logic array (PLA)

Dawit Hiluf [1]

Ben-Gurion University of Negev department of Physical Chemistry, Be'er-Sheva 84105, Israel
Physics Department, Mekelle University, P.O.Box 231, Mekelle, Ethiopia.

E-mail: `dawit@post.bgu.ac.il`

Abstract. A programmable logic array (PLA) is an integrated circuit (IC) logic device that can be reconfigured to implement various kinds of combinational logic circuits. The device has a number of AND and OR gates which are linked together to give output or further combined with more gates or logic circuits. This work presents the realization of PLAs via the physics of a three level system interacting with light. A programmable logic array is designed such that a number of different logical functions can be combined as a sum-of-product or product-of-sum form. We present an all optical PLAs with the aid of laser light and observables of quantum systems, where encoded information can be considered as memory chip. The dynamics of the physical system is investigated using Lie algebra approach.

1. Introduction

The recent success of the computing industry coupled with limitations put forth by Moore's law has opened to a search of a paradigm shift in either the way we manufacture chips, transistors and/or in the way we do computation. This range of research routes include quantum computing [1], molecular computing [2, 3, 4], ternary valued logic [5, 6, 7], parallel computation [8, 9], and multivariable logic beyond binary [10, 11].

A programmable logic array (PLA) is a type of reconfigurable logic device that can be tailored to implement the desired combinational logic circuits. The device has a number of AND and OR gates which are linked together to produce output. PLAs are advantageous as they are compact thereby, consequently, doesn't take much space in designing complex circuits, this is more so pronounced in logic circuit such as feedback and control where one takes into account several variables in search of efficient logic design and implementation.

Nowadays PLAs are readily obtainable from the market and reconfigure (i.e "program") them to provide desired Boolean functions. The most commonly used method of 'programming' places fuses between the inputs to a gate and the gate itself. In this case by a fuse we meant a breakable, by applying high current, electrical connection. By analyzing the desired output's Boolean expression, one readily identifies which fuses to blow and which to be left, thereby clearing connections that are not required to implement the Boolean function. It is this act of disconnection that we refer as programming, wherein all the connections are available before reconfiguring.

[1] Present address: Department of Chemistry, Sungkyunkwan University, Suwon 440-746, Korea

IOP Conf. Series: Journal of Physics: Conf. Series **987** (2018) 012033 doi:10.1088/1742-6596/987/1/012033

2. The system

The physical system we consider is a very well studied, adiabatic population transfer in solids, specifically speaking on Rare-earth-metal ions doped in solids. Recently the authors of [12, 13, 14] have demonstrated the implementation of population transfer by stimulated Raman adiabatic passage (STIRAP) in Praseodymium ions (Pr^{3+}), doped into a Y_2SiO_5 crystal. Rare-earth ions doped into an inorganic crystal exhibit optical transitions that do not occur in free ions. These are due to the interaction of the ions with the crystal field. The Energy level scheme in Pr^{3+}:Y_2SiO_5, hereafter (Pr:YSO) is shown in Fig.(1). The optical transitions in Pr:YSO exhibit very narrow homogenous line widths width in the order of a few kHz. However due to crystal impurities, different ions see slightly different crystal fields and the optical line is inhomogeneously broadened by several GHz. Due to this inhomogeneous broadening, a laser field with single frequency will always address several ions in the in homogeneously broadened line.

The benefit of using Rare-earth materials is that they exhibit sharp optical transitions and long decoherence times. This in turn facilitates the implementation of coherent adiabatic processes. It is worth noting that all the three ground states can be coupled to all three excited states, and altogether, nine transitions are possible. The hyperfine states in Pr:YSO are spaced between 4.6 MHz and 17.3 MHz from each other. The system we consider is a three level system shown in Fig.(1) in a Λ scheme quantum system of states $|0\rangle$, $|1\rangle$, & $|2\rangle$ with energies $\hbar\omega_0$, $\hbar\omega_1$, and $\hbar\omega_2$ respectively. These three levels of the system are coupled by a two lasers. A pump pulse drives the transition between states $|0\rangle$ and $|1\rangle$. A second pulse, the Stokes pulse, drives the transition between $|1\rangle$ and $|2\rangle$. That is, we have transitions between states $|0\rangle$ and $|1\rangle$, and between states $|1\rangle$ and $|2\rangle$, but no transition between states $|0\rangle$ & $|2\rangle$, it is dipole forbidden. [15, 16] Stimulated Raman adiabatic passage (STIRAP) is an adiabatic process which provides

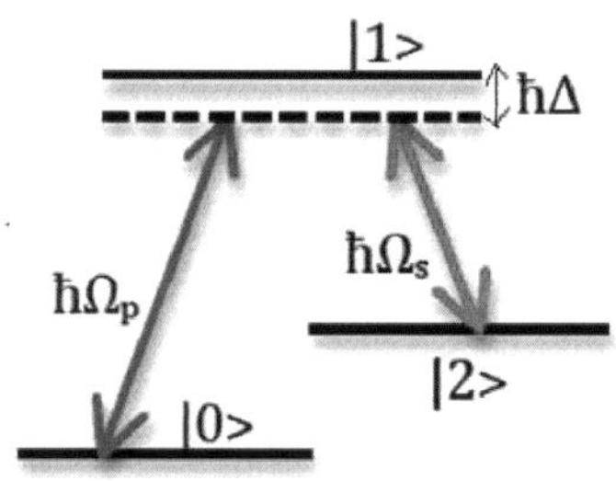

Figure 1. Three level Λ scheme, with coupling lasers. Blue (Ω_P) pump laser, and red (Ω_S) Stokes laser, along with the corresponding energy levels and the laser frequencies. Detuning $\hbar\Delta_{p/s} = \hbar\omega_{jk} - \hbar\omega_{p/s}$ where $jk = 01, 12$. We consider the case $\Delta_p = \Delta_s = \Delta$

complete coherent transfer in a Λ-type quantum system. By preparing the system to be initially on the ground state $|0\rangle$, one can transfer 100% of the population to state $|2\rangle$ [15]. To do so we couple the states, via the pulses, counter intuitively, that is we first drive the states $|1\rangle$ and $|2\rangle$ using the Stokes pulse, and later we couple states $|0\rangle$ & $|1\rangle$, using pump laser. STIRAP is robust in that it is insensitive to the decaying intermediate state $|1\rangle$. Our purpose here is to go beyond STIRAP.

3. The Pulse Sequence

Fractional STIRAP (FSTIRAP) is variant of STIRAP that can be used to create a superposition of states [15, 17]. We aim to create coherent superposition between two hyperfine levels, instead of transferring 100% population between the states. To do so we have to interrupt the STIRAP evolution. In FSTIRAP scheme, like that of STIRAP, the Stokes pulse precede the Pump pulse, but the two pulses vanishes smoothly and simultaneously. If, for example, all the population is initially prepared to be in the ground state: we have Stokes pulse applied when $t \to -\infty$ but the pump pulse is not yet applied therefore $\frac{\Omega_P(-\infty)}{\Omega_S(-\infty)} \to 0$. After a time, i.e when $t \to +\infty$, we interrupt the pulses in such a manner as to get a constant from their ratio. The condition can mathematically be put as Eq.(1)

$$\lim_{t \to -\infty} \frac{\Omega_P(t)}{\Omega_S(t)} = 0, \quad \lim_{t \to +\infty} \frac{\Omega_P(t)}{\Omega_S(t)} = \tan\alpha, \tag{1}$$

where α is a constant mixing angle $\left(0 \leq \alpha \leq \frac{\pi}{2}\right)$. A convenient realisation of the condition Eq.(1), using three pulses, a pump pulse and two Stokes pulses (one with the same time dependence as the pump and another coming earlier) is provided by Vitanov et al. to be [17]

$$\Omega_P(t) = \Omega_0 \sin\alpha \; e^{\frac{-(t-\tau)^2}{\sigma^2}} \tag{2a}$$

$$\Omega_S(t) = \Omega_0 e^{\frac{-(t+\tau)^2}{\sigma^2}} + \Omega_0 \cos\alpha \; e^{\frac{-(t-\tau)^2}{\sigma^2}} \tag{2b}$$

where $\tau, \sigma, \&\ \Omega_0$ are respectively defined to be the centre, width, and the peak of the pump pulse. Note that, of the two Stokes pulses one has same time dependence as the pump and another coming earlier. Moreover Ω_0 is also the peak of each of these two Stokes pulses.

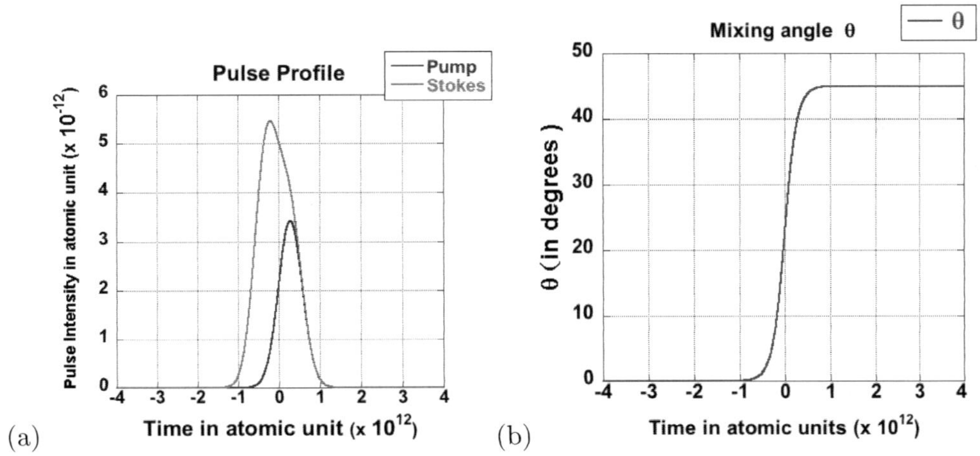

Figure 2. (a) Pulse Profile, blue (Ω_P) Pump laser, and red (Ω_S) Stokes laser. (b) Mixing angle θ, defined by the ratio of the pulse profiles, i.e. $\tan\theta = \frac{\Omega_P(t)}{\Omega_S(t)}$

3.1. Hamiltonian of the system

For quantum systems with distinct states, external perturbations change the state of the system. The three states of the system, i.e. $|0\rangle$, $|1\rangle$, & $|2\rangle$ respectively are the eigenstates of the the unperturbed part of the Hamiltonian $\hat{H}_0$, and $\hat{H}_I$ is the part of the Hamiltonian representing the interaction of the system with the laser field. Therefore, the total Hamiltonian can be expressed as the sum of the two parts [16].

$$\hat{H} = \hat{H}_0 + \hat{H}_I \tag{3}$$

where the unperturbed part is, $\hat{H}_0 = \hbar\omega_0|0\rangle\langle 0| + \hbar\omega_1|1\rangle\langle 1| + \hbar\omega_2|2\rangle\langle 2|$, and the interaction part of the Hamiltonian is $\hat{H}_I = \hat{V}_P + \hat{V}_S$, where $\hat{V}_j = -\mu_j \cdot E_j \cos\omega_t$ with $j = P, S$. With the aid of which and considering a rotating frame, i.e. $e^{i\omega_0 t}|0\rangle\langle 0| + e^{i(\omega_1-\Delta)t}|1\rangle\langle 1| + e^{i\omega_2 t}|2\rangle\langle 2|$, the total Hamiltonian in the interaction picture and in RWA can be rewritten as

$$\hat{H} = \frac{-h}{2}\begin{pmatrix} 0 & \Omega_P(t) & 0 \\ \Omega_P(t) & 2\Delta_p & \Omega_S(t) \\ 0 & \Omega_S(t) & 2(\Delta_p - \Delta_s) \end{pmatrix} \tag{4}$$

Although the presence of the single detuning does not prevent population transfer, it is important that the two-photon resonance condition apply, i.e $\hbar(\Delta_p - \Delta_s) = 0$. Because this condition is crucial for the creation of the dark state [15, 16].

3.2. Adiabaticity

The system we consider is shown Fig.(1). The population transfer mechanics, FSTIRAP, is easily understood in Hilbert space whose coordinate basis vectors are instantaneous eigenstates of the time varying Hamiltonian of Eq.(4). In section (3.1) we have obtained the Hamiltonian of a three level system whose eigenstates are the energy level of the system, these states were labeled as $|0\rangle$, $|1\rangle$, and $|2\rangle$, we refer to such basis the diabatic basis. But it turns out, it is helpful to transform into different basis to get a complete understanding of the system at hand. To this end it will be insightful if we work in the adiabatic basis. When the two-photon resonance condition $\hbar(\Delta_p - \Delta_s) = 0$ fulfilled, Eq.(4) reduces to

$$\hat{H}(t) = \frac{-h}{2}\begin{pmatrix} 0 & \Omega_P(t) & 0 \\ \Omega_P(t) & 2\Delta & \Omega_S(t) \\ 0 & \Omega_S(t) & 0 \end{pmatrix} \tag{5}$$

It is worth recalling that for the three level system depicted in Fig.(1), the transition between levels $|0\rangle$ and $|2\rangle$ is dipole forbidden, level $|1\rangle$ can be off-resonance by certain detuning. Introducing an adiabatic $\{|a_+\rangle, |a_0\rangle, |a_-\rangle\}$ and diabatic basis $\{|0\rangle, |1\rangle, |2\rangle\}$ (also known as the bare states) respectively, we can make a unitary transformation $|\Phi\rangle = U|\Psi\rangle$ [18]. Following which we make the expansion of the wave function in terms of the time dependent probability amplitude corresponding to each basis as:

$$|\Psi\rangle = \sum_{j=0}^{2} C_j(t)\,|j\rangle, \qquad |\Phi\rangle = \sum_{k=\pm,0} a_k(t)\,|a_k\rangle \tag{6}$$

The unitary matrix U diagonalises the Hamiltonian Eq.(5) into $D = U^\dagger H U$, adiabatic Hamiltonian, where the transformation matrix U is given by [18]

$$U = \begin{pmatrix} \sin\phi\sin\theta & \cos\phi & \sin\phi\cos\theta \\ \cos\theta & 0 & -\sin\theta \\ \sin\theta\cos\phi & -\sin\phi & \cos\phi\cos\theta \end{pmatrix} \tag{7}$$

After diagonalising the Hamiltonian Eq.(5) we find that the new basis vectors $\{|a_+\rangle, |a_0\rangle, |a_-\rangle\}$ corresponds to the eigenvalues $\{\lambda_+, \lambda_0, \lambda_-\}$ respectively where

$$\lambda_+ = \frac{1}{2}\left(\Delta + \sqrt{\Delta^2 + \Omega^2}\right) = \Omega\cot\phi, \tag{8a}$$

$$\lambda_0 = 0, \tag{8b}$$

$$\lambda_- = \frac{1}{2}\left(\Delta - \sqrt{\Delta^2 + \Omega^2}\right) = -\Omega\tan\phi \tag{8c}$$

IOP Conf. Series: Journal of Physics: Conf. Series **987** (2018) 012033 doi:10.1088/1742-6596/987/1/012033

with $\Omega^2 = \Omega_P^2 + \Omega_S^2$ where the mixing angles are defined to be

$$\tan\theta = \frac{\Omega_P(t)}{\Omega_S(t)}, \tag{9a}$$

$$\tan 2\phi = \frac{\Omega(t)}{\Delta} \tag{9b}$$

The Hamiltonian matrix element for non-adiabatic coupling between state $|a_0\rangle$ and either one of $|a_+\rangle$ or $|a_-\rangle$ is given by $\langle a_\pm|\dot{a}_0\rangle$. Non-adiabatic coupling is small if this matrix element is small compared to the field induced splitting $|\lambda_\pm - \lambda_0|$ of the energies of these states, that is

$$|\langle a_\pm|\dot{a}_0\rangle| \ll |\lambda_\pm - \lambda_0| \tag{10}$$

Using Eq.(7) and the transformation $|\Phi\rangle = U|\Psi\rangle$ one readily obtain

$$|a_+\rangle = \sin\theta\sin\phi|0\rangle + \cos\phi|1\rangle + \cos\theta\sin\phi|2\rangle \tag{11a}$$

$$|a_0\rangle = \cos\theta|0\rangle - \sin\theta|2\rangle \tag{11b}$$

$$|a_-\rangle = \sin\theta\cos\phi|0\rangle - \sin\phi|1\rangle + \cos\theta\cos\phi|2\rangle \tag{11c}$$

Notice that the adiabatic state $|a_0\rangle$ is independent of state $|1\rangle$, which is the leaky state. Therefore following the adiabatic state $|a_0\rangle$ we can achieve the complete population transfer from the ground state $|0\rangle$ to the target state $|2\rangle$. For appreciable non-adiabatic transition to occur, the following conditions has to be met

$$\left|\dot{\theta}(t)\right| \geq \frac{1}{2}\Omega(t) \tag{12a}$$

$$\left|\dot{\theta}(t)\right| \geq \frac{1}{\sigma} \tag{12b}$$

The first condition means the adiabatic condition has to be violated, that is the non-adiabatic coupling has to be comparable or larger than the eigenvalues splitting [17]. To summarize, so far we have described our physical system which is a Rare-Earth metal ion doped in Inorganic crystal, specifically Praseodymium ions (Pr^{3+}), doped into a Y_2SiO_5 crystal. The ion's electronic states are split into so called "crystal field states"; the crystal field states are split into hyperfine states. The hyperfine interaction consists of hyperfine interaction from the ions and magnetic hyperfine interaction with the crystal field. The hyperfine interaction splits the crystal field states further into doubly degenerate levels. We have identified, based on the experimental measurement, three level lambda system out of the nine possible transitions. Working in the adiabatic basis has revealed that we have two important controlling parameters: time-dependent mixing angles θ and ϕ. With these angles we can manipulate the population transfer amongst the states of the atomic system. To avoid, for example, complete transfer of population between the initial and the target state, as one would have in STIRAP scheme, we demanded the ratio of our pulse profiles (of pump and Stokes) to be constant after interaction. This demand means our pulses after long time decay smoothly and simultaneously. Such scheme is referred to as Fractional STIRAP (FSP); with such pulse profile we managed to create superposition of the two hyperfine ground states $|0\rangle$ and $|2\rangle$. Furthermore we have obtained the Hamiltonian under two-photon resonance, RWA (Rotating Wave Approximation), in the interaction picture.

4. Dynamics of the system

The Quantum dynamic are calculated by the density matrix formalism [19]. In this method a statistical average over the assembly of the expectation values for the individual ions is included.

This in turn yields information about observable macroscopic quantities. The formalism is advantageous in that it greatly simplifies the calculations since it allows one to handle a large number of dynamical variables in a systematic way. The physical significance of the individual density matrix elements depends on the representation with which the matrix is calculated. In a representation where H_0 is diagonal, the elements of the density matrix are defined as $\rho_{jk} = \langle\psi_j|\hat{\rho}|\psi_k\rangle$ where $\hat{\rho}$ is the density operator.

In this representation, the diagonal elements ρ_{jj} are the ensemble averaged occupation probabilities for the state $|\psi_j\rangle$. The off-diagonal elements ρ_{jk} give information about the ensemble averaged phases of the quantum states, these matrix elements are commonly known as coherences. The matrix ρ is Hermitian. The motion of the populations ρ_{jj} is coupled to the coherences ρ_{jk}, therefore we should deal the density matrix as a whole. As our density matrix ρ is a 3×3 Hermitian matrix, it contains nine distinct real variables. Because of conservation of probability, only eight of these variables are independent.

Lie Algebra approach

The dynamical evolution of an N-level atomic system can be expressed in terms of density matrix $\hat{\rho}$ which satisfies the Liouville equation [20]

$$i\hbar\frac{\partial}{\partial t}\hat{\rho} = \left[\hat{H}, \hat{\rho}\right] \tag{13}$$

Because $\langle\hat{P}_{mn}(t)\rangle = Tr\left(\hat{P}_{mn}\hat{\rho}(t)\right) = \rho_{nm}(t)$ the dynamics can be equally explored by studying $\hat{P}_{mn}$, where $\hat{P}_{mn} = |m\rangle\langle n|$. In this section we will obtain the equation of motion for the eight dimensional coherence vector, we do so by using the Alhassid-Levine formalism [21]. One can also obtain the equation of motions using the Hioe-Eberly [22, 23, 24] formalism; In Hioe-Eberly approach, the state of a three-level system is represented by pseudo-spin vector $\vec{S}$ having 8 real components, 9 if we do not impose normalization. This description provides an elegant geometrical framework. The Hamiltonian under RWA, in the interaction picture, and two-photon resonance is expressed as

$$\hat{H} = \hbar\begin{pmatrix} 0 & \alpha(t) & 0 \\ \alpha(t) & \Delta & \beta(t) \\ 0 & \beta(t) & 0 \end{pmatrix} \tag{14}$$

where the half Rabi frequency is defined by $\alpha(t) = \frac{1}{2}\Omega_P(t)$ and $\beta(t) = \frac{1}{2}\Omega_S(t)$.
The $\hat{G}_\alpha$s are our generators, which will be expressed in terms of the operators $|\alpha\rangle\langle\beta|$. Using this generators we will expand our Hamiltonian and density matrix as

$$\hat{\rho}(t) = \frac{\hat{I}}{3} + \frac{1}{2}\sum_{\alpha=0}^{8}\langle G_\alpha\rangle\hat{G}_\alpha \tag{15a}$$

$$\hat{H}(t) = \frac{\hbar}{2}\sum_{\alpha=0}^{8} h_\alpha(t)\hat{G}_\alpha \tag{15b}$$

with traceless generators $\hat{G}_\alpha$ and $\hat{I}$ is the identity operator. We stress here that Eq.(15a) is always true, but Eq.(15b) is a special case that is valid for our problem.
The coefficients $\langle G_\alpha\rangle$ and $h_\alpha(t)$ are given by

$$\langle G_\alpha\rangle = Tr\left(\hat{\rho}(t)\hat{G}_\alpha\right) \tag{16a}$$

$$\hbar h_\alpha(t) = Tr\left(\hat{H}(t)\hat{G}_\alpha\right) \tag{16b}$$

Furthermore the generators $\hat{G}_\alpha$ have the following properties

$$Tr\left(\hat{G}_\alpha\hat{G}_\beta\right) = 2\delta_{\alpha\beta} \tag{17a}$$

$$\left[\hat{G}_\alpha, \hat{G}_\beta\right] = 2if_{\alpha\beta\gamma}\hat{G}_\gamma \tag{17b}$$

where the repeated indexes are summed up from 1 to 8 and $f_{\alpha\beta\gamma}$ is the completely antisymmetric structure constants of the SU(3) group is given in Table A1. The Identity operator is given by $|0\rangle\langle 0| + |1\rangle\langle 1| + |2\rangle\langle 2|$. We note that Eq.(17a) tells us that the trace of the product of the generators is orthogonal, and Eq.(17b) is the closure property.
The generators $\hat{G}$ in turn are closed under the commutation relation Eq.(17b)

$$\left[\hat{H}, \hat{G}_\alpha\right] = \imath\hbar \sum_{\beta=0}^{8} \hat{G}_\beta g_{\beta\alpha} \tag{18}$$

where the $g_{\beta\alpha}$ are (possibly complex) numerical coefficients.

In their pioneering work on the information theoretic approach, Alhassid and Levine starting from Eq.(19), the equation of motion for observable $\hat{G}$ in the Heisenberg picture, replaced Eq.(13) by a set of coupled equations for the mean value of the observables as given by Eq.(20)

$$i\hbar\frac{d\hat{G}}{dt} = \left[\hat{G}, \hat{H}\right] \tag{19}$$

In doing so they demand that the commutation between the Hamiltonian and the generators be closed under the Lie algebra as Eq.(18) [21, 25]. The closure means that the commutation between the Hamiltonian and the generators is a linear combination of the generators [26]. Moreover the Hamiltonian is expanded in terms of the generators $\hat{G}$ via Eq.(15b). The equation of motion for the expectation values of the observables $\langle G_\alpha\rangle$, takes the form

$$\frac{d\langle G_\alpha\rangle}{dt} = -\sum_\beta \langle G_\beta\rangle g_{\beta\alpha} \tag{20}$$

Equation(20) is a set of coupled linear differential equations that completely determine the time evolution of the expectation values $\langle G_\alpha\rangle$, provided that one knows the corresponding initial values.

So as to exploit the $SU(3)$ Lie algebra let us define three groups of generators as

$$\begin{aligned} \hat{s}_m &= |j\rangle\langle k| + |k\rangle\langle j| \\ \hat{s}_n &= -i\left(|j\rangle\langle k| - |k\rangle\langle j|\right) \\ \hat{s}_p &= -\sqrt{\frac{2}{l\,(l+1)}}\left(j\rangle\langle j| - l|l+1\rangle\langle l+1|\right) \end{aligned} \tag{21}$$

where $0 \le j \le k \le 2$, $0 \le l \le 2$, $m = 1, 2, 3$, $n = 4, 5, 6$, and $p = 7, 8$.

Using Eqs.(14),(16b), and the generators Eq.(21) we note that one can readily express the Hamiltonian (to within an addition of multiple of an Identity matrix) in terms of the generators as

$$\hat{H}\,(t) = \frac{\bar{h}}{2}\left[\Omega_P\hat{s}_1 + \Omega_S\hat{s}_2 + \Delta\hat{s}_7 - \frac{\Delta}{\sqrt{3}}\hat{s}_8\right] + \text{const} \tag{22}$$

where const $= \frac{-\hbar}{2}\left(\frac{2}{3}\sum_{j=0}^{2}\omega_j\right)\hat{s}_0$–commutes with the Hamiltonian– where $\hbar\omega_j$ is energy of level j. Making use of the commutation relation between the Hamiltonian Eq.(22) and the generators Eq.(21), and then equate both sides of Eq.(15b) we find the complete surviving elements $g_{\alpha\beta}$, in a matrix form is

$$g = \begin{pmatrix} 0 & 0 & 0 & \Delta & 0 & \beta & 0 & 0 \\ 0 & 0 & 0 & 0 & -\Delta & -\alpha & 0 & 0 \\ 0 & 0 & 0 & \beta & -\alpha & 0 & 0 & 0 \\ -\Delta & 0 & -\beta & 0 & 0 & 0 & 2\alpha & 0 \\ 0 & \Delta & \alpha & 0 & 0 & 0 & -\beta & \sqrt{3}\beta \\ -\beta & \alpha & 0 & 0 & 0 & 0 & 0 & 0 \\ 0 & 0 & 0 & -2\alpha & \beta & 0 & 0 & 0 \\ 0 & 0 & 0 & 0 & -\sqrt{3}\beta & 0 & 0 & 0 \end{pmatrix} \tag{23}$$

where the half-Rabi frequency is defined by $\alpha(t) = \frac{1}{2}\Omega_p$, $\beta(t) = \frac{1}{2}\Omega_s$, and note that g is antisymmetric.

We now solve our equation of motion given by Eq.(20) numerically. It is worth pointing out that the equation of motion is exact. Recall we have a three level lambda system whose population is initially prepared to be on the ground state $|0\rangle$. We then apply FSTIRAP pulse profile and follow the evolution of the observable vector. By arranging the expectation values of the generators we can form a vector of the observables as $\vec{S} = \left(\langle s_1\rangle, \langle s_2\rangle, \ldots \langle s_8\rangle\right)^T$ the equation of motion Eq.(20), can now be rewritten as

$$\frac{d}{dt}\vec{S} = g\vec{S} \tag{24}$$

We first solve Eq.(24) numerically for a very small value of detuning and the FSTIRAP pulse profile; we recall we prepare the system to be on the ground state $|0\rangle$. Of course it goes without saying that the results will vary when we take different initial state, if we take a different pulse profile, STIRAP case for example. The consequence of change in the initial state, change in the input parameters leading to different output, in digital logic term, basically means we have a Finite State Machine (FSM) where the output of the system in not only dependent on the input applied but also is dependent on the initial state of the system [27].

5. Logic Machine

The discussion in this paper focuses on isolated systems, however, it has to be pointed out that the inclusion of noise destroys most of the observables (coherences) and hence we refrain from making logic out of such observables. Yet we can design our pulse profile to yield only the desired observables that withstand the effect of noise on the system, long enough to do computation. These observables are the population on each state along with the coherence between the two lower levels (i.e. between states $|0\rangle$ and $|2\rangle$). Using the results we obtain by such scheme we will try to show here the implementation of logic machines: a programmable logic array (PLA) it can be depicted as parallel outputted logic (endowed with Memory).

Before proceeding it is worth differentiating a ROM (Read Only Memory) and PLA (Programmable Logic Array) for the sake of clarity. Both of these used to implement logic functions, and they both use the 'Sum of Products' logic configuration, which consists of a primary array of AND gates and a secondary array of OR gates. The OR function (Sum) is applied to outputs of AND (product) arrays. ROM is made of an AND gates array and OR gates array. AND array provides all the combinations of inputs, and OR array is used to select the necessary combinations. Therefore, AND array is always fixed. For example, in a three input system, AND array produces all the combinations (product terms) of $ABC, AB\bar{C}, A\bar{B}C, A\bar{B}\bar{C}, \bar{A}BC, \bar{A}B\bar{C}, \bar{A}\bar{B}C, \bar{A}\bar{B}\bar{C}$ where $\bar{X}$ implies the complement (NOT) of

X. Then an OR gate can be used to select the necessary product terms to implement the given logic function. Any logic function of A, B, C can be implemented using those product terms. For example

$$\begin{aligned} f(A,B,C) =& \bar{A}\bar{B} + \bar{B}\bar{C} \\ =& \bar{A}\bar{B}\bar{C} + A\bar{B}\bar{C} + \bar{A}\bar{B}C + A\bar{B}\bar{C} \end{aligned} \tag{25}$$

In like manner an array of OR gates can implement an array of logic functions. Therefore ROM is used to store programs. Programming the ROM means configuring those OR array by selecting the necessary products. PLA too is made of two OR and AND arrays, but both the arrays are configurable unlike in ROM. This also provides a 'Sum of Products' term, but in a different way. Since terms for AND gates are also possible, it can give more product terms like $AB, \bar{C}, C$, etc. Therefore it is much easier to implement logic functions compared to ROM.

For example, $\bar{A}\bar{B} + \bar{B}\bar{C}$ can be directly implemented by selecting $\bar{A}, \bar{B}$ for one AND gate, $\bar{B}, \bar{C}$ for another AND gate and making outputs of those AND gates to inputs of an OR gate.

A PLA provides a bank of AND gates, followed by a bank of OR gates, and optionally a flip-flop per output. It also will have an array of programmable "fuses", the "fuses" are short-circuits, so each input of each AND is connectable to any of the chip inputs or outputs, or their complements. By programming we mean fusing connections. When we program the PLA, the programmer blows the fuses to leave only the connections we want. Then, the OR array similarly can be programmed to OR together various combinations of the AND outputs. The fuse technology incurs minimal delay, so propagation delay from input pin to output pin can be as small as two gate delays or up to four, depending on how many inverted signals are needed, which automatically beats the usual ROM topology for speed. In a PLA the "OR array" is fixed and the "AND array" is programmable. Each output must be formed from a restricted number of minterms but those minterms can each cover multiple input combinations.

On top of this, a read-only memory (ROM) is a logic circuit that can generate all of the possible minterms of its inputs. With the same numbers of inputs and outputs, a ROM is capable of more general logic, in that we can burn any truth table we like into the ROM. However, as a logic device, the ROM is not optimal. If we attempt to burn several independent little functions into a ROM, we will find we have a lot of 'don't care' cases that must nonetheless be programmed into the ROM to get the desired output. The PLA is better for logic because we don't have to deal with the don't-cares, but the trade-off is that we can't program completely arbitrary functions of the bits. However, they are nonetheless sufficiently flexible to be useful for many common 'glue-logic' tasks, and they are generally faster at producing output.

What we aim to show here is how the system yields logic in parallel and serves as PLA, for which we will obtain the corresponding Boolean equations. In what follows we describe how we implement the PLA. To this end we apply a sequence of three pulses separated by some waiting time t. This means that after the system interacts with the first pulse we have some time window t_1 before applying the second pulse. During this time the observable vector propagates freely. Then we apply the second pulse and wait another time t_2 before applying the last pulse. Each of the pulses can take either the STIRAP or FSTIRAP pulse profile, here after **SP** and **FSP** respectively. We can follow the evolution of the observables before, during, and after each of the pulses are applied, but what we take as an output is the observable after the third pulse is applied. We assign logic **0**, when we apply the **SP** pulse and logic **1** for **FSP** pulse profile. Because we have three inputs (the triple pulse sequence), and because each can be either **0** or **1**, we have $2^3 = 8$ possible combination of inputs as seen in the truth table (1). The three pulse sequences are labeled as input variable **ABC**, where **A** stands for the first pulse and **C** is the last pulse in the sequence, **B** being the intermediate pulse. We use logic assignment for the populations to be logic **1** if we have reading of populations ≥ 0.2 else it is **0**; and for the coherences it is logic **1** if the absolute value of the coherences is ≥ 0.1, else **0**. With these logic assignments the possible input combinations, along with their outputs is shown in Table (1).

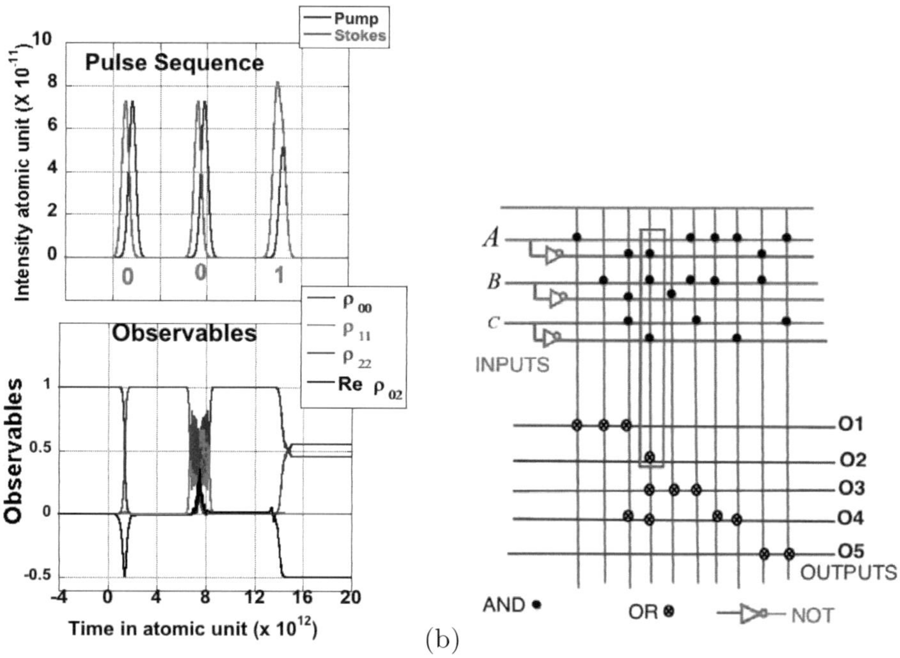

(a) (b)

Figure 3. (a)Plots of observables for pulse sequence 001(see text). (b) Programmable logic array (PLA). In this case programming means fusing connections

Table 1. Truth table for the triple input pulses labeled as **ABC**, the outputs **O1, O2, O3** are populations of state $|k\rangle$, $k = 0, 1, 2$, and **O4, O5** are the real and imaginary part of the coherence

ABC	O1	O2	O3	O4	O5
000	0	0	1	0	0
001	1	0	1	1	0
010	1	1	1	1	1
011	1	0	0	0	1
100	1	0	1	1	0
101	1	0	1	0	1
110	1	0	0	1	0
111	1	0	1	1	1

We show in Fig.(3a), the numerical result of the observable vector, for the input sequence **001** (**SP-SP-FSP**), which is also shown in the table via red ink–second row. We can see from the truth Table (1) we have obtained five logic outputs simultaneously read after the system interacts with the last (i.e. third) pulse. This means we can store information in each of these outputs. Moreover, if we have many, say **N**, independent three level systems we can have a massive parallel logic memory, 5^N, where we can store information. Each of these three level systems could be arranged to have different initial condition, set on different detuning, and apply different pulse sequence to store a variety of information. From the truth table we can find an expression representing each of the five output logic functions. Such logic function is derived from the truth table by **OR**-ing (logic **OR**), logic OR is read as A OR B, is expressed as A+B, all the terms that correspond to those input lines for which the output assumes the logic value **1**. Each term is an **AND** (logic **AND**), logic AND is read as A AND B, is expressed as AB, of the input variables. The input variable appears in uncomplemented form in the product (i.e. logic **AND**) if it has logic value **1** in the corresponding input line, and it appears in complemented form if it has logic value **0**. For example, the output logic function **O2** has only one logic value **1** (row 3 in Table (1)) corresponding to input line **010**. Hence the product term (i.e. logic **AND**) that corresponds to this input line is $\bar{A}B\bar{C}$. Such product term contains each of the 3 input variables in either complemented or uncomplemented form, and is called minterm. If, however, we have logic value **1**, for more than one input line combinations; then we **OR** all of such minterms to find the logic function of the output. An expression of this type is known as a canonical sum of products (**SOP**). Therefore using the **SOP** for the output logic functions shown in the truth table we get

$$\begin{aligned} O1 =& \bar{A}\bar{B}C + \bar{A}B\bar{C} + \bar{A}BC + A\bar{B}\bar{C} + A\bar{B}C + AB\bar{C} + ABC \\ =& \bar{A}\bar{B}C + (A \oplus B) + AB \\ O1 =& \bar{A}\bar{B}C + A + B \end{aligned} \tag{26}$$

where we have used $(\bar{X} + X) = 1, A \oplus B = \bar{A}B + A\bar{B}$ for simplification, where X stands A, B, and C, also $\oplus$ represent logic Exclusive OR (**XOR**). In like manner we obtain the logic function of the other outputs

$$\begin{aligned} O2 =& \bar{A}B\bar{C} \\ O3 =& \bar{B} + B(A \odot C) \\ O4 =& \bar{A}(B \oplus C) + A\bar{C} + AB \\ O5 =& \bar{A}B + AC \end{aligned} \tag{27}$$

where $(B \oplus C) = \bar{B}C + B\bar{C}$ and $(A \odot C) = (\bar{A}\bar{C} + AC) = \overline{(A \oplus C)}$ and also $\odot$ represent logic Exclusive NOR (**XNOR**). Therefore we have here, expressed by Eqs.(26) and (27), logic output functions in parallel. As the laser pulse is very long, and as the observables are long lived, these output functions serve as memory. We can form an Integrated Circuit (IC) that implements these five output functions as shown in Fig.(3b). It comprises of an **AND** plane where they form input lines and are shown by the red lines, these input lines represent both the complemented and the uncomplemented form; and an **OR** plane where we form the outputs functions and are shown by the blue lines. The dots, in the input lines, tell us what form of the input variable we take (i.e. complemented or uncomplemented). To obtain the minterms that is present in the in the logic output function, we take the product of each of the input variables down the line (see the green box in Fig.(3b), for example, it is formed by taking the product of the input variables $\bar{A}, B$, and $\bar{C}$ i.e. $\bar{A}B\bar{C}$). To obtain the output of the logic function we **OR** the minterms we formed for each logic functions. After we have done that we get the following Integrated Circuit (IC) for the five output functions.

We stress that the very act of applying the pulses 'fuses' the input-output connections in our system, this in turn yields fixed input combinations (an AND array in the PLA) and outputs (an OR array of PLA) as given by Eqs.(26) and (27) thereby allowing the possibility of implementing PLA using three level dynamics.

The logic functions we get will differ if we redefine our logic assignment for the input pulses. If we now assign logic value **1** for when we use the **SP** pulse profile and logic **0** for **FSP** pulse profile we obtain the following logic equations.

$$\begin{aligned}
O1 =&\bar{A} + A\bar{B} + AB\bar{C} \\
O2 =&A\bar{B}C \\
O3 =&\bar{A}\bar{C} + C(A \oplus B) + AB \\
O4 =&\bar{A}\bar{B} + C(A \oplus B) + AB\bar{C} \\
O5 =&\bar{A}\bar{C} + A\bar{B}
\end{aligned} \tag{28}$$

We can also change threshold value we took for the purpose of the logic assignments for the observables and thereby obtaining different logic functions for the observables, the five outputs. Moreover, we can also change the initial state of the system leading to different logic functions; hence we have a dynamic memory. By dynamic we mean the memory changes whenever we change either of: the initial state of the system, the inputs, redefine input logic assignments, and change the logic assignments of the observables by re binning the threshold value.

6. Conclusion

In a nut shell, in this paper we have explored the dynamics of the physical system, optically addressed three-level system, using Alhassid-Levine scheme, where we have obtained equation of motion for the average values of the observables which we numerically solve for different cases of inputs (i.e. Pulse profile, detuning, and initial state of the system). By applying sequence of three pulses we managed to propose a Memory Logic that yields its output in parallel and serves as Programmable logic array (PLA) where by programmable we mean fusing connections. Moreover we expressed the logic function of each output in terms of the input variables (and their complemented form). The execution of logic functions in parallel and storing them has advantage of yielding storing vast data in parallel, the size and complexity of the stored data increases as we increase the physical system to N-level system. Although contact with the environment washes away the information stored, we argue that we have ample time to retrieve the stored information before it goes away.

Acknowledgments

The author thanks Prof. Raphael D. Levine of Hebrew University of Jerusalem for introducing him to Lie algebra and molecular logic. This work was supported by the European Commission Future and Emerging Technologies Proactive Project MULTI (317707).

Appendix

The generators chosen to fit the GellMann SU(3) matrices are [24]

$$\begin{aligned}
&\hat{G}_1 = \begin{pmatrix} 0 & 1 & 0 \\ 1 & 0 & 0 \\ 0 & 0 & 0 \end{pmatrix}, \quad \hat{G}_2 = \begin{pmatrix} 0 & 0 & 0 \\ 0 & 0 & 1 \\ 0 & 1 & 0 \end{pmatrix}, \quad \hat{G}_3 = \begin{pmatrix} 0 & 0 & 1 \\ 0 & 0 & 0 \\ 1 & 0 & 0 \end{pmatrix}, \\
&\hat{G}_4 = \begin{pmatrix} 0 & -i & 0 \\ i & 0 & 0 \\ 0 & 0 & 0 \end{pmatrix}, \quad \hat{G}_5 = \begin{pmatrix} 0 & 0 & 0 \\ 0 & 0 & -i \\ 0 & i & 0 \end{pmatrix}, \quad \hat{G}_6 = \begin{pmatrix} 0 & 0 & -i \\ 0 & 0 & 0 \\ i & 0 & 0 \end{pmatrix}, \\
&\hat{s}_7 = \begin{pmatrix} -1 & 0 & 0 \\ 0 & 1 & 0 \\ 0 & 0 & 0 \end{pmatrix}, \quad \hat{G}_8 = \frac{1}{\sqrt{3}} \begin{pmatrix} -1 & 0 & 0 \\ 0 & -1 & 0 \\ 0 & 0 & 2 \end{pmatrix}
\end{aligned} \tag{A.1}$$

where the corresponding surviving structure constants f_{ijk} are given below

Table A1. Values of the surviving structure constants

ijk	147	135	126	432	465	736	752	368	258
f_{ijk}	-1	$\frac{1}{2}$	$\frac{1}{2}$	$\frac{1}{2}$	$-\frac{1}{2}$	$-\frac{1}{2}$	$-\frac{1}{2}$	$-\frac{\sqrt{3}}{2}$	$-\frac{\sqrt{3}}{2}$

References

[1] Michael A Nielsen, Isaac L Chuang 2010 *Quantum Computation and Quantum Information* (Cambridge University Press)
[2] Remacle F, Levine R D 2001 *J. Chem. Phys.* **114** pp 10239–46
[3] Beil F, Halfmann T, Remacle F and Levine, R D 2011 *Phys. Rev. A* **83** p 033421
[4] Remacle F, Weinkauf R and Levine R D 2006 *J Phys Chem A* **110** pp 177-84
[5] Klein M, Rogge S, Remacle F and Levine, R D 2007 *Nano letters* **7** pp 2795–99
[6] Klein M, Mol JA, Verduijn J, Lansbergen GP, Rogge S, Levine R D and Remacle F 2010 *Appl. Phys. Lett.* **96** p 043107
[7] Mol JA, Van der Heijden J, Verduijn J, Klein M, Remacle F and Rogge, S 2011 *Appl. Phys. Lett.* **99** p 263109
[8] Fresch B, Hiluf D, Collini E, Levine R D and Remacle F *Proc. Natl. Acad. Sci. U. S. A* **110** pp 17183–88
[9] Remacle F, Levine R D 2007 *Eur. Phys. J. D* **42** pp 49–59
[10] Fresch B, Bocquel J, Hiluf D, Rogge S, Levine R D and Remacle F 2017 *ChemPhysChem 18* pp 1790-97
[11] Fresch B, Cipolloni M, Yan T, Collini E, Levine R D and Remacle F 2015 *J. Phys. Chem. Lett. 6* pp 1714-18
[12] Klein J, Beil F and Halfmann T 2008 *Phys. Rev. A* **78** p 033416
[13] Beil F, Klein J, Nikoghosyan G and Halfmann T 2008 *J. Phys. B* **41** p 074001
[14] Klein J, Beil F and Halfmann T 2007 *Phys. Rev. Lett.* **99** p 113003
[15] Vitanov N V, Halfmann T, Shore B W and Bergmann K 2001 *Annu. Rev. Phys. Chem* **52** pp 763–809
[16] Shore B W 2008 *Acta Phys. Slovaca* **58** pp 243–486
[17] Vitanov N V, Suominen K A and Shore B W 1999 *J. Phys. B* **32** p 4535
[18] Vitanov N V , Stenholm, S 1999 *Phys. Rev. A.* **60** pp 3820–32
[19] Fano U 1957 *Rev. Mod. Phys.* **29** pp 74–93
[20] Levine R D 2011*Quantum mechanics of molecular rate processes* (Courier Dover Publications)
[21] Alhassid Y, Levine R D 1977 *J. Chem. Phys.* **67** pp 4321–39
[22] Hioe F T, Eberly J H 1981 *Phys. Rev. Lett.* **47** pp 838–41
[23] Hioe F T, Eberly J H 1982 *Phys. Rev. A.* **25** p 2168
[24] Hioe F T 1983 *Phys. Rev. A.* **28** p 879
[25] Alhassid Y, Levine R D 1978 *Phys. Rev. A.* **18** p 89
[26] Robert G 2006 *Lie Groups, Lie Algebras, and Some of Their Applications* (Dover Publisher)
[27] Kohavi Z, Jha N K 2010 *Switching and finite automata theory* (Cambridge University Press)

IOP Conf. Series: Journal of Physics: Conf. Series **987** (2018) 012034 doi:10.1088/1742-6596/987/1/012034

Graphene like ZnO 2D structure from semiconductor to conductor behaviour by doping F

Ahmad Yazdani*, Behrad Barakati and Asiye Shokri

Department of Physics, Tarbiat Modares University, P.O. Box 14115-175, Tehran, Iran

*Yazdania@modares.ac.ir

Abstract. 2D ZnO honeycomb structure was recently fabricated. This new semiconductor material with band gap about 1.67 ev is great for applications which needs this gap and such dimension. This material has a great electronic structure, which was investigated in this article by details for the first time by this view. Duo to this especial electronic structure here we found that this material could be change its behaviour from semiconductor to metal material with different Fermi surfaces which could be used for different applications by doping F elements on the surface . The first section in your paper

1. Introduction

In recent years many 2D structures were fabricated such as graphene [1], transition metal dichalcogenides (TMDs)[2], hexagonal boron nitride (hBN) [3] etc. with thicknesses ranging from a single to a few atomic layers which, open new area of research in physics. These materials have great electronic structures that is different from their bulk structures [4]. In modern electronic devices area, low dimension material is going to use for miniaturization of electronic devices. 2D crystals are great of interest to understand in fundamental physics at reduced dimension because of their unique and functional properties. These 2D materials exhibit charming and great physical properties, from conductor to semiconductor range of materials, which could be used in the wide range of applications in nano scale electronic and photoelectronic devices [5-9].

In this situation, founding new 2D Structure that could be stable in the nature would be great of interest; actually, if it could be tunable from semiconductor to conductor material. One of the structures that fabricated in recent years is 2D ZnO graphene like honeycomb structure [10, 11].

Bulk zinc oxides structures are widely used in different studies in environmental and energy-related applications such as solar cell devices [12], sensors[13], photo catalysis[14] etc.. However, the new structure (2D ZnO with honeycomb structure) has more interested structure because of less band gap and his unique electronic structure, which is investigate in this article with more details. Because of the open surface area that is duo to monolayer structure and its double chemical bond between Zn and O elements this structure could have tunable electronic structure.

2. Calculation methods

First-principles calculations within density functional theory (DFT) is used for achieving to our results by using Wien2k and FHI-aims packages. Generalized gradient approximation GGA functional use as exchange-correlation potential. All structures have been treated within supercell geometry using the periodic boundary conditions. The interaction between ZnO monolayers in adjacent supercells examine as a function of their spacing. Since we used the spacing of 100 Å which these monolayers have not interaction with each other in the calculations. In the self-consistent potential and total-energy calculations, the Brillouin zone is sampled by 12*12*1 special k points for 2D honeycomb ZnO and also for doping with F element. All structure was relaxed and the structure stability was checked out.

3. Result and discussion

At first the electronic structure of single ZnO layer was investigated. As could be seen in the DOS results (see picture 1) the band gap is 1.67 ev which is have great agreement with Topsakal et all results [10]. On the other hand as could be seen we have 2 different type of hybridization in this material:

1- In the XY plane the hybridization is included of S, Px and Py orbitals of Oxygen ; S, Px,Py,dX^2Y^2 and dXY orbitals of Zinc and also dZ^2 orbital of Zinc because as you know this orbital have Electronic Cloud in XY plane. This hybridization create σ chemical bond.
2- Duo to Z direction the hybridization is include of Pz orbital of Oxygen and Pz, dXZ, dYZ and dZ^2 orbitals of Zinc. This hybridization create Π chemical bond. Which the second type band is located in the middle of the first band duo to the energy and located in out of XY plane in the Z direction duo to the space.

The Zn-O bond length in this structure was obtained 1.90 A.

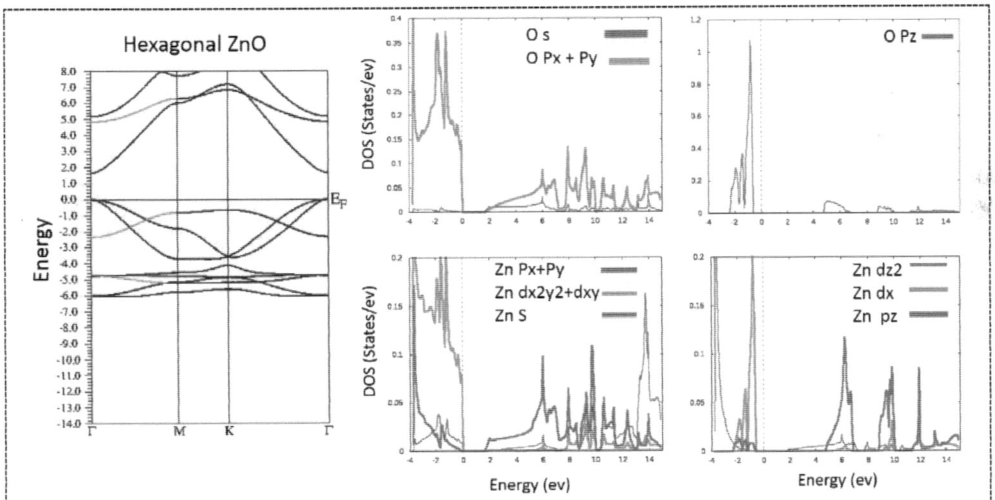

Figure 1. This pictures shows the band structure and pDOS of ZnO honeycomb structure in 2 way of in plane and out of plane direction.

Therefore, because of this electronic structure by doping hole in the second type orbitals (the orbitals which are located out of XY plane) metallic characteristic behavior could be induced to this material by broken Π bond and could construct dangling bond. In this situation, we doping F elements to the surface of ZnO sheet by different concentration (One, two and three F elements for each three ZnO hexagon). The results show all of the structures could be metallic (see figure 2). These three structures have different Fermi surfaces, which could be used for different applications (see figure 2 around fermi level).

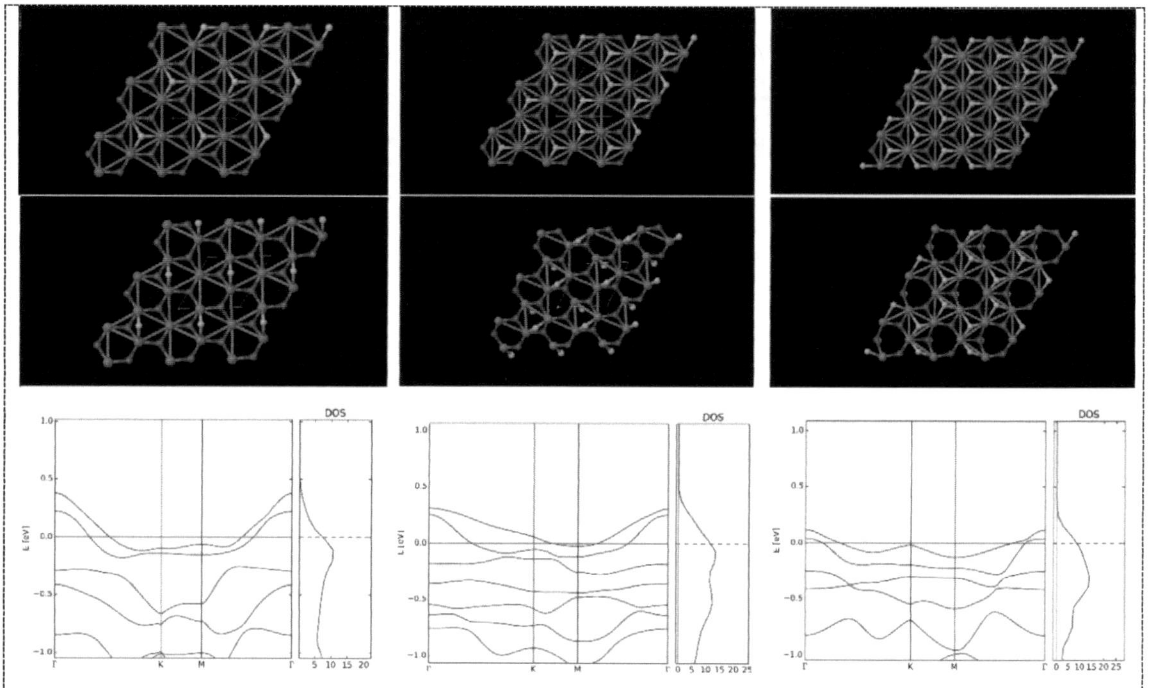

Figure 2. In this pictures the first row shows the selected structure and second row shows the optimize structure with different F doping concentrations and the last row shows band structure and Dos of this three state.

4. Conclusion:

Therefore, 2D ZnO honeycomb structure was fabricated. This material has a great electronic structure which was investigated in this article by details. This material has semiconductor electronic character with 1.67 ev band gap and chemical double bond between Zn and O elements. Duo to this especial electronic structure we found that this material could be change its electronic characteristic from semiconductor to metal with F elements doping by break the Π bonds.

References

[1] Phaedon Avouris, Christos Dimitrakopoulos. 2012 Materials Today. 15(3) 86-97
[2] Bao Zhang, Xiao Ji, Kui Xu, Chi Chen, Xiong Xiong, Junlin Xiong, Yifeng Yao, Ling Miao, Jianjun Jiang. 2016 Electrochimica Acta. 217 1-8
[3] P.C.Mende, Q.Gao, A.Ismach, H.Chou, M.Widom, R.Ruoff, L.Colombo and R.M.Feenstra. 2017 Surface Science. 659 31-42
[4] Akhilesh Pandey, Shankar Dutta, Anand Kumar, R. Raman, Ashok K. Kapoor, R Muralidhran. 2016 Adv. Mater. Lett. 7(10) 777-782
[5] Xia Y N, Yang P D, Sun Y G, Wu Y Y, Mayers B, Gates B, Yin Y D, Kim F, Yan H Q . 2003 Adv. Mater. 15 353-389
[6] Xue X, Nie Y, He B, Xing L, Zhang Y, Wang Z L 2013 Nanotechnology 24 225501.
[7] Yao J, Yan H, Lieber CM. 2013 Nat. Nanotechnol. 8 329-335.
[8] Su Y, Liu C, Brittman S, Tang J, Fu A, Kornienko, Kong Q, Yang PD. 2016 Nat. Nanotechnol. 11 609-612.
[9] Cao J L, Wang Z X, Zhan X Y, Wang Q S, Safdar M, Wang Y J, He J 2014 Nanotechnology. 25 105705.

[10] M. Topsakal, S. Cahangirov, E. Bekaroglu, and S. Ciraci. 2009 Phys. Rev. B. 80 235119
[11] Huimin Wang, Guangzhao Qin, Guojian Li, Qiang Wang and Ming Hu. 2017 Phys. Chem. Chem. Phys.19 12882-12889
[12] Qingzhi An, Paul Fassl, Yvonne J. Hofstetter, David Becker-Koch, Alexandra Bausch, Paul E. Hopkinson, Yana Vaynzof. 2017 Nano Energy.39 400-408
[13] Ekasiddh Wongrat, Narong Chanlek, Channarong Chueaiarrom, Wichittra Thupthimchun, Benjarong Samransuksamer, Supab Choopun. 2017 Ceramics International. 43 S557-S566.
[14] Jothi Ramalingam R., Arun K. Shukla, K. Kombaiah, J. Judith Vijaya, Ahmed M. Tawfeek. 2017 Optik - International Journal for Light and Electron Optics. 148 325-331.

IOP Conf. Series: Journal of Physics: Conf. Series **987** (2018) 012035 doi:10.1088/1742-6596/987/1/012035

Investigation of the phase formation from Ni-modified nanostructured silicon

O V Volovlikova, Y I Shilyaeva*, A Y Berezkina, D I Smirnov and S A Gavrilov

National Research University of Electronic Technology - MIET, 1 Shokin Square, Zelenograd, Moscow, 124498, Russia

*shyliaeva@gmail.com

Abstract. The phase formation behavior during the thermally induced solid-state reaction of the Ni-modified nanostructured Si has been investigated using a combination of differential scanning calorimetry and x-ray diffraction analysis. The effect of nanostructured interface on the silicides formation temperatures is discussed. Special attention is paid to the methodical aspects of the characterization of the thermal behavior of the Ni-modified nanostructured Si. The role of the thermal instability of nanostructured Si in the calorimetric studies of the silicides formation is also reviewed.

1. Introduction

Transition metal silicides are widely used as metal contact materials for multilayer metallization in modern silicon microelectronics. As the trend of device-miniaturization continues, the nanostructured silicides come into use in different applications such as field emitters, nanoscale interconnects, biomedical cell separation systems, as high-performance catalysts and others [1-3]. A preferred use of Ni silicides is explained by several reasons, in particular, relatively low formation temperatures, low specific resistivity and a small amount of silicon consumed during silicidation.

The formation process of the nickel silicide by solid-state interaction usually involves applying metal coating to the silicon surface using various techniques (evaporation, sputtering, and electrochemical deposition) and subsequent heat treatment. Mechanism of silicide growth during solid-state interaction have been fairly well studied for thin-film systems. Studies [4-6] devoted to questions of phase identification, stability and formation sequence during the solid-state reaction at the metal/Si interface. The influence of nature of alloy components and impurities on silicidation mechanism is discussed in [7]. Formation of Mn, Fe and Co silicides in thin-film systems are reviewed in [8-10], etc.

The key challenge in this research area is to reduce the temperature of silicide formation. Realization of solid-state synthesis in nanoscale is a promising solution that could significantly reduce the silicide formation temperature. However, as is known according to classical thermodynamic concepts, the role of the Si/metal interface increases in the case of nanostructured systems. In this regard, changes in thermal behavior can be observed. In particular, the temperature characteristics of silicidation processes, stability of phases and the phase formation sequence during solid-state reaction can change. The use of nanostructured Si as templates (matrices) for the synthesis of Si/metal nanocomposites allows to vary the geometric parameters of structures and opens new possibilities for studying the silicide formation processes. In addition, in this case, nanostructured Si is an active matrix directly involved in the solid-state interaction.

In this work, the phase formation behavior during the thermally induced solid-state reaction of the Ni-modified nanostructured Si has been investigated using a combination of differential scanning calorimetry and x-ray diffraction analysis. The effect of nanostructured interface on the silicides formation temperatures is discussed. Special attention is paid to the methodical aspects of the characterization of the thermal behavior of the Ni-modified nanostructured Si. The role of the thermal instability of nanostructured Si in the calorimetric studies of the silicides formation is also reviewed.

2. Experimental details

2.1. Sample preparation
Single crystal silicon wafers (n-type) of (111) orientation and 0,001 Ω·cm resistivity were used for fabricating Si/Ni samples. Si wafers were pre-treated consistently in a mixture of H_2SO_4 and H_2O_2 (1:1) and HF:H_2O (1:4), rinsed with deionized water and isopropyl alcohol.

Figure 1 shows the flowchart of the fabrication of membrane-based and thinned substrate-supported Si/Ni samples. Nanostructured porous Si layers were prepared by electrochemical anodizing [11] of pre-treated Si wafers in a polytetrafluoroethylene (PTFE) cell in a mixture of HF and CH_3COOH (7:4) in the galvanostatic mode, the cathode material – Pt. The current density was 15 mA/cm^2 and the anodizing time – 2 hours. In order to obtain Si membrane the separation of the porous Si layer was performed by gradually increasing the current density of the anodizing by 10-15 mA/cm^2 every 3-5 min. The maximum current density at detachment was 62.2 mA/cm^2. The conductive Ni layer on the backside of Si membranes was applied by deposition of a 200 nm thick Ni layer using magnetron sputtering. Thinned Si substrates after membrane detachment were reused to form substrate-supported nanostructured porous Si layers (Figure 1).

Figure 1. Schema of the fabrication of membrane-based and thinned substrate-supported Si/Ni samples.

Si membranes with conductive layer and substrate-supported nanostructured porous Si were subjected to Ni electrodeposition after etching a possible thin native oxide layer for 10 s in diluted HF solution. Cathodic Ni deposition was carried out from an aqueous solution containing 213 g/l $NiSO_4 \cdot 7H_2O$, 25 g/l $NiCl_2 \cdot 6H_2O$, 25 g/l H_3BO_3 and 3 g/l of saccharin (pH = 2 – 3). The current density was 3.5 mA/cm^2. Ni was used as the anode. The deposition time was about 2 hours.

2.2. Structural characterization
The structure of the samples was characterized by means of x-ray diffraction analysis (XRD). Measurements were carried out using the "RIKOR-8" x-ray measuring system ("IRO", Moscow, Russia). Samples were studied in the 2θ scan mode with a wavelength $\lambda(CuK_\alpha)$ = 1,541 Å using Bragg–Brentano focusing. A Ni filter was used for the K_β-line stripping, scanning time of 800 s in a point. The investigation of the samples' morphology was carried out by means of scanning electron microscope (SEM), JEOL JSM-6010PLUS/LA.

IOP Conf. Series: Journal of Physics: Conf. Series **987** (2018) 012035 doi:10.1088/1742-6596/987/1/012035

2.3. Differential scanning calorimetry

The investigation of the thermal behaviour of samples was carried out using the precalibrated differential scanning calorimeter Netzsch DSC 204 F1 Phoenix (Netzsch-Geratebau GmbH, Germany). The heating of the samples of 10 – 15 mg was conducted in press-fitted aluminium crucibles in an argon atmosphere with four different heating rates (5, 10, 20, and 30 °C/min). The lid of the crucible was punched before sealing. Empty aluminium crucible was used as a reference. A sample fragment was cut in the shape of a disc and placed on the bottom of the crucible, then pressed with the lid with the upturned convex side down to provide good contact between the material and the bottom of the crucible.

3. Results and discussion

From the point of view of calorimetric investigations, it is important to limit the influence of the substrate on the DSC signal. For this purpose, membrane-based and thinned substrate-supported Si/Ni samples were prepared for DSC analysis. Porous Si layer prepared according to the procedure described in [11] has nanometer-sized pores and when coated with Ni it allows to fabricate a nanostructured Si/Ni interface.

The results of electron microscopy (Figure 2) reveal that the thickness of nanostructured porous Si layer was of the order of 80 μm. A SEM images of the membrane-based Si/Ni sample illustrate a dense Ni film on the surface of nanostructured porous Si layer.

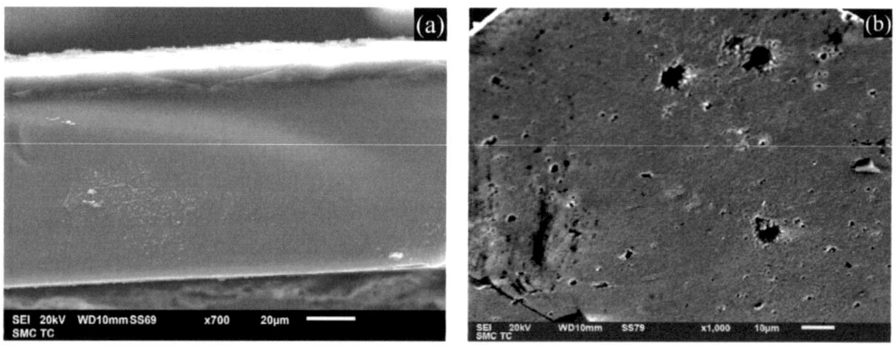

Figure 2. SEM image of membrane-based Si/Ni samples: (a) cross-section and (b) top view.

Also, it is important to distinguish which effects relate to the Si/Ni composite structure as a whole and which are due to the nature of their components taken separately. It is therefore necessary to have reliable information about the thermal behavior of the nanostructured porous Si layer. It is known that porous silicon is unstable during heat treatment and is prone to oxidation. Figure 3a shows the DSC scans of Si membrane. During the first heating up to 500 °C, two thermal effects are observed that are not reproducible upon cooling and reheating. Weakly expressed endothermic effect displayed at 80 – 110 °C is caused by loss of residual water and alcohol used in the fabrication process. This is evidenced by the fact that the loss of the samples' weight after the first heating was 1 – 3%.

In the range of 250 – 400 °C, the exothermic effect of a complex shape is observed, presumably because of the low-temperature oxidation (preoxidation) [12] in an atmosphere of non-dehydrated argon. This effect is also weakly expressed due to a small amount of oxidizing agents in the purge gas and is observed only at relatively high heating rates (30 °C/min).

Figure 3b presents DSC signal of Si membrane with 200 nm thick conductive Ni layer deposited using magnetron sputtering. In the range of 80 – 110 °C, the distinctive nonlinearity of the DSC signal is observed, because of loss of residual water and alcohol. The exothermic bend of the DSC curve in the range of 150 – 450 °C can be explained by the low-temperature oxidation of Si membrane and by the solid-state reaction between Si membrane and conductive Ni layer. This fact should be taken into account when analyzing membrane-based Si/Ni samples.

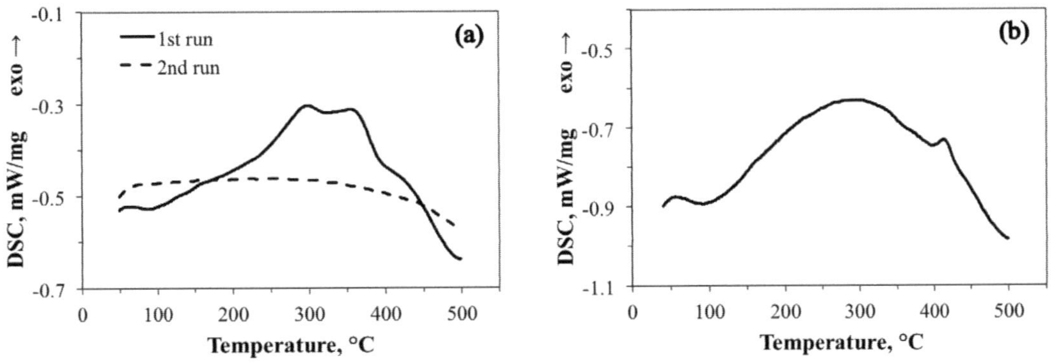

Figure 3. DSC scans of as-prepared Si membrane (a) and Si membrane with conductive Ni layer (b).

Figure 4 presents DSC curves, showing similar behavior of the membrane-based and thinned substrate-supported Si/Ni samples at the specified temperature range. In the range of 80 – 110 °C, the distinctive nonlinearity of the DSC signal is observed, because of loss of residual water and alcohol. Also, in both cases the exothermic effect is observed in the range of 200 – 300 °C non-reproducible upon cooling and reheating. Evidently, this exo-effect relates to the Si/Ni nanocomposite structure as a whole.

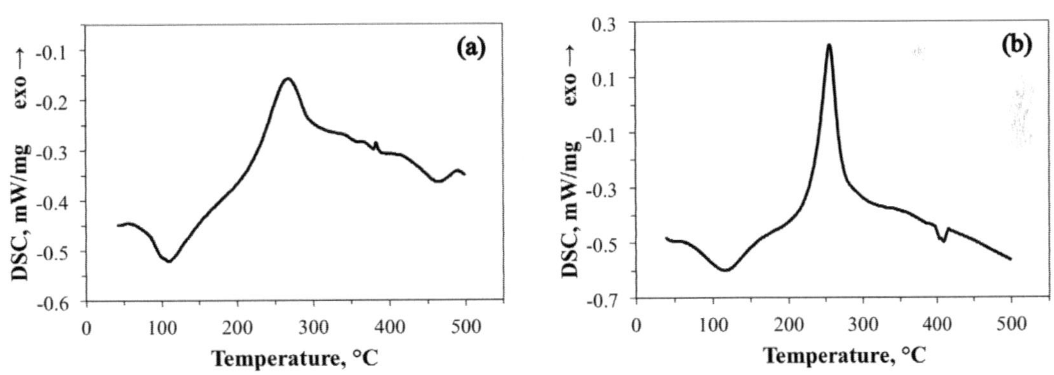

Figure 4. DSC scans of the membrane-based (a) and thinned substrate-supported (b) Si/Ni samples.

Combining both XRD (Figure 5) and DSC measurements it was established that this effect is caused by solid-state interaction at the nanostructured Si/Ni interface. According to the data obtained earlier, Ni_2Si and NiSi reflections appears on XRD spectra after heat treatment at 200 °C [13]. After heat treatment at 500 °C, XRD shows the presence of Ni and NiSi phase. It is also necessary to note, that exo-effect is more clear and intensive in the case of Si/Ni sample on thinned Si substrate, while the DSC signal of membrane-based Si/Ni sample has a complicated shape presumably due to a contribute of reaction between Si membrane and Ni conductive layer.

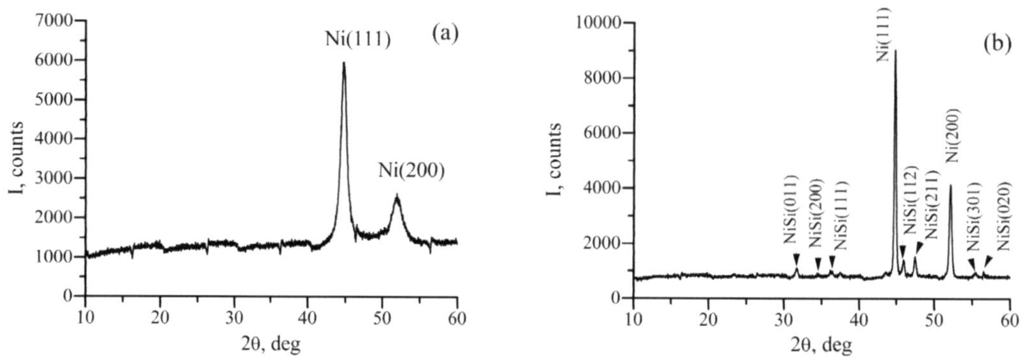

Figure 5. X-ray diffractograms of thinned substrate-supported Si/Ni samples before (a) and after (b) heating up to 500 °C

Endothermic effect on the DSC curve of the thinned substrate-supported Si/Ni sample can be caused by the loss of surface hydrogen from porous silicon. [14].

Figure 6a exemplifies the DSC curves of the thinned substrate-supported Si/Ni samples corresponding to different heating rates. As it follows from the data in this figure, the extrapolated peak onset temperature (T_{onset}) of the solid-state reaction at the nanostructured Si/Ni interface decreased with decreasing heating rate.

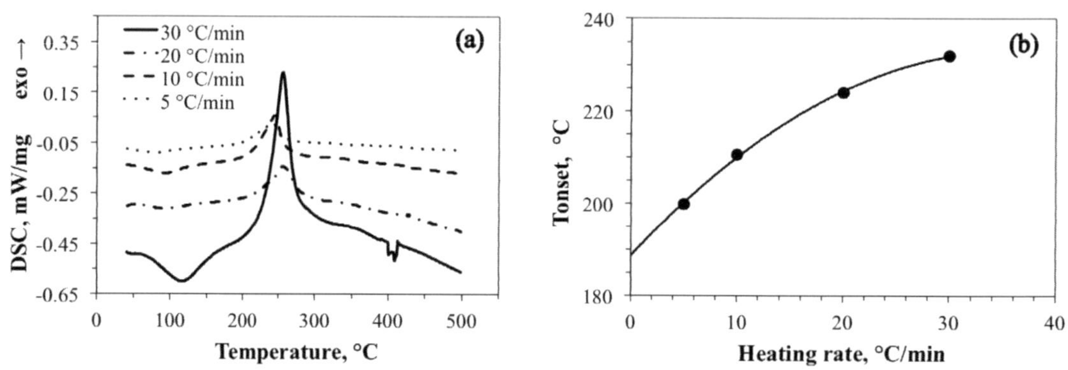

Figure 6. Set of DSC curves of the thinned substrate-supported Si/Ni samples for different heating rates (a) and (b) observed values of T_{onset} as a function of heating rate.

The true temperature of the solid-state reaction at the nanostructured Si/Ni interface estimated by the extrapolation of observed values of T_{onset} to zero heating rate (Figure 6b) is 188.7 °C. This temperature is lower than that for similar thin-film systems [15] and is much lower than the equilibrium temperatures of the formation of silicide phases on the Ni-Si phase diagram [16].

4. Conclusions

In our work, the membrane-based and thinned substrate-supported Si/Ni samples with nanostructured interface have been electrochemically prepared and investigated using a combination of DSC and XRD analysis. DSC analysis in combination with XRD reveal that the thermal behavior of Ni-modified nanostructured Si is caused both by instability of nanostructured Si during heating and solid-state interaction at the nanostructured Si/Ni interface. The estimated true temperature of the solid-state reaction at the nanostructured Si/Ni interface (188.7 °C) is lower than that for similar thin-film systems and is much lower than the equilibrium temperatures of the silicides formation on the Ni-Si phase diagram. It was experimentally shown, that the atmosphere as well as presence of secondary components (such as metal conductive layer) contribute to the DSC signal and this fact should be taken into account when analyzing nanostructured Si-based systems.

Acknowledgment
The reported study was funded by RFBR according to the research project No. 16-33-00712 мол_а.

References

[1] Lin J-Y, Hsu H-M and Lu K-C 2015 *Cryst. Eng. Comm.* **17** 1911–6
[2] Kim D-J, Seol J-K, Lee M-R, Hyung J-H, Kim G-S, Ohgai T and Lee S-K 2012 *Appl. Phys. Lett.* **100** 163703
[3] Zhao A, Zhang X, Chen X, Guan J and Liang C 2010 *J. Phys. Chem.* **114** 3962–7
[4] Beyers R and Sinclair R 1985 *J. Appl. Phys.* **57** 5240–5
[5] Gromov D G, Mochalov A I, Pugachevich V P, Sorokin I N 2000 *Appl. Phys. A* **70** 333–40
[6] Baglin J E, d'Heurle F M, Petersson C S 1980 *Appl. Phys. Lett.* **36** 594–6
[7] Mangelinck D, Hoummada K, Panciera F, Kousseifi M E, Blum I, Descoins M, Bertoglio M, Portavoce A, Perrin C and Putero M 2014 *Phys. Status Solidi A* **211** 152–65
[8] Gomoyunova M V, Malygin D E, Pronin I I 2006 *Phys. Solid State* **48** 2016-20
[9] Varnakov S N, Zabluda V N, Ovchinnikov S G, Gomoyunova M V, Grebenyuk G S, Pronin I I 2014 *Phys. Solid State* **56** 812-5
[10] Gomoyunova M V, Grebenyuk G S, Pronin I I 2014 *Tech. Phys.: The Russian Journal of Applied Physics* **59** 1492-8
[11] Solanki C S, Bilyalov R R, Poortmans J, Celis J P, Nijs J and Mertens R 2004 *J. Electrochem. Soc.* **151** C307-14
[12] Herino R, Perio A, Barla K and Bomchil G 1984 *Materials Lett.* **2** 519–23
[13] Shilyaeva Y I, Pyatilova O V, Berezkina A Y, Sysa A V, Dudin A A, Smirnov D I and Gavrilov S A 2016 *Proc. SPIE.* **10224**, Int. Conf. on Micro- and Nano-Electronics 102240E
[14] Bomchil G, Halimaoui A and Herino R 1988 *Microelectron. eng.* **8** 293-310
[15] Liew K P, Li Y, Yeadon M, Bernstein R, Thompson C V 2003 *Adv. Mater. for Micro- and Nano-Systems* 2003-01
[16] Lyakishev N P 1996 *Phase Diagrams of Binary Metallic Systems: reference book in 3 vol.* (Moscow: Mashinostroenie) **1** 89-90

IOP Conf. Series: Journal of Physics: Conf. Series **987** (2018) 012036 doi:10.1088/1742-6596/987/1/012036

Magnetic Thermometer: Thermal effect on the Agglomeration of Magnetic Nanoparticles by Magnetic field

Daeseong Jin and Hackjin Kim*

Department of Chemistry, Chungnam National University, Taejeon 34134 Korea
*E-mail: hackjin@cnu.ac.kr

Abstract. We have investigated the agglomeration of magnetite nanoparticles in the aqueous solution under magnetic field by measuring temporal change of magnetic weight. The magnetic weight corresponds to the force due to the magnetization of magnetic materials. Superparamagnetic magnetite nanoparticles are synthesized and used in this work. When the aqueous solution of magnetite nanoparticle is placed under magnetic field, the magnetic weight of the sample jumps instantaneously by Neel and Brown mechanisms and thereafter increases steadily following a stretched exponential function as the nanoparticles agglomerate, which results from the distribution of energy barriers involved in the dynamics. Thermal motions of nanoparticles in the agglomerate perturb the ordered structure of the agglomerate to reduce the magnetic weight. Fluctuation of the structural order of the agglomerate by temperature change is much faster than the formation of agglomerate and explained well with the Boltzmann distribution, which suggests that the magnetic weight of the agglomerate works as a magnetic thermometer.

1. Introduction

Magnetic nanoparticles have been studied extensively for their wide applications in various fields.[1-3] Magnetite nanoparticles have been used as magnetic recording medium as well as friction-reducing ferrofluid. Hyperthermia, drug delivery and MRI contrast enhancement are recent important examples of medical applications. Modifications of magnetic nanoparticle surface play important roles in the recycle of catalysts. Magnetic nanoparticles reveal very complicated behaviors under magnetic field.[4] Detailed knowledge about the magnetic nanoparticles under magnetic field will be instrumental in improving many application processes.

We have studied the dynamics of agglomeration of magnetite nanoparticles by monitoring the magnetic weight with an electronic balance. The magnetic weight W given by Eq. (1) corresponds to the vertical magnetic force $\mathbf{F_z}$ that the magnetic nanoparticles experience under magnetic field,

$$W = \mathbf{F_z} = [VM(\mathbf{\nabla B})]_z \quad (1)$$

where V and M are the volume and the magnetization of the sample, and $\mathbf{\nabla B}$ is the magnetic field gradient. The magnetic weight can be measured easily as in the Gouy balance. The magnetite nanoparticles are superparamagnetic in the solution before applying the magnetic field. As the nanoparticles agglomerate and the ordered structure develops inside the agglomerate under magnetic field, the magnetization as well as the magnetic of the agglomerate weight is growing. The ordered structure of the agglomerate is assigned to ferromagnetic-like state. The magnetic weight of the agglomerate is sensitive to the temperature. Some particles of the agglomerate that do not contribute to the magnetic weight due to the

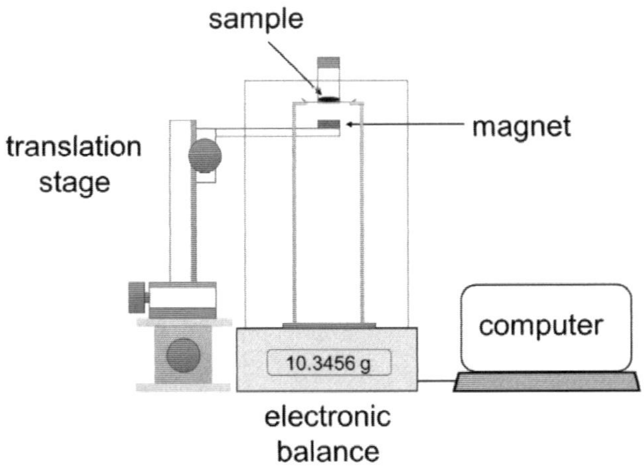

Figure 1. Experimental setup. A conventional electronic balance connected to a computer is used to monitor the magnetic weight change.

thermal motions are considered to be in the superparamagnetic-like state. The temperature effect on the magnetic weight of the agglomerate is explained well with the Boltzmann distribution between the superparamagnetic-like state and the ferromagnetic-like state of the agglomerate, which suggests that the magnetic weight of the agglomerate is a good indicator of temperature, that is, the magnetic thermometer.

2. Experimental

Figure 1 shows the experimental setup for monitoring the temporal change of the magnetic weight.[5] The magnetic field is applied in the direction parallel to the gravity with the NdFeB disc magnet of 2 cm diameter and 1 cm thick. The strength of the magnetic field can be varied by translating the magnet vertically and measured with a magnetometer. The magnetic field strength is in the order of a few tens T m^{-1} which is regarded as the low gradient range.[1] The magnetic weight of the sample is recorded every minute and the laboratory temperature is also monitored every minute with a digital thermometer.

The aqueous solution of 1.5 wt % magnetite nanoparticle is used in this work. The magnetite nanoparticles are synthesized by following a well-known procedure.[6] Tetramethylammonium hydroxide is used as the surfactant for magnetite nanoparticles of ~11 nm diameter.[5] High concentration solution of the synthesized nanoparticles reveals the characteristic behaviors of ferrofluid under magnetic field.[4] The magnetite nanoparticles are stable to keep the superparamagnetic state in the aqueous solution for several months or longer.

3. Results and Discussion

3.1. Agglomeration of nanoparticles under magnetic field

Figure 2 shows the agglomeration of magnetite nanoparticle under magnetic field. The magnet of Figure 2 is smaller than the magnet used in the experiment of Figure 1. Under no magnetic field, the superparamagnetic nanoparticles are well dispersed to form a homogeneous solution. Under magnetic field, the nanoparticles agglomerate at the bottom of the cuvette. It takes several hours to form the apparent agglomerate. As the particles agglomerate, the solution turns pale. The agglomerate does not change apparently thereafter even though the magnetic weight of the sample increases steadily, which indicates that the structural relaxation of the agglomerate undergoes continuously. When the magnetic field is removed without disturbance, the agglomerate remains as it is for a while but the nanoparticles of the agglomerate disperse and the original homogeneous solution is recovered in a short time.

3.2. Temporal change of magnetic weight

Figure 3 shows a temporal change of the magnetic weight of the sample at 30 T m^{-1}. The magnetic weight is set to zero when no magnetic field is applied. The magnetic weight of the sample jumps at the

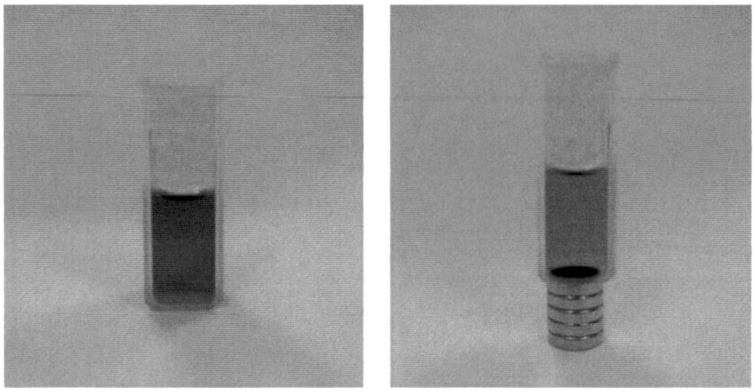

Figure 2. Agglomeration of nanoparticles under magnetic field. It takes several hours for the nanoparticles to agglomerate at the bottom of a cuvette.

instant of placing the sample on the balance. This jump is due to the Neel and Brown mechanisms[7] whose response times are much shorter than that of the electronic balance, a few msec. The slow growth of the magnetic weight after the jump W(t) fits well to the stretched exponential function as following

$$W(t) = W(\infty) + [W(0) - W(\infty)] \exp[-(t/\tau)^{\beta}] \quad (2)$$

where $W(\infty)$ is the equilibrium magnetic weight and W(0) corresponds to the magnetic weight by the Neel and Brown mechanisms. The exponential function with $0 < \beta < 1$ is called a stretched exponential. The fitting curve given as the red line in Figure 3 has the parameters of W(0) = 1.090 g, $W(\infty)$ = 2.454 g, τ = 802 min and β = 0.802. The magnetization calculated from the $W(\infty)$ is 1.38 x 10^5 A m^{-1} which corresponds to 29% of the bulk magnetization of magnetite, 4.7 x 10^5 A m^{-1}.[1] Deviation of the fitting curve from the measured magnetic weight results mainly from the temperature fluctuation, which is observed clearly when the laboratory temperature changes very much. The temperature effect on the magnetic weight is discussed below.

The stretched exponential behavior of the dynamics is observed when the energy barrier involved in the dynamics is not a single value but has some distribution, which is a characteristics of complicated processes.[8-10] The spectrum of energy barrier is wider for smaller β.[11] With a proper expression of the rate constant such as the Arrhenius relation, the energy barrier distribution function can be determined by the inverse Laplace transform of the temporal behavior of the dynamics.[8,11] The data of Figure 3 gives the negatively skewed function for the energy barrier with the energy maximum of 52.3 kJ mol^{-1} and the full width at the half maximum of 1.6 kJ mol^{-1}. Details of the process of determining the energy barrier distribution function is discussed in ref. 11.

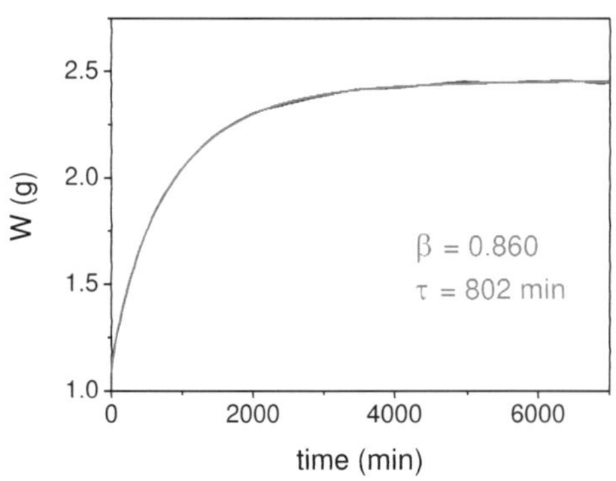

Figure 3. Temporal change of magnetic weight with the fitting curve of a stretched exponential function (red line). Two data overlap so well that it is not easy to see the difference.

IOP Conf. Series: Journal of Physics: Conf. Series **987** (2018) 012036 doi:10.1088/1742-6596/987/1/012036

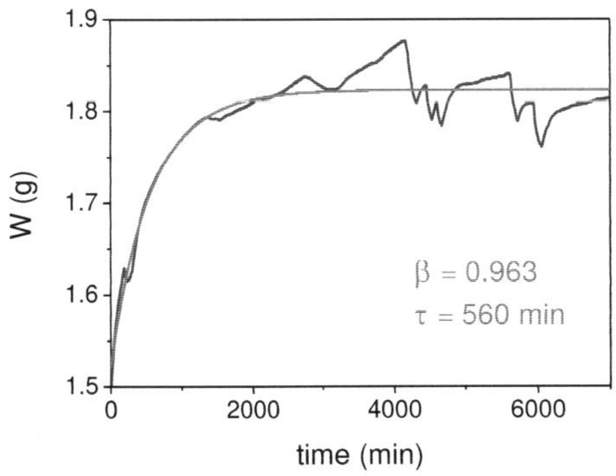

Figure 4. Fluctuation of the magnetic weight growth by temperature variation. The stretched exponential fitting curve is given as the red line.

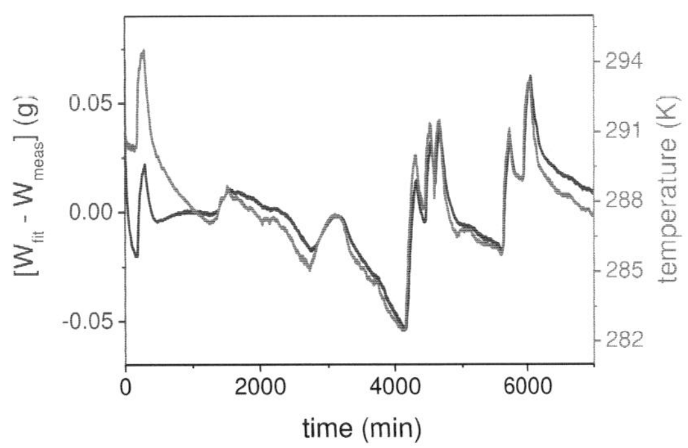

Figure 5. Laboratory temperature (red line) and difference of the fitting curve and the measured magnetic weight of Figure 4, W_{fit} - W_{meas} (blue line).

3.3. Effect of temperature fluctuation on the magnetic weight

Thermal motions of the nanoparticles of the agglomerate perturb the structural order of the agglomerate, which results in the decrease of magnetic weight. Figure 4 shows the magnetic weight of the sample measured in the cold winter season when the heating of the laboratory building is off and the laboratory temperature goes down at night time. Change of the magnetic weight by temperature fluctuation can be seen as a result of interconversion between the superparamagnetic-like state and the ferromagnetic-like state of the agglomerate. The fitting parameters of the data of Figure 4 are $W(0) = 1.520$ g, $W(\infty) = 1.823$ g, $\tau = 560$ min and $\beta = 0.963$. The magnetization estimated with the $W(\infty)$ is 1.03×10^5 A m^{-1} corresponding to 22 % of the bulk magnetite magnetization. The magnetic weight increases faster and $W(\infty)$ is smaller but W(0) is larger in Figure 4 compared with the data of Figure 3. The larger β of Figure 4 indicates that the energy barrier distribution is narrower in Figure 4.

The fitting curve of Figure 4 is considered as the magnetic weight at the average temperature. The magnetic weight greater than the fitting curve corresponds to the agglomerate whose temperature is lower than the average. As the heating of the laboratory building is on in the morning and the laboratory temperature begins to rise, the magnetic weight as well as the magnetization of the agglomerate decreases to be smaller than those at the average temperature. The enhanced thermal motions of nanoparticles may interrupt and block complicated pathways to the formation of the ordered agglomerate to retard the growth of the magnetic weight.

Figure 5 show the difference of the fitting curve and the measured magnetic weight $\Delta W = W_{fit} - W_{meas}$, and the laboratory temperature. The laboratory temperature and the magnetic weight difference change synchronously, which suggests that the magnetic weight of the agglomerate can work as a thermometer. The temperature and the weight difference deviate somewhat at the beginning of agglomeration where the magnetite nanoparticles are dispersed in the whole solution and do not agglomerate at the bottom of the cuvette yet. Diffusion of nanoparticles along the magnetic field gradient is dominant at the early stage of agglomeration. After the agglomerate forms at the bottom and the magnetic weight reaches the almost flat slow increasing stage, the structural relaxation inside the agglomerate is a main cause for the magnetic weight growth. At this stage, the magnetic weight of the agglomerate becomes more sensitive to the temperature, that is, more synchronous with the temperature than at the early stage of agglomeration. As the agglomerate becomes mature, the deviation of the temperature and the magnetic weight reappear at shown in Figure 5. This deviation is different in the direction from that of the beginning of the dynamics, which suggests that the origins of the deviations are different., The agglomerate is not formed yet at the beginning of the dynamics while the internal structure of the agglomerate changes progressively after the formation.

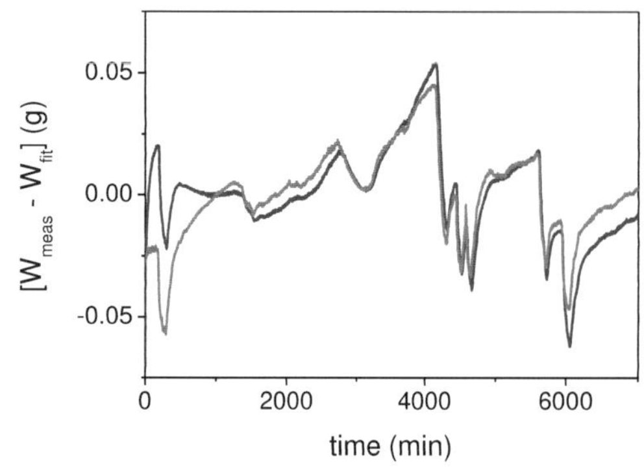

Figure 6. Difference of the measured magnetic weight and the fitting curve of Figure 4 (blue line) and the magnetic weight difference simulated with energy difference E_b of Eqs. (3)-(5) (red line).

3.4. Analysis of the magnetic weight fluctuation with Boltzmann distribution

In order to analyse the magnetic weight fluctuation with the Boltzmann distribution, it is assumed that the magnetic weight results from only the ferromagnetic-like state of the agglomerate and that thermal motions of the nanoparticles cause the transition to the superparamagnetic-like state which reduces the magnetic weight. The magnetic weight is assumed to be proportional to the relative ratio of the ferromagnetic-like state.

Eq. (3) shows the Boltzmann relation of the magnetic weight.[12] The nanoparticles of the agglomerate are assumed to be in the two-level system. The W_o corresponds to the magnetic weight when all the nanoparticles of the agglomerate are assumed to be at the ferromagnetic-like state. The W_o is the magnetic weight of the agglomerate at 0K, and the W_{meas} would approach one half of W_o when the temperature goes up. The W_{meas} is the magnetic weight of the agglomerate when the nanoparticles are distributed between the two states. Eqs. (4) and (5) are modifications of Eq. (3). As mentioned, the W_{fit} is the magnetic weight when the agglomerate is assumed to be at the average temperature, T_a.

$$\frac{W_o - W_{meas}}{W_{meas}} = \exp(-\frac{E_b}{RT}) \tag{3}$$

$$W_{meas} = W_o\,[1 + \exp(-\frac{E_b}{RT})]^{-1} \tag{4}$$

$$W_{fit} = W_o\,[1 + \exp(-\frac{E_b}{RT_a})]^{-1} \tag{5}$$

where the E_b represents the energy difference between the two states. This energy difference is interpreted as the interparticle interaction energy in the agglomerate.

We have three measured parameters involved in the picture of the Boltzmann distribution, W_{meas}, T and T_a, and the W_{fit} from the fitting with the stretched exponential. From the ratio of W_{meas} and W_{fit}, we can simulate the magnetic weight difference, (W_{meas} - W_{fit}) by varying E_b. Figure 6 shows the measured and the simulated magnetic weight difference. The best fitting E_b is 4.1 ± 0.1 kJ mol^{-1}, which is similar as the weak hydrogen bond energy. This energy difference is smaller than the previously reported value of 7.6 kJ mol^{-1}.[12] The variation of E_b reveals the complexity of the agglomerate. Both E_b's are much smaller than the maximum of the energy barrier involved in the formation of the agglomerate.

In order to improve the agreement of the measured and the simulated magnetic weight at the beginning and the end of Figure 6, the E_b should change in opposite directions. The E_b should be raised at the beginning of the dynamics while it should be lowered at the end, which confirms that the characters of the dynamics are changing as the agglomeration goes on. At the beginning of the dynamics, the diffusion is dominant and loosely bonded agglomerate is formed. As the structural relaxation of the agglomerate goes on, the stable structure of the agglomerate is developing eventually.

4. Concluding Remarks

It is not easy to carry out accurately the measurement of magnetic weight in this work. Manual placement of the sample on the balance causes the variation of the magnetic field strength. Slow microscopic transformation and inhomogeneity of the nanoparticles in the solution sample could prevent the reproducibility of the experiments. However, the main obstacle of this work lies in the complexity of the agglomeration of nanoparticles itself. The agglomeration is not a single step process and there are so many different agglomerates with the same macroscopic properties and so many pathways to reach a specific agglomerate. Therefore, the energy landscape of the agglomeration[13] should be employed to understand the details of agglomeration. There are many big energy barriers for the formation of the agglomerate and small barriers for the stabilization of the agglomerate.

Nonetheless the two-level system works well for the explanation of thermal fluctuation of the magnetic weight. Instantaneous variation of the magnetic weight following the temperature change makes the magnetic weight act as a thermometer. Small interparticle interaction energy inside the agglomerate is important for the magnetic thermometer working at room temperature.

5. References

[1] Yavuz CT, Mayo JT, Yu WW, Prakash A, Falkner JC, Yean S, Cong L, Shipley HJ, Kan A, Tomson M, Natelson D and Colvin VL *Science* 2006 **314** 964
[2] Wang D, Deraedt C, Ruiz J and Astruc D *Acc. Chem. Res.* 2015 **4** 1871
[3] Revia RA and Zhang M *Materials Today* 2016 **19** 157
[4] Rosensweig RE *Ferrohydrodynamics* 2003 Cambridge University
[5] Jin D and Kim H *Bull. Korean Chem. Soc.* 2013 **34** 1715
[6] Kang Y, Risbud S, Rabolt JF and Stroeve P *Chem. Mater.* 1996 **8** 2209
[7] Bedanta S and Kleemann W *J. Phys. D: Appl. Phys.* 2009 **42** 013001
[8] Austin RH, Beeson K, Eisenstein L, Frauenfelder H, Gunsalus IC and Marshall VP *Phys. Rev. Lett.* 1974 **32** 403
[9] Lindsey CP and Patterson GD *J. Chem. Phys.* 1980 **73** 3348
[10] Palmer RG, Stein DL, Abrahams E and Anderson PW *Phys. Rev. Lett.* 1984 **53** 958
[11] Jin D and Kim H *Bull. Korean Chem. Soc.* 2015 **36** 424
[12] Jin D and Kim H *Bull. Korean Chem. Soc.* 2016 **37** 962
[13] Wales DJ *Energy Landscapes* 2003 Cambridge University

Acknowledgments

This study was financially supported by research fund of Chungnam National University in 2017.

IOP Conf. Series: Journal of Physics: Conf. Series **987** (2018) 012037 doi:10.1088/1742-6596/987/1/012037

Determination of permeability of ultra-fine cupric oxide aerosol through military filters and protective filters

E Kellnerová[1,3,4], Z Večeřa[2,5], J Kellner[1], T Zeman[1] and J Navrátil[1]

[1]University of Defence, Kounicova 65, 602 00 Brno, Czech Republic
[2]Institute of Analytical Chemistry, Czech Academy of Sciences, Veveří 967/97, 602 00 Brno, Czech Republic

Email: eva.kellnerova@unob.cz

Abstract. The paper evaluates the filtration and sorption efficiency of selected types of military combined filters and protective filters. The testing was carried out with the use of ultra-fine aerosol containing cupric oxide nanoparticles ranging in size from 7.6 nm to 299.6 nm. The measurements of nanoparticles were carried out using a scanning mobility particle sizer before and after the passage through the filter and a developed sampling device at the level of particle number concentration approximately 750000 particles·cm^{-3}. The basic parameters of permeability of ultra-fine aerosol passing through the tested material were evaluated, in particular particle size, efficiency of nanoparticle capture by filter, permeability coefficient and overall filtration efficiency. Results indicate that the military filter and particle filters exhibited the highest aerosol permeability especially in the nanoparticle size range between 100–200 nm, while the MOF filters had the highest permeability in the range of 200 to 300 nm. The Filter Nuclear and the Health and Safety filter had 100% nanoparticle capture efficiency and were therefore the most effective. The obtained measurement results have shown that the filtration efficiency over the entire measured range of nanoparticles was sufficient; however, it was different for particular particle sizes.

1. Introduction

Nanomaterial is defined as a natural, incidental or manufactured material containing particles, in an unbound state or as an aggregate or as agglomerate and where 50% or more of the particles with one or more external dimensions are in the size range 1-100 nm [1]. Under certain circumstances, solid and liquid particles in the air adversely affect the health status of people. These are primarily aerosols made up of ultra-fine particles with a mean aerodynamic diameter in the range of several μm to nm. Such inhaled particles have a wide range of effects and most commonly damage the cardiovascular and respiratory tract [2]. The toxicity of nanoparticles has been clearly demonstrated in some biological systems, but this attribute depends largely on the size, shape, surface and reactivity of the particles. The effect may also be influenced by the susceptibility of an individual [3].

---

[3]Author to whom any correspondence should be addressed.
[4]The financial support of the University of Defence of the Czech Republic by the research project SV16-FVL-K106-KELL is gratefully acknowledged.
[5]This work was supported by the grant project GAP503/11/2315 of the Institute of Analytical Chemistry of the Czech Academy of Sciences.

1.1. Military environment
Ultra-fine particles (UFP) are ubiquitous in workplaces, even in urban environments and pose a significant health hazard to humans. Repeated smog situations are monitored when the legal limits of airborne dust and selected chemical pollutants in the air are exceeded. Due to the expected introduction of specialized nanotechnologies to the Army of the Czech Republic (ACR), it is necessary to be involved in the UFP monitoring process. Another reason is that military nanotechnology has its own specific features over normal industrial nanoproducts, which must be taken into account when estimating their health risks. In the ACR, the most common threats are military explosives and chemical weapons emitted into the air in the form of aerosols [4]. This applies not only to the training of combat units but also to selected departments of military units, repair workrooms and workplaces carrying out delaboration and ecological disposal of ammunition and explosives, where the staff can be exposed to their influence.

1.2. Protective military filters
Respiratory protective devices in the ACR are most often filter respirators. Filter respirator consist of a capsule filter, the functional part, which is stored in a plastic or metal housing. In accordance with the Czech technical standard 132 (CSN EN 132) [5], the particulate filter is a separate filter or is usually part of combined filters that are capable of entrapping dispersed solid and liquid particles and designated gases and vapors from the passing air. Particle filters can be of three classes, with increasing efficiency marked 1 to 3 (P1, P2, P3) and they are marked with a white color code.

Most modern filters are combined (including anti-particle and anti-gas layers) [6]. Such filters usually have to comply with a number of technical standards, such as CSN EN 143 [7], CSN EN 12942 [8], CSN EN 14387 [9] in the Czech Republic. Until now, testing of military and fire protection filters as a protection against inhalation of nanoparticles from the air is not mandatory. Therefore, the effectiveness of these filters against nanoparticles is not officially confirmed. The technical requirements for military means of individual respiratory protection, military protective masks, names and definitions are set by Czech Defense Standard 841503 (CDS 841503) [10].

1.3. Evaluation of filtration
From the point of view of the filtration mechanism, we divide filtration into a surface and depth filtration. The principle of surface filtration is the mechanical capture of each particle, which is larger than the space between the filter fibers. On the other hand, in depth filtration, more forces are applied simultaneously - Van der Waals forces, electrostatic forces and forces induced by surface tension. Total efficiency of capture of ultrafine aerosols (E) [%] is a combination of all capture principles and their share in overall efficiency varies with the size and shape of each particle, where G1 is the amount of dispersion after the filter, G2 is the total amount of dispersion before the filter and the proportion G1 / G2 is the penetration (P) [11]:

$$E = \left(1 - \frac{G_1}{G_2}\right) \cdot 100 \tag{1}$$

For calculation the permeability coefficient (K_p) [%], the calculation given in CDS 841503, where the maximum permissible paraffin oil aerosol permeability coefficient of protective filters for all army protection masks is set at a maximum of 0.001% [10] and a formula (2) is used:

$$K_P = \frac{G_1}{G_2} \cdot 100 \tag{2}$$

2. Materials and methods
Cupric oxide nanoparticles (CuONP) were continuously synthesized in an accredited laboratories of the Institute of Analytical Chemistry of the Czech Academy of Sciences.

2.1. Preparation of cupric oxide (CuO) nanoparticles

CuONP were synthesized by aerosol route in a flow tube reactor by thermal decomposition of the organic precursor of copper acetylacetonate (Aldrich) and subsequently oxidized in Carbolite TZF 15/50/610, Carbolite Limited, UK at 700 °C in the presence of 15.1% vol. of oxygen. Copper acetylacetonate vapours were obtained from the solid form of copper acetylacetonate in a saturation vessel at 148.2 °C. The released vapours were transported with nitrogen (flow rate 0.8 L·min^{-1}, purity 5.5) to the flow reactor. The total nitrogen / oxygen mixture flow rate was 2 L·min^{-1}. The CuONP nitrogen stream was mixed with air flow (96 L·min^{-1}) at the reactor outlet and used for both commercial and army filter experiments in a specially developed sampling device (see figure 1).

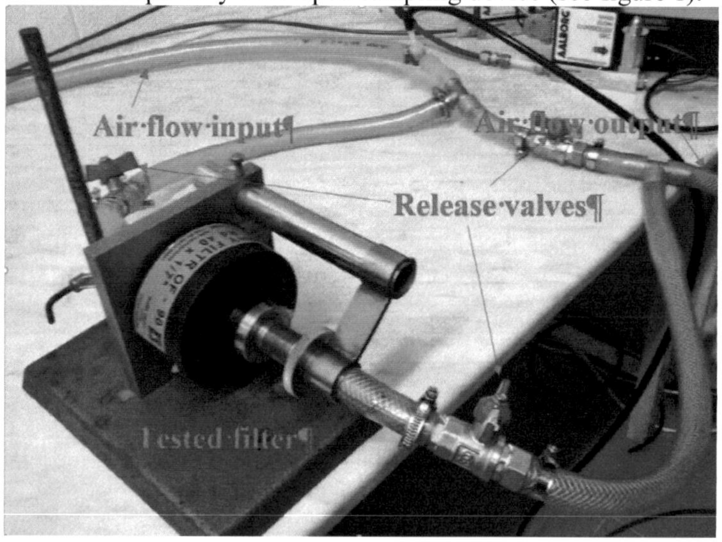

Figure 1 Sampling device (source:own).

Total particle number concentrations of produced CuONP (C_N) after mixing with air were 4.88 x 10^7 particles·cm^{-3} on average. The size of generated CuO nanoparticles ranged from 7.6 to 299.6 nm. The nanoparticle concentration and particle size distribution in the inhalation chamber were continuously measured by scanning mobility particle sizer (SMPS, model 3936L72, TSI, USA). The counts of nanoparticles at the input of the filter are shown in figure 2. Figure 2 below illustrates number concentrations of particular sizes of nanoparticles in measured range. The maximum amount of nanoparticles is entering the tested filters in the range of particle sizes between 30–40 nm. Samples were measured sequentially, four times at five-minute intervals (total ten to twenty-five minutes).

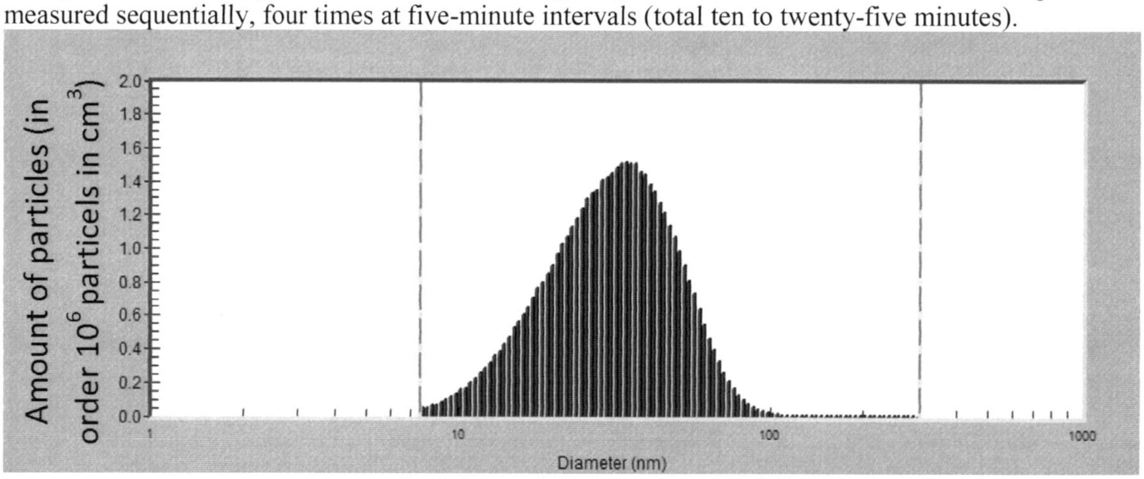

Figure 2 Number size distribution of particles before entering filters (source:own).

2.2. *Method for evaluation of acquired results*

Acquired results were reviewed using Aerosol Instrument Manager software that exports data of normalized particle number concentrations in analysed sample (NC_i). NC_i was calculated by the equation (3):

$$NC_i = \frac{C_i}{\lg D_{i+1} - \lg D_{i-1}} \quad (3)$$

where C_i [N / cm^3] is particle number concentration for i-th channel and D_i is the diameter of particles measured in i-th channel. The decadic logarithms of widths of channels are identical with the distinction of 64 channels per decade:

$$C_i = \frac{NC_i}{64} \quad (4)$$

Total particle number concentration (C_N) was calculated by the equation, where n means total number of channels:

$$C_N = \sum_{i=1}^{n} C_i \quad (5)$$

2.3. *Characterization of generated CuO nanoparticles and tested filters*

With regard to particle concentration, nanoparticle distribution was measured directly by SMPS [12]. The nanoparticles were measured using SMPS before and after passing the filters at a particle number concentration before entering the filter (C_i) of about 7.5 x 10^5 particles·cm^{-3}.

In our measurement, the military filter OF-90, which belongs to the standard equipment of every soldier in ACR, was included. In addition, another type of protective filters may be used depending on the environment and pollutant levels in the ambient air. Such filters are usually used by the Fire Rescue Brigade (FRB), but occasionally by the ACR as well. That is why our research included a wide range of protective filters, used by the Czech FRB. All tested filters and their detailed specifications are defined in table 1.

Table 1 Tested filters (source:own).

Filter	Diameter	Height	Weight	Usage against	Maximal K_p [%]
OF-90	110 mm	90 mm	260 g	Solid and liquid substances and aerosols of poisonous substances according to CSN EN 143 class P3, biological aerosols and radioactive dusts	0.001 [10]
Filter P3	110 mm	50 mm	90 g		0.05 [7]
Filter P3 with two threads 40 × 1 / 7´´	110 mm	65 mm	130 g		0.05 [7]
Filter MOF-2	110 mm	90 mm	260 g	Chemical warfare agents in the form of gases and vapours, solid and liquid aerosols of poisonous substances according to CSN EN 143 class P3, biological aerosols and radioactive dusts	0.05 [7]
Filter MOF-3	110 mm	90 mm	260 g		0.05 [7]
Filter MOF-4	110 mm	90 mm	260 g		0.05 [7]
Filter MOF-5	110 mm	90 mm	260 g		0.05 [7]

Filter	Diameter	Height	Weight	Usage against	Maximal K_p [%]
Scott Health and safety CFR 22 CBRN A2B2E1K1-P3 RD	40 mm	/	300 g	Gases and vapours from organic compounds with a boiling point above 65 °C, inorganic gases and vapours, e.g. chlorine, hydrogen sulphide, hydrogen cyanide, acid gases and vapours e.g. sulphur dioxide, ammonia and organic ammonia derivatives (EN 14387), other gases and vapours tested: chloropicrin, cyanogen chloride, sarin, hydrogen cyanide, mustard gas and phosphine as well as CS and CN, Arsine, DMMP, particulates: solid and liquid toxic and radioactive particles and micro-organisms, e.g. bacteria and viruses	0.05 [7]
FN (Nuclear ULPA filter AX6650HS)	/	/	72 g·m^{-2}	Heavy industry environment, chemical industry, laboratories, pharmaceutical industry, remediation activities	0.05 [7]

3. Results
Particle number concentrations (C_i) measured before and after passage through a particulate or combined filter were performed at initial concentration of an aerosol of 7.5 x 10^5 particles cm^{-3}.

3.1. Measured values
Four measurements were made at 5 minute intervals (at 10, 15, 20 and 25 minute) and the average particle concentration was determined. Concentration values were then recalculated to particle penetration by the filter, respectively the permeability coefficient K_p. The resulting values measured for individual filters ranged from 7.6 nm to 299.6 nm. Acquired results are shown in table 2 and figure 3.

Table 2 Average permeability and efficiency of all tested filters [%] (source:own).

	E [%]	K_p [%]
FN	100.000	0.000
P3 with two threads	100.000	$9.561 \cdot 10^{-5}$
P3	100.000	$7.136 \cdot 10^{-5}$
MOF-2	99.999	$7.028 \cdot 10^{-4}$
MOF-3	100.000	$3.576 \cdot 10^{-6}$
MOF-4	100.000	$7.517 \cdot 10^{-6}$
MOF-5	100.000	$5.063 \cdot 10^{-5}$
Scott Health and safety	100.000	0.000
OF-90	100.000	$3.671 \cdot 10^{-6}$

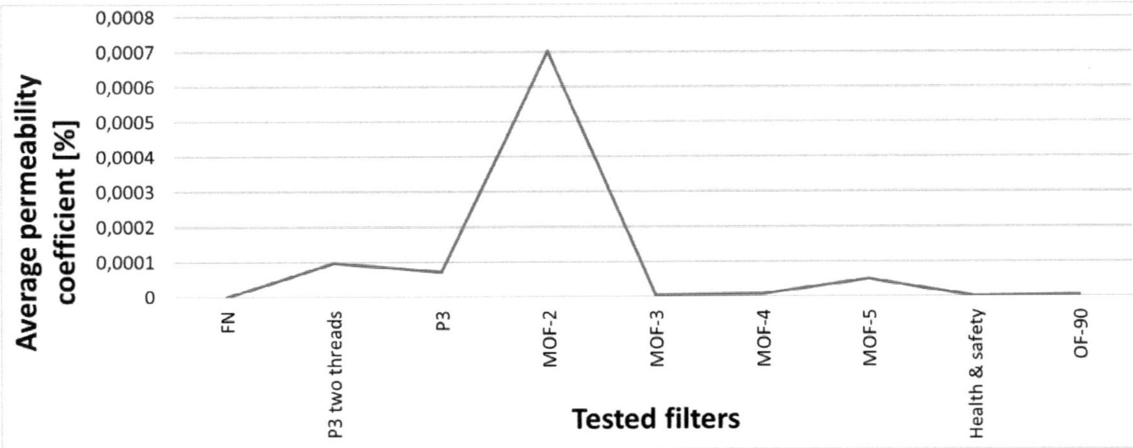

Figure 3 Average permeability coefficient of tested filters (source:own).

In past years, very high tactical and technical requirements have been set for the coefficient of permeability of the MOF protection filters [13]. According to CSN EN 143, the maximum permeability coefficient K_p of the test aerosol of paraffin oil and sodium chloride 0.05% [7] is determined. Similarly to military filters, according to CDS 841503 K_p 0.001% is applied [10] and implemented according to the methodology CSN EN 14387 [9]. The requirements of the standardized technical norms [5; 7; 8; 9] and the Czech defense standard [10] have been met in all tested filters.

3.2. *Permeability of particular sizes of nanoparticles in measured range*

The K_p value has been measured and calculated in the whole range of our measurements for all tested filters.

3.2.1. Filter OF-90. Regarding the military filter, the highest permeability of the filter was measured for nanoparticle sizes from 100 to 200 nm, where the filter efficiency was 99.995%. The corresponding K_p value of 0.005% in this range slightly exceeds the limit of 0.001% given by the standard [10]. This can be documented at the values shown in figure 4:

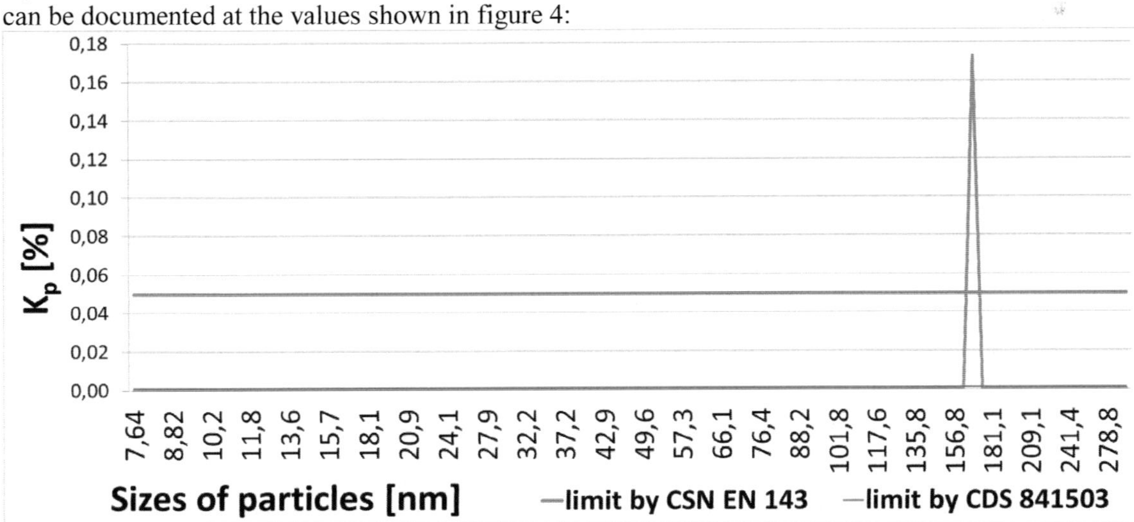

Figure 4 Average permeability of particles in different sizes of particles – filter OF-90 (source:own)

3.2.2. Particle filters P3. For both particle filters, the highest permeability was also observed in the nanoparticle size range from 100 nm to 200 nm (see figure 5):

- the P3 filter had the lowest efficiency in this range at 99.995% (K_p 0.005%),
- the P3 filter with two threads had the lowest efficiency at 99.997% (K_p 0.003%).

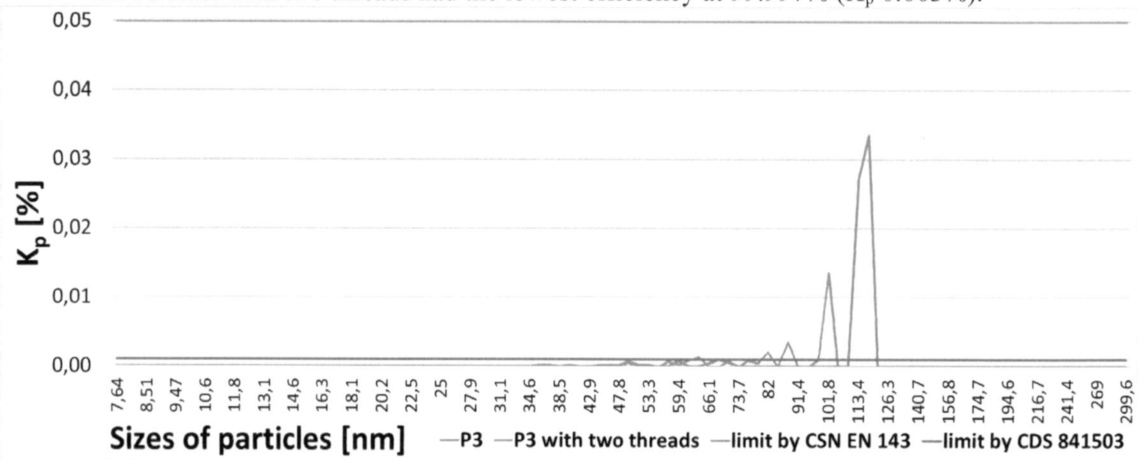

Figure 5 Average permeability of particles in different sizes of particles – particle filters (source:own)

3.2.3. Filters MOF. The MOF-3 filter had the highest filter permeability for nanoparticles from 100 to 200 nm where the filter efficiency was 99.996% (K_p 0.004%). In contrast, the MOF-2, MOF-4 and MOF-5 filters had the highest throughput in nanoparticles ranging from 200 nm to 300 nm (see figure 6):

- the MOF-2 filter had the lowest efficiency in this range at 99.901% (K_p 0.030%),
- the MOF-4 filter had the lowest efficiency at 99.919% (K_p 0.081%, the value exceeds the limit given by technical norm [7]),
- the MOF-5 filter had the lowest efficiency at 99.930% (K_p 0.070%, this value exceeds the limit given by technical norm [7]).

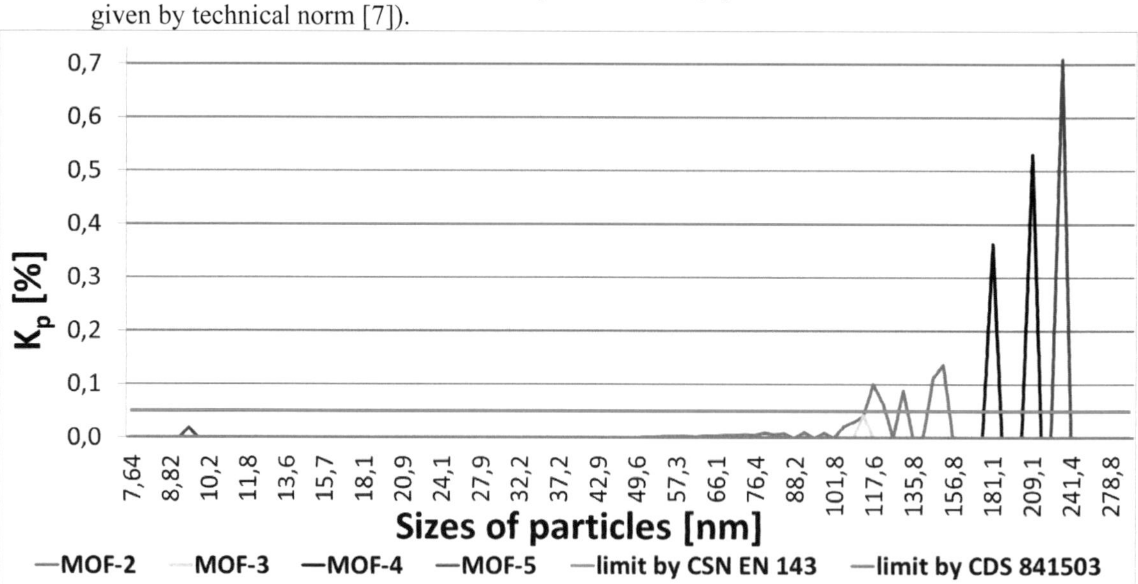

Figure 6 Average permeability of particles in different sizes of particles – MOF filters (source:own)

IOP Conf. Series: Journal of Physics: Conf. Series **987** (2018) 012037 doi:10.1088/1742-6596/987/1/012037

3.2.4. Filter FN a Health and safety filter. Protective filters providing a high degree of penetration efficiency against nanoparticle permeability have been highly effective. Both of these filters had zero nanoparticle penetration and the capture efficiency was 100% in both cases.

4. Conclusion

In this research, the efficiency of the antiparticle filters introduced in the ACR and the protection of the population was carried out using ultra-fine aerosol of cupric oxide with a particle size in the range of 7.6 to 299.6 nm. The testing was carried out in the laboratories of the Institute of Analytical Chemistry of the Academy of Sciences of the Czech Republic. The test showed that although the experiment was carried out under modified conditions using an ultrafine cupric oxide aerosol, compared to the aerosol and the measuring range specified by standards [7; 10], the required maximum aerosol permeability coefficients for any of the tested filters were not exceeded. The recorded results indicate that the military filter and particle filters exhibited the highest aerosol permeability through the filters, especially in the nanoparticle size range between 100–200 nm, while the MOF protective filters had the highest permeability in the range of 200 to 300 nm. On the other hand, the FN filter and the Health and Safety filter, did not release CuONP particles at all and had 100% nanoparticle capture efficiency and were therefore the most effective. The results confirm that the worst efficiency of filtration is usually at the interface of applying different proportions of surface and depth filtration. The range of particle sizes that begin to be low for surface filtration and too high to be applied for the depth filtration mechanism is critical. The obtained results of the experiment indicate that the particles most subject to this effect are most often in the range of 100 to 200 nm.

References

[1] EC 2011 Official Journal of the European Union Recommendations on the definition of nanomaterial. Available from URL: http://eur-lex.europa.eu/LexUriServ/LexUriServ.do?uri=OJ:L:2011:275:0038:0040:EN:PDF
[2] Information about health risks related to the quality of the air in 2015 (Prague: Ministry of the Environment of the Czech Republic) p 25
[3] Kendall M and Holgate S 2012 Health impact and toxicological effects of nanomaterials in the lung *Respirology* **17** 743 – 758
[4] Ščurek R 2008 *Selected Technical Means of Detection and Pyrotechnical Protection on the Airport* (Ostrava: VŠB – Technical Univesity of Ostrava) p 62
[5] CSN EN 132 *Respiratory Protective Devices* 2000 (Prague: Bulletin of the Office for Standardization, Metrology and State Examination of the Czech Republic) p 24
[6] Hylák Č and Pivovarník J 2016 *Individual and Collective Civil Protection of the Czech Republic* (Prague: Ministry of the Interior of the Czech Republic) p 194
[7] CSN EN 143 *Respiratory protective devices* 2000 (Prague: Bulletin of the Office for Standardization, Metrology and State Examination of the Czech Republic) p 30
[8] CSN EN 12942 *Respiratory protective devices* 1999 (Prague: Bulletin of the Office for Standardization, Metrology and State Examination of the Czech Republic) p 47
[9] CSN EN 14387 *Respiratory protective devices* 2004 (Prague: Bulletin of the Office for Standardization, Metrology and State Examination of the Czech Republic) p 20
[10] Czech Defence Standard 181503 2007 *Respiratory devices for individual protection* (Prague: Office for Defense Standardization, Cataloging and State Quality Assurance) p 20
[11] Brown R C 1993 *Air Filtration* (Oxford: Pergamon Press) p 272
[12] Kellnerová E, Večeřa Z, Kellner J and Zeman T 2017 Evaluation of selected pollutants and measures to reduce their impacts on health in military workplaces *Chem Engineer Trans* **57** 433 – 438
[13] Kroupa M 2003 *Means of Individual Protection* (Prague: Ministry of the Interior of the Czech Republic – Directorate General of Fire Rescue Service of the Czech Republic) p 44

IOP Conf. Series: Journal of Physics: Conf. Series **987** (2018) 012038 doi:10.1088/1742-6596/987/1/012038

Magnetic, thermal and luminescence properties in room-temperature nanosecond electron-irradiated various metal oxide nanopowders

S Yu Sokovnin [1,2], M E Balezin [2], V G Il`ves [2]

[1] Ural Federal University, Yekaterinburg, Russia
[2] Institute of Electrophysics UB RAS, Yekaterinburg, Russia

E-mail: sokovnin@iep.uran.ru

Abstract. By means of pulsed electron beam evaporation in vacuum of targets non-magnetic, in bulk state, Al_2O_3 and YSZ (ZrO_2-8% Y_2O_3) oxides, magnetic nanopowders (NPs) with a high specific surface were produced. The NPs were subsequently irradiated in air by electrons with energy of 700 keV, using a URT-1 accelerator for 15 and 30 minutes. The magnetic, thermal, and pulsed cathodoluminescence (PCL) characteristics of NPs were measured before and after irradiation. It was established that the electron irradiation non-monotonically changes the magnetization of the pristine samples. To the contrary, a clear correlation between the intensity of PCL and the irradiation doses is found in the oxides. There was a decrease in the intensity of PCL after irradiation. Luminescent and thermal properties reflect the transformation of structural defects in NPs more strongly after the exposure to a pulsed electron beam in comparison with corresponding changes of the NPs magnetic response.

1. Introduction

The magnetic properties of nanoparticles (NPles) depend on many factors - the chemical composition, type of the crystal lattice, degree of its deficiency, the shape and the size, morphology, interaction of particles with the matrix and neighboring particles, phase structure and type of structural defects, etc. [1]. By changing the size, shape, composition and structure of NPles, it is possible, within limits, to control the magnetic characteristics of nanomaterials.

The room temperature ferromagnetism (RTFM) is associated with lattice defects and / or added ions, whereas the added ions can be non magnetic. This phenomenon is called defect-induced magnetism [2]. The identification of RTFM in non-magnetic dopant-doped oxide also attracts a lot of attention in connection with possible use of the oxides in semiconductor spintronics. Therefore, the search for a suitable non-magnetic additive, which makes non-magnetic oxides magnetic, is very important [3].

The study of the relationship between magnetic and luminescent characteristics (and the luminescence is a good tool for studying structural defects in solids) makes it possible to reveal the nature of magnetism in DMS, more deeply, in particular, in DMS based on nonmagnetic oxides. Presently, engineering of nanostructures using electronic irradiation has been widely used [4].

The possibility of improving properties by controlling the particle size can lead to potentially useful technological applications. There studies on the use of electron beam irradiation for controlling the urface properties of NPles are very rare. There are few works related to the effect of electron irradiation on structural, optical and thermal properties of NPles [5].

In this work the magnetic, thermal, and luminescent characteristics of oxide NPs obtained by pulsed electron evaporation in vacuum were studied before and after their exposure to the high-energy electron beam.

2. Experimental

Nanopowders of pristine Al_2O_3 [6, 7] and Y_2O_3-doped ZrO_2 (YSZ) oxides [8, 9] were previously obtained using the method of evaporation by a pulsed electron beam in a vacuum on NANOBEAM-2 apparatus [10]. NPs were deposited on the windowpanes of a large area placed around the target at a distance of 10-15 cm. The residual pressure in the evaporation chamber was 4 Pa. Next, the samples were irradiated on an electron accelerator URT-1M (700 keV, ~100 ns, 35 pps) [11]. The irradiation time was 15 minutes (1st series), then another 15 minutes (2nd series). The samples were placed on an aluminum plate cooled by fans. The samples irradiated at adsorbed dose (AD) of 0, 31.5 and 63 MGy are labeled as **1**, **2** and **3**, respectively.

A detailed description of the preparation of specific oxide targets for electron evaporation is given in [6-10]. The following diagnostics of NPs were used in the experiments were used the following diagnostics of NPs. The XRD was made on diffractometers XRD 7000 Shimadzu (Japan) and D8 Discover (Holland). The morphology and the sizes of NPAs were observed by means of the on JEM 2100 microscope. Magnetic measurements were carried by Faraday's scales (sensitivity of scales ~ 10^{-5}emu, the range of measurements on the magnetic field – to 12 kOe). The PCL spectra were recorded at room temperature on a KLAVI-1 spectral analyzer. TG and DSC measurements were performed on a Netzsch thermal analyzer at a heating rate of 10 °C min^{-1} in air atmosphere. N2 adsorption–desorption isotherms were obtained on a TriStar 3000 V6.03 apparatus at 77 K under continuous adsorption condition. BET and BJH analyses were used to determine the surface area, pore size distribution, and pore volume.

3. Results and discussion

3.1. Properties of NPs before irradiation

Some physicochemical characteristics of oxide NPs before irradiation are given in Table 1. All NPs, without exception, contained an amorphous fraction that was formed during the crystallization of NPles on a cold substrate. X-ray diffraction patterns of unirradiated oxide NPs and their analysis were performed in [6-10]. Measurement of adsorption isotherms showed that all powders obtained by the PEBE method belong to mesoporous NPs with disordered pores, a high pore volume and a relatively large pore diameter in the range of 15 to 20 nm.

Table 1. Some physical parameters of oxide NPs obtained by a PEBE, before irradiation

Parameters	Al_2O_3	**YSZ**
S_{BET} (m^2/g)	*124*	*98.7*
Phase composition	*α, δ & amorphous*	*amorphous/cubic*
Crystallinity (%)	*not determine*	*29*
BJH Desorption average pore diameter (4V/A)	-	*20.4*

3.2. Al_2O_3

The evolution of the specific magnetization (σ) and the PCL spectrum of the multiphase Al_2O_3 NP from the irradiation time are shown in Fig. 1 and 2.

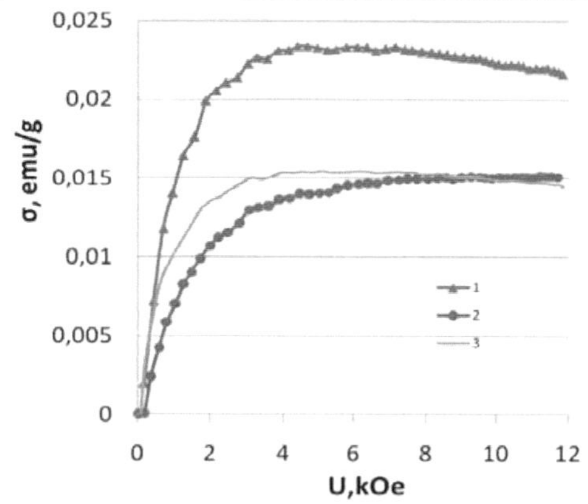

Figure 1. Dependences of the change in the magnetization of the Al_2O_3 NP.

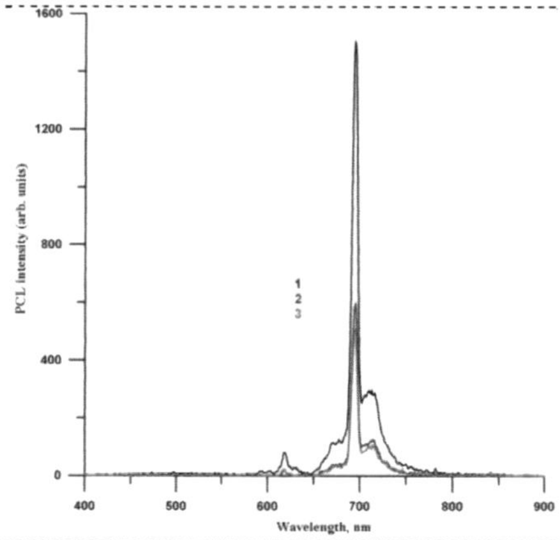

Figure 2. PCL spectra of pristine and irradiated samples Al_2O_3.

The PCL spectrum is an important tool for obtaining fundamental information about structural defects in the lattice. The unirradiated Al_2O_3 sample showed (Figure 2) the maximum PCL intensity compared to samples irradiated with 700 MeV electrons. The PCL spectra of all samples are completely identical and have a similar morphology. All spectra consist of two bands in the red region of the spectrum, centered at ~ 620 and 720 nm. The narrow peak at 694.3 nm -R line, is from Cr^{3+} chromium impurity ions. Preservation of the PCL form of the non-irradiated sample Al_2O_3 after exposure to large doses of radiation indicates that there is no possible phase transformation in the sample with an electron irradiation. The absence of a peak in the defective green band at a wavelength of ~ 500 nm indicates the absence of oxygen vacancies (F centers) and/or copper ions in the Al_2O_3 sample.

The behavior of the dependences of the specific magnetization (σ) on the AD is more complicated. In the magnetization curves of the first arbitrary selection of NP, a decrease in σ with an increase in the AD was observed, which correlates with a change in the PCL spectrum.

3.3. Y_2O_3-doped ZrO_2

Fig. 3 shows that with the increase in the AD, the intensity of PCL emission of YSZ NP was reduced. The intensity is due to the formation of non-emission by-products or defects on the surface of NPles [12]. A decrease in the intensity of PCL indicates a decrease in the rate of recombination between electrons and holes. The location of the main peaks of the PCL spectra did not change significantly with increasing AD, which indicates the preservation of the pristine phase composition of the YSZ NP at the given irradiation levels.

The magnetization saturation (Fig. 4) of the YSZ sample increased by almost 20% after 15 minutes of irradiation and did not change later with an increase in the AD. This sample could contain insignificant magnetic impurities (the YSZ NP was collected from glass substrates using a steel razor), as indicated by the relatively high specific magnetization of the unirradiated sample, which is an order of magnitude higher than the magnetization of the YSZ sample from our early work [9]. It is possible that the increase in the magnetization of the YSZ sample was due to the effect of electron irradiation on the impurity magnetic iron particles. The magnetization of strong ferromagnets such as Fe and its oxides, as shown above, in the analysis of the effect of electron irradiation on particles with different types of magnetism, usually increases with the AD increase. The presence of magnetic impurities in the YSZ sample indirectly confirms the reach of saturation by all the magnetization curves at the same magnetic field and the uniformity of the magnetization curves, independently of the AD. The nature of the magnetic ordering in the YSZ NP requires further studies, but it is obvious that the irradiation changed the concentration of defects in the NP, which was reflected in the growth of the magnetization of the NP.

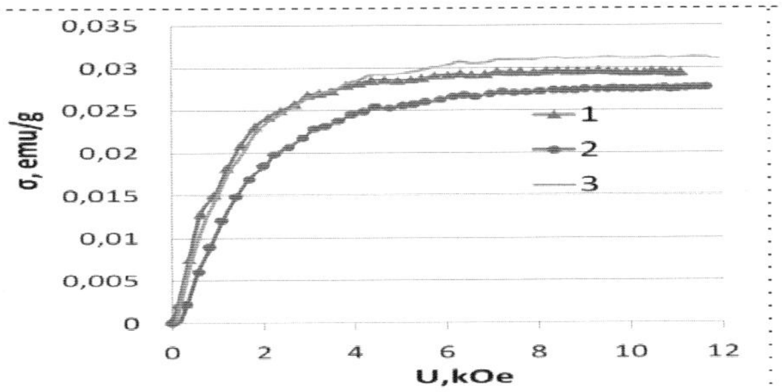

Figure 3. Dependences of the change in the magnetization of the YSZ NP.

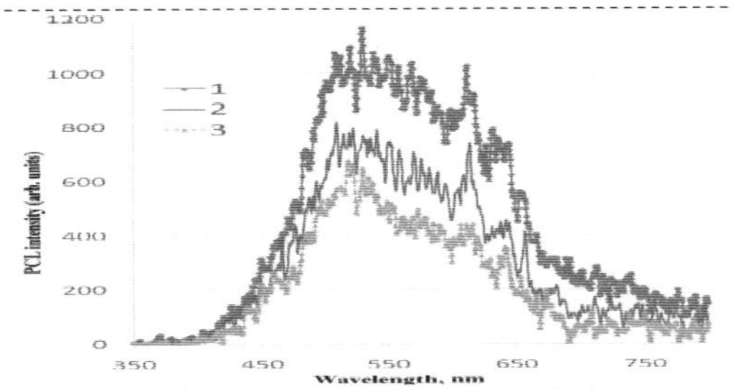

Figure 4. PCL spectra of pristine and irradiated samples of the YSZ NP.

The results of the thermal analysis of the YSZ NP by the synchronous DSC-TG method are shown in Fig. 5. On the TG curves of all samples in the temperature range from 100 to 430 ° C, water molecules and carbon-containing compounds adsorbed on the surface of mesoporous YSZ NPles were removed.

On the TG curve of the non-irradiated sample (Fig. 5), two temperature ranges are clearly visible: **I** = 70-250 ^{0}C and **II** = 250-430 ^{0}C, corresponding to evaporation of water and desorption of carbon-containing compounds, respectively. On the TG curves of irradiated samples, range **II** is barely noticeable. The decrease in the endothermic peaks of the **II** band in irradiated samples is caused by the influence of a strong oxidant-ozone on carbon-containing compounds adsorbed on the surface and in the pores of the samples (it is well known that ozone reacts with carbon at normal temperature to form carbon dioxide: $2C+ 2O_3 \rightarrow 2CO_2+O$). In the **III** temperature range from 430 to ~ 600 ^{0}C thermal peaks on the DSC curves were absent in all samples, with simultaneous, linear mass growth.

It is important to note that the increased mass growth (increase in the steepness of the TG curves) for all samples began at the same temperature around 430 °C and continued uninterruptedly up to 1400 °C. That indicates the same nature of the transformation starting at a temperature of 430 °C in all samples. It is likely that the growth of mass in the temperature range **III** was mainly caused by thermal annealing of oxygen vacancies. In the **IV** range from 430 °C to 1030 °C, crystallization of the amorphous fraction in the monoclinic ZrO_2 phase occurred in a wide temperature range in all samples. Synchronous growth of the TG and DSC curves for all samples in the V temperature range from 1030 to 1130 °C shows that the completion of the crystallization of the amorphous fraction in the **IV** range promotes the acceleration of the annealing of oxygen vacancies.

In the **VI** range, the final polymorphic transformation of the monoclinic phase to tetragonal phase occurs at the same temperature of about 1130 °C in all samples. (The region of thermal stability of the bulk monoclinic phase is from 273 to 1160 °C [13]). The decrease in the temperature of the polymorphic transformation by 30 °C is exclusively due to the influence of the size factor and is not connected with the influence of the electron beam on the samples.

In the **VII** range, the growth of the mass of the samples in the TG curves, as a result of the annealing of the oxygen vacancies, continued until the end of the heating to 1400 °C (At the same time, there were no thermal transformations in the DSC curves).

The change in the mass of the samples at their heating from 430 to 1400 °C increased nonlinearly with an increase in the irradiation dose in the sequence 6.8 → 7.3 → 12.3%. An almost twofold increase in the mass of the sample at maximum irradiation clearly indicates that irradiation by electrons substantially changes the concentration of oxygen vacancies at the surface and inside of the volume of the YSZ mesoporous nanopowder. The decrease in the concentration of oxygen vacancies on the surface of irradiated samples as a result of radiation annealing confirms a consistent decrease in the PCL intensity of the samples (Fig. 3) induced by electrons with an energy of 180 keV. In turn, the increase in the continuous growth of the mass of irradiated samples indicates an increase in the concentration of oxygen vacancies in irradiated samples as compared to the pristine concentration of vacancies in the unirradiated sample.

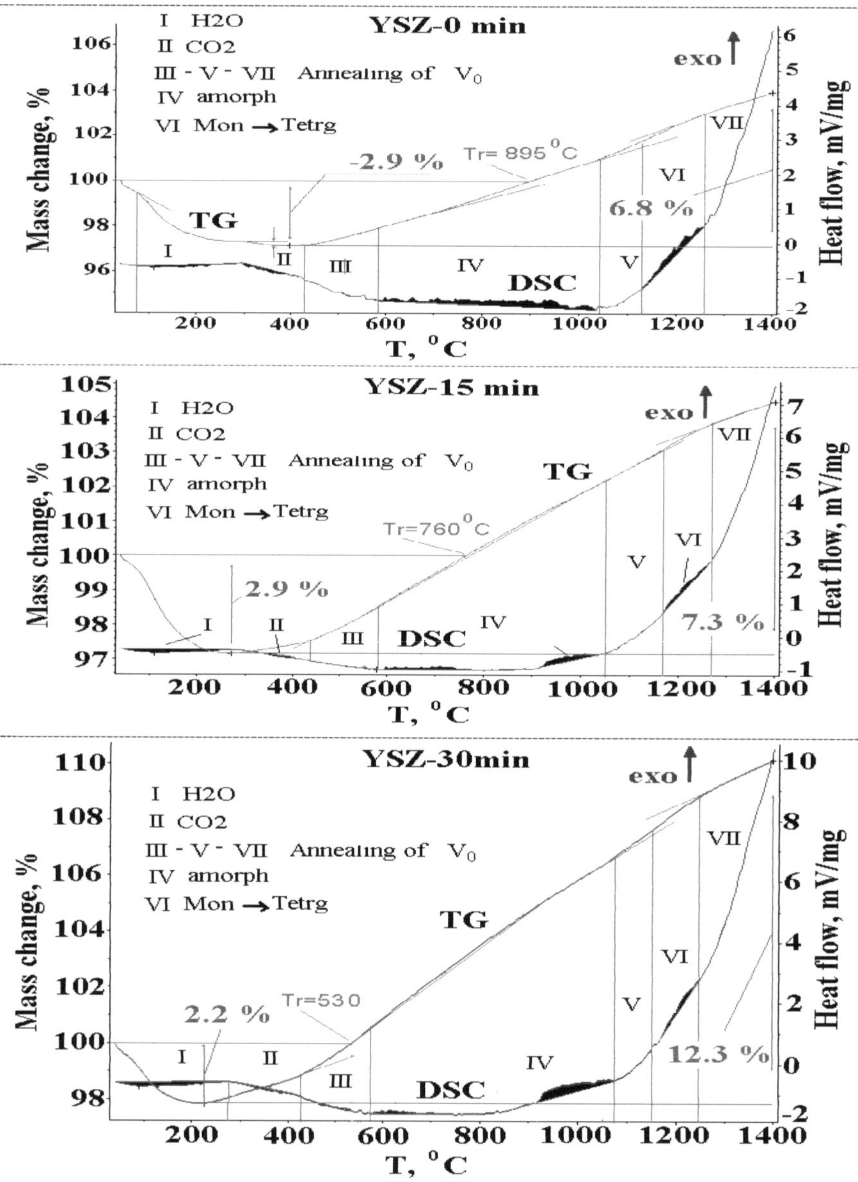

Figure 5. Simultaneous DSC-TG curves of pristine and irradiated samples YSZ .

Thus, the use of a complex approach to study the nature of defects induced by irradiation in multiphase mesoporous NPs allows a qualitative estimation of the change in the concentration of oxygen vacancies after exposure. From the obtained data, the following conclusions can be drawn:

- irradiation with high doses of high-energy electrons significantly changes the concentration of oxygen and, possibly, cationic vacancies, which clearly demonstrates the continuous increase in the mass of irradiated samples at heating in comparison with the change in the mass of the unirradiated sample;

- the impact of electrons does not change the temperature of the various phase transformations (crystallization temperature of the amorphous phase, the temperature of the polymorphic transformations), which indirectly indicates the absence of a significant change in the original size of the nanocrystallites after irradiation;
- the kinetics of the sample mass transformation (mass reversal) significantly changes after irradiation. The temperature at which the mass of the sample, after the initial weight loss, returns to its original level (Tr), successively decreases 895 →760 → 530 °C with the dose increase, which confirms that the irradiated samples contain significantly less amount of adsorbed water and carbonaceous impurities;
- the unchanged location of the PCL peaks indicates the absence of phase transformations caused by irradiation.

4. Conclusions

Thus, the preliminary investigation of the effect of a pulsed electron beam with a high electron energy on the NPs of some nonmagnetic metal oxides showed the following:

- the effect of electron irradiation on the magnetic properties of NPs is less pronounced than the effect on structurally sensitive PCL spectra and thermal properties;
- the specific magnetization of NPles of non-magnetic oxides varies nonmonotonically from the AD of irradiation and varies arbitrarily for each particular sample;
- there is a general tendency to decrease the PCL intensity of NP of various metal oxides with an increase in the irradiation dose, probably caused by a decrease in the concentration of structural defects;
- synchronous thermo-gravimetric analysis of the behavior of the samples before and after irradiation by the electrons significantly expands the idea of the formation and change of the defect structure of NP samples of nonmagnetic oxides as a result of their irradiation with low-energy electrons, in the synthesis process by the PEBE method, and with further irradiation by high-energy 700 keV electrons.

This work was performed within the subject of the state task [0389-2014-0005]; was partial supported by the Russian Foundation for Basic Research [15-08-01381]

References

[1] Gubin S, Koksharov Y, Khomutov G, Yurkov G 2005 *Russ. Chem. Rev.* **74** 489
[2] Peng H, Xiang H, Wei S, Li S, Xia J, Li J 2009 *Phys. Rev. Lett.* **102** 017201
[3] Hsu H, Tung Y, Chen Y, Chen M, Lee J, Sun J 2011 *Phys. Status Solidi* RRL **5** 12 447
[4] Gonzalez-Martinez I, Bachmatiuk A, Bezugly V, Kunstmann J, Gemming T, Liu Z, Cuniberti G, Rummeli M 2016 *Nanoscale* **8** 11340.
[5] Nagase T, Umakoshi Y 2005 *Materials Transactions* **46** 616
[6] Il`ves V, Sokovnin S, Zuev M 2016 *J.Lumin.* **179** 474
[7] Sokovnin S, Il'ves V, Surdo A, Mil'man I, Vlasov M 2013 *Nanotechnologies in Russia* **8** 466
[8] Sokovnin S, Il'ves V 2009 *Tech. Phys. Lett.* **35** 1026
[9] Sokovnin S, Il'ves V 2012 *ISRN Nanomaterials* Article ID 504634 11
[10] Sokovnin S, Il'ves V 2012 *J. Ferroelectrics* **436** 101
[11] Sokovnin S, Balezin M 2016 *Radiation Physics and Chemistry* S0969-806X(17)30918-0
[12] Chethan P, Joshi P, Raj Mohan S, Dhami T, Jayakrishna K, Koteshwar Rao K, Kukreja L, Ganesh S 2013 *Adv. Mat. Lett.* **4** 454
[13] Kumar S, Ojha A, 2016 *Materials Chemistry and Physics* **169** 13.

IOP Conf. Series: Journal of Physics: Conf. Series **987** (2018) 012039 doi:10.1088/1742-6596/987/1/012039

Investigation of the Morphological Evolution and Etching Kinetics of black Silicon During Ni-Assisted Chemical Etching

OV Volovlikova[a]*, G O Silakov[a], S A Gavrilov[a], A A Dudin[b], G O Diudbin[b], Y I Shilyaeva[a]

[a] National Research University of Electronic Technology - MIET, Zelenograd, Moscow, Russia
[b] Institute of Nanotechnology of Microelectronics, Russian Academy of Sciences, Moscow, Russia
*e-mail: VolovlikovaOV@org.miet.ru

Abstract. Kinetics of contact angle of solutions contained HF (40%):H_2O_2(30%):C_2H_5OH:H_2O on Si/Ni has been investigated. The effect of the ethanol concentration on the morphology of the porous silicon formed by Ni-assisted chemical etching is discussed. The influence of ethanol concentration on the morphology of the porous silicon was established by water contact angle and roughness factor.

1. Introduction

Superhydrophobic surfaces are a key technology for their ability to reduce friction losses in microchannels and their self cleaning properties [1-4]. Hydrophobicity is a property of a solid surface, which plays important roles in many industrial applications. Material with wide application is porous silicon [5-7]. The past decade has seen the rapid development of porous silicon (*p*-Si) in many applications including microelectronics, implantation, drug delivery, and biosensor [8]. Silicon surface chemistry is of fundamental importance because of the very relevant role of silicon in modern technology. The morphology of silicon wafer is modified by an electrochemical wet etching [9-11], galvanic etching [12-14], metal-assisted chemical etching (MACE) [15-17] to produce pores with controlled size and distribution and coated with a silane hydrophobic layer.

Perspective method for porous silicon formation is MACE using thin film of Ni [18]. There is no residual Ni on the surface of the por-Si after MACE with Ni films [19]. The effect of non-wetting the porous surface is achieved by creating a stable Cassie state on the surface. At this state there are air gaps between the liquid and the surface [20]. It is possible to obtain both hydrophilic and hydrophobic surfaces by changing the geometric parameters of the porous layer. Wetability of the surface is characterized by the water contact angle (CA). CA can be increased by increasing the surface roughness factor or the pore diameter [21]. One of the methods to change the surface roughness is to change the solutions composition for MACE contained HF (40%), H_2O_2(30%), and H_2O. The addition of alcohol to the solution promotes an increase in the wettability of the surface and decrease in the roughness. The effect of the ethanol concentration on the morphology of the porous silicon formed by Ni-assisted chemical etching is not investigated.

Thus, the aim of the work is studying the kinetics of contact angle of solution contains ethanol on the initial surface Si/Ni. The effect of the ethanol concentration on the morphology of the porous silicon formed by Ni-assisted chemical etching was investigated.

2. Experimental details

Wafers of *p*-type Si ((100), 0.01 Ω·cm) are used as substrates. Preliminary to be etched single-crystalline wafers were washed with piranha etch (mixture of sulfuric acid H_2SO_4 (98%) and hydrogen peroxide H_2O_2 (30%) (1:2 in volume) at 130°C during 10 min. Then they were rinsed in deionized (DI) water and dried by jet of isopropyl alcohol vapor. Si wafers were rinsed in HF-H_2O solution (1:4 in volumes) to remove native SiO_2. Subsequently, a 50 nm thick Ni thin film (99.99 % purity) was deposited by magnetron sputtering (Si/Ni samples). Then wafers were cut into 2×2 cm^2 pieces and used as samples. All solutions contained HF (40%):H_2O_2(30%):C_2H_5OH:H_2O. There were 14 solutions with various component concentrations (table 1).

Table 1 – Solutions composition

Solution number	volume part HF	volume part H_2O_2	volume part C_2H_5OH	volume part H_2O	molar ratio HF/H_2O_2	molar ratio HF/C_2H_5OH
I	8	8	4	40	1.60	2.56
II	8	8	8	40	1.60	1.28
III	8	8	12	40	1.60	0.85
IV	8	12	4	40	1.06	2.56
V	8	12	8	40	1.06	1.28
VI	8	12	12	40	1.06	0.85
VII	8	8	0	40	1.60	10^8 (∝)
VIII	8	8.5	4	40	1.50	2.56
IX	8	9	8	40	1.42	1.28
X	8	10	12	40	1.28	0.85
XI	8	12	0	40	1.06	10^8 (∝)
XII	8	13	4	40	0.98	2.56
XIII	8	14	8	40	0.91	1.28
XIV	8	15	12	40	0.85	0.85

The solution was dripping on the Si/Ni surface for kinetics of contact angle investigation. Solution drop profile began to change immediately. We use a Fujifilm Finepix S4000 camera, to record images of millimeter size sessile solution drops, at rest over horizontal samples. Samples Si/Ni were immersed in the solutions presented in Table1 for black silicon morphology investigation. Etching duration was 20, 40 and 60 min. Then samples were rinsed in ethanol and dried in air. Morphology of the samples surface were investigated by scanning electron microscopy (SEM) using a JSM-6010PLUS/LA (JEOL company) and Helios NanoLab 650. An atomic force Dimension Icon microscope (Bruker) was used for the morphological characterization of the samples by employing tapping mode AFM. We adopt the sessile drop method to measure the contact angles (CA) over the porous silicon samples. We also used a Goniometer LK-1 [22] to record images of sessile DI water (18.2 Ω) drops. Drop volume was 10 ml. We use freely available software, DropShape to process images and measure contact angles.

3. Results and Discussion

3.1 Kinetics of contact angle

Contact angles have been obtained from Si/Ni samples. The initial CA of VII, XI, XII and XIV solutions is 40°, 22°, 11°, and 1° respectively. Mean observed water CA evolution over 60 seconds of monitoring is presented in Figure 1.

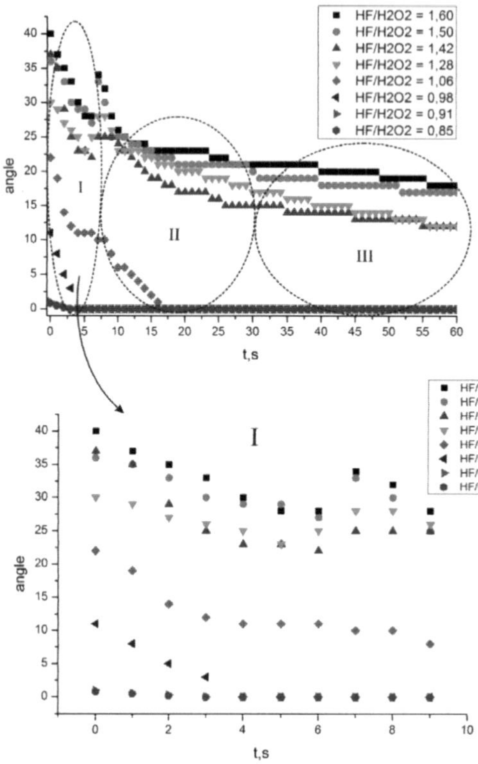

Figure 1. CA vs. monitoring duration.

The maximum value of the contact angle 40° is observed on the Si/Ni surface when an investigated solution contained no alcohol and minimum value of H_2O_2 (VII solution). The minimum value of the contact angle 1° is observed on the Si/Ni surface when an investigated solution contained 12 volume part of C_2H_5OH and 15 volume part of H_2O_2 (XIV solution).

Contact angle decreases with molar ratio HF/H_2O_2 decreasing. It can be explained by H_2O_2 increasing. The ratio of HF/H_2O_2 decreases due to a decrease in the concentration of HF and growth of the concentration of C_2H_5OH. Water has a high surface tension because it is strongly attracted to itself. However, at the adding alcohol, instead of water interacting only with other water molecules, it now interacts (less strongly) with alcohol, and the surface tension of the mixture will be lower (Figure 2) [9]. Extreme hydrophilicity has been observed for molar ratio $HF/H_2O_2 = 0.91$ and 0.85.

The kinetic curves of the solutions contact angle of the Si/Ni surface (Figure 1) can be divided into 3 sections. Each section characterizes the corresponding process takes place on the Si/Ni surface during etching with solutions of I-XIV (Figure 2).

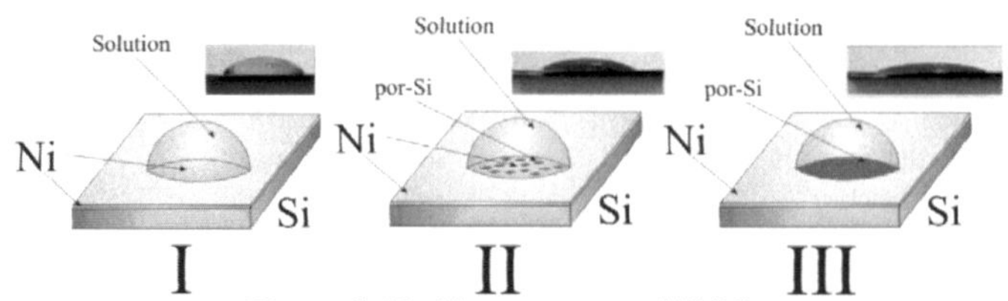

Figure 2. Etching process of Si/Ni.

Table 2 shows the CA value and the rate of CA decrease.

Table 2 – Decreacing contact angle (CA) rate vs. duration

I	HF/H_2O_2	1.6	1.5	1.42	1.28	1.06	0.98	0.91	0.85
	CA (1s), °	40	36	37	30	22	11	1	0.8
	CA (6 s), °	28	27	22	23	11	0	0	0
	CA, °/sec	2.1	1.5	2.6	1.03	1.85	1.92	0.15	0.12
II	HF/H_2O_2	1.6	1.5	1.42	1.28	1.06	0.98	0.91	0.85
	CA (7 s), °	34	33	25	28	10	0	0	0
	CA (30 s), °	21	21	15	18	0	0	0	0
	CA, °/sec	1.2	1.27	0.98	0.86	0.14	0	0	0
III	HF/H_2O_2	1.6	1.5	1.42	1.28	1.06	0.98	0.91	0.85
	CA (31 s), °	21	20	15	17	0	0	0	0
	CA (60 s), °	18	17	12	12	0	0	0	0
	CA, °/sec	0.1	0.09	0.1	0.18	0	0	0	0

The behavior of a drop solution of each section is determined by the following processes:

1 section:

- duration up to 6 seconds;
- the contact angle depends on the surface energy of Si/Ni;
- H_2O_2 – constant value equals to 1.75 mol/L for molar ratio HF/H_2O_2 in the rage of 1,28-1,6 and 2.5 mol/L for HF/H_2O_2 in the rage of 0.85-1.06;

2 section.

- duration from 7 to 30 seconds;
- the increasing of the CA is noted at 7 seconds. The obtained results indicate a change in the surface tension of the solution. This can be due to the dissolution of the metal film and the formation of porous silicon.
- the rate of decrease of the CA in section II is lower than in section I.

3 section.

- duration over 30 seconds;
- characterized by the absence of Ni and the formation of porous silicon;
- the rate of decrease of CA is practically equal to zero.

Ethanol concentration influence on the CA can be demonstrated in Figure 3.

The increasing of HF/C_2H_5OH and decreasing the C_2H_5OH concentration respectively in the solution leads to increasing the CA throughout the duration of the treatment. At the initial time of monitoring (Figure 3a) the difference between the contact angles for HF/H_2O_2 = 1.6 and 1.06 is ~ 7-8°. The difference in CA is reduced to 2-3° at 30 seconds of monitoring. The difference in CA is reduced to 1-2° at 60 seconds of monitoring.

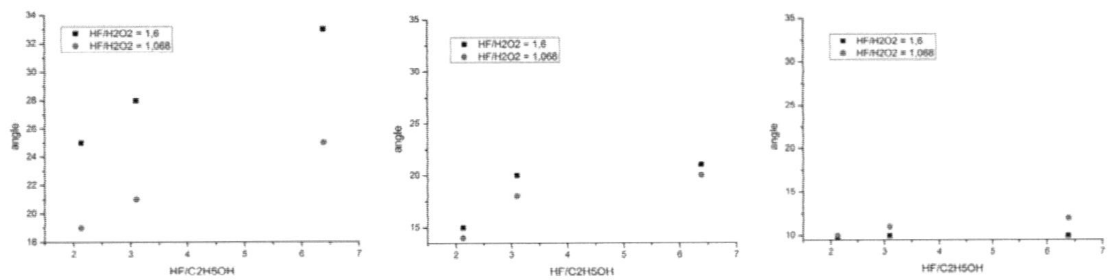

a b c

Figure 3. Initial contact angle (a), during 20 sec (b) and during 60 sec (c).

Thus, it has been found that the addition of C_2H_5OH to the etching solution results in a decreasing in CA and a surface tension of the solution. We investigate XIII and XIV solutions as completely wetting solutions. In addition alcohol-containing solution, promotes the detachment of H_2 from Si during the formation of the porous layer. Such an etchant composition improves wetting of the surface during etching and leads to the uniformity of the modified Si.

3.2 Morphological Evolution

Figure 4 shows SEM images of the *por*-Si after 40 minute duration of treatment in HF (40%):H_2O_2(30%):C_2H_5OH:H_2O.

Figure 4. SEM images of por-Si after treatment in solution with HF/H_2O_2 and HF/C_2H_5OH: 1.6 and no ethanol (a), 1.6 and 0.85 (b), 1.06 and no ethanol (c), 1.06 and 0.85 (d).

The lower the concentration of H_2O_2 in the solution, the more uniform porous layer is formed (Figure 4 a, c). Macropores begin to form on the surface of porous silicon after etching with solution containing C_2H_5OH (Figure 4 b, d).

The porous silicon samples after 40 min treatment were characterized via Atomic Force Microscopy (AFM). The results obtained are shown in Figure 5.

IOP Conf. Series: Journal of Physics: Conf. Series **987** (2018) 012039 doi:10.1088/1742-6596/987/1/012039

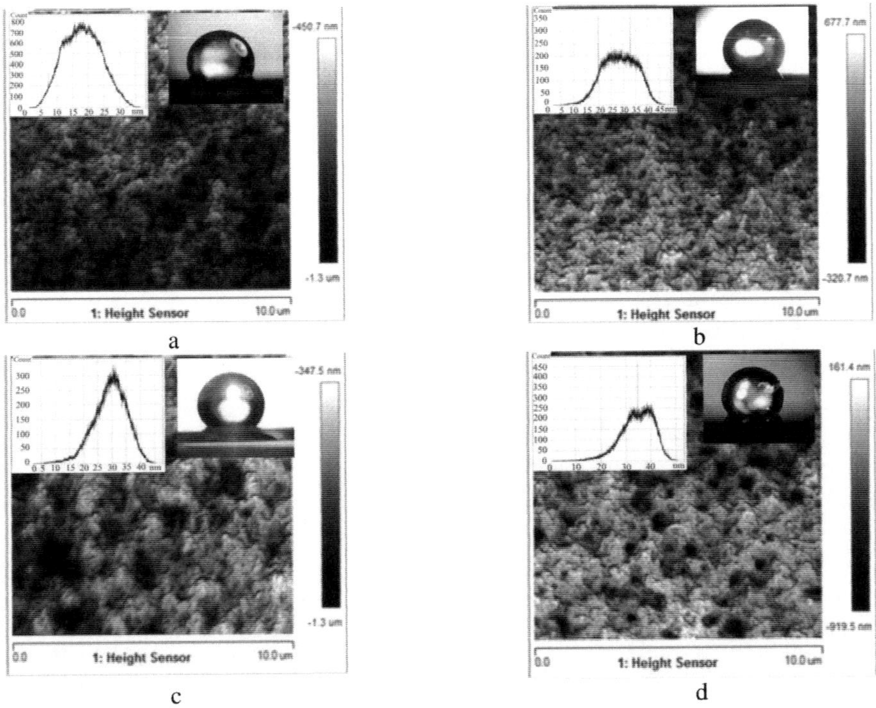

Figure 5. AFM images for the formation of pores in Si in a solution of $HF/C_2H_5OH=2.56$ (b), $HF/C_2H_5OH=1.28$ (c), $HF/C_2H_5OH=0.85$ (d), image zoom of (a-d) 10×10 μm. The inset of Figure shows the silicon nanocrystal size distribution on the surface (left), water contact angle (right).

From Figure 5, we observe that the formation of macropores is feasible after etching in solution $HF/C_2H_5OH=1.28$ и $HF/C_2H_5OH=0.85$. Through section analysis with the AFM software, the diameter of the pores was obtained with values between 300 and 900 nm. Open pores contacting with the external environment leads to decreasing the solid-liquid contact in the Cassie state. Therefore the contact angle of wetting is increased. Water CA is 115° in the case of absence of ethanol in etching solution, at $HF/C_2H_5OH=2.56$ water CA is 134°, at $HF/C_2H_5OH=1.28$ water CA is 136°, at $HF/C_2H_5OH=0.85$ water CA is 138° (Figure 5 a, d). Thus, the higher the ethanol concentration in the solution, the higher the water CA on porous Si, because of pores formation on the surface. It is known that CA is one of the main techniques to indicate surface roughness. Roughness is determined by composition of a solution. Cassie–Baxter equation describes CA (θ_{CB}) of roughness surface [23].

$$\cos\theta_{CB} = f\cos\theta + f - 1,$$

$$\cos\theta_{CB} = -\frac{1-f}{r-f},$$

where θ_{CB} is the liquid–solid contact angle on the textured surface, θ is the static contact angle on the flat surface, and f is the fraction of the liquid–solid contact area, r - roughness ratio factor.
The results of calculations are shown in Figure 6.

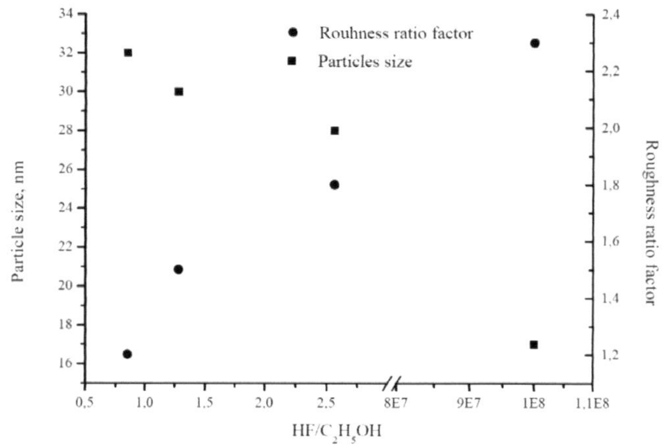

Figure 6. Particle size and roughness ratio factor vs. ratio HF/C_2H_5OH.

It was established that the value of the roughness calculated from Cassie–Baxter equation has an inverse dependence on the ethanol concentration in the solution. Through roughness analysis with the AFM software, the distribution histograms of the silicon particles were obtained (Figure 6). It has been established that the smaller the particle size of the porous layer, the greater its roughness.

The chemical composition of the porous surface has influence on the CA in addition to roughness and pores. Chemically bonded hydrogen atoms forms during the formation of pores in contact with the HF-containing electrolyte. Porous silicon acquires hydrophobic properties. Thus, water CA on as-etched porous silicon is from 135° to 144° depending on the composition of the electrolyte. Due to the oxidation of the samples in air, the water CA is decreased due to formation of the hydrophilic Si-OH groups [24]. For the samples stored in air for 1 week, water CA is 117°-142°, depending on the composition of the electrolyte; for 3 weeks, water CA is 111°-136°, for 8 weeks, water CA does not exceed 125°. Since we assume that the value of the CA depends on the Si-H groups, it will be necessary to carry out an additional study.

4. Conclusion.

The addition of ethanol to the etching solution contained $HF+H_2O_2$ during Ni-assisted chemical etching leads to decreasing of echtant CA on Si/Ni. CA of etchtant contained HF and H_2O_2 on Si/Ni is 40°, at $HF/C_2H_5OH = 0.85$ etchant CA is 1°. The lower the concentration of H_2O_2 in the solution, the more uniformly porous layer is formed. Macropores begin to form on the surface of porous silicon after etching with solution containing C_2H_5OH. We observe that the formation of macropores with diameter 300 - 900 nm is feasible after etching in solution $HF/C_2H_5OH=1.28$ и $HF/C_2H_5OH=0.85$.

Water CA on the porous silicon is 115° in the case of absence of ethanol in etching solution, at $HF/C_2H_5OH=2.56$ water CA is 134°, at $HF/C_2H_5OH=1.28$ water CA is 136°, at $HF/C_2H_5OH=0.85$ water CA is 138°. Thus, the higher the ethanol concentration in the solution, the higher the water CA on porous Si, because of pores formation on the surface. It was established that the value of the roughness ratio factor calculated from Cassie–Baxter equation has an inverse dependence on the CA and on the ethanol concentration in the solution, respectively. It has been established that the smaller the particle size of the porous layer, the greater its roughness.

Acknowledgment
This study was supported by the Russian Science Foundation, project no. 16-19-10625.

[1] Furstner R and Barthlott W 2005 *Langmuir* **21** 956-61.
[2] Zhang M, Feng Sh, Wang L, Zheng Y 2016 *Biotribology* **5** 31-43.
[3] Yang H, Jiang P 2010 *Langmuir* **26** 12598-04.
[4] Bhushan B, Jung Y C, Koch K 2009 *Langmuir* **25**, 3240-8.
[5] Gullis G, Canham L T, Calcott P D J 1997 *J. Appl. Phys.* **82**, 909-65.
[6] Schmidt V, Riel H, Senz S, Karg S, Riess W, Gösele U. 2006 *Small* **2** 85-8.
[7] Goldberger J, Hochbaum A I, Fan R, Yang P 2006 *Nano Lett.* **6** 973-7.
[8] Foll H, Christophersen M, Carstensen J, and Hasse G 2002 *Materials Science and Engineering R: Reports* **39** 93–41.
[9] Barillaro G, Nannini A, Piotto M 2002 *Sensors and Actuators A* **102** 195–01.
[10] Lehmann V and Full H 1990 *J. Electrochem. Soc.* **137** 653-9.
[11] Gavrilov S A, Karavanskii V A, Sorokin I N 1999 *Russian Journal of Electrochemistry* **35** 729-34.
[12] Ashruf C M A, French P J, Bressers P M M C, Sarro P M, and Kelly J J 1998 *Sens. Actuators A: Phys.* **66** 284-91.
[13] Kolasinski K W 2014 *Nanoscale Res. Lett.* **9** 432-40.
[14] Pyatilova O V, Gavrilov S A, Shilyaeva Yu I, Pavlov A A, Shaman Yu P, and Dudin A A 2017 *Semiconductors* **51** 173–7.
[15] Qu Y, Liao L, Li Yu, Zhang H, Huang Yu, and Duan X 2009 *Nano Lett.* **9** 4539-43.
[16] Lee Ch-L, Tsujino K, Kanda Yu, Ikeda Sh, and Matsumura M 2008 *J. Mater. Chem.* **18** 1015-20.
[17] Huang Z, Geyer N, Werner P, Boor J, and Gösele U 2011 *Adv. Mater.* **23** 285-08.
[18] Volovlikova O V, Gavrilov S A, Sysa A V, Savitskiy A I, Berezkina A Yu 2017 *ElConRus* 1213-6.
[19] Yue Z, Shen H, Jiang Y et al. 2014 *J Mater Sci: Mater Electron* **25** 1559-63.
[20] Bracco G, Holst B 2013 Surface Science Techniques (Springer) 1 chapter 3-34.
[21] Yilei Zh 2007 Retrospective Theses and Dissertations. 15934.
[22] http://www.openscience.ru/chem/index.php?page=wetting&item=003.
[23] Zhang T, Zhang P, Li Sh, Li W, Wu Zh and Jiang Y *Nanoscale Research Letters* 2013 **8** 351-6.
[24] Jiang L, Li S, Wang J, Yang L, Sun Q, and Li Zh 2014 *Journal of Nanomaterials*, ID 526149, 6 pages.

Synthesis of nanocrystalline α-Fe2O3 by using thermal oxidation of Fe Films

G FORTAS[1,2*], I Saidoun[3], H Abboud[3], N Gabouze[2], N Haine[1], A Manseri[2], M Zergoug[4], H Menari[2], S Sam[2], H Cheraga[2] and I Bozetine[2]

[1]Faculté de Physique, Université des Sciences et de la Technologie Houari Boumediene, BP 32 El Alia 16111 Bab Ezzouar Alger, Algérie.
[2]Centre de Recherche en Technologie des Semi-conducteurs pour l'Energétique, 02 bd Frantz Fanon, Alger, Algérie.
[3]Faculté de Génie des Procédés et Mécaniques, Université des Sciences et de la Technologie Houari Boumediene, BP 32 El Alia 16111 Bab Ezzouar Alger, Algérie.
[3]Centre de Recherche en Technologie Industrielles, Route de Dely Ibrahim, BP 6416002 Cheraga, Algérie

E-mail: fortasghania@crtse.dz

Abstract. α-Fe_2O_3 hematite films were prepared by thermal oxidation from Fe films electroplated on silicon. Electrodeposition of Fe thin films was carried out from a sulfate bath containing an ammonium chloride complexing agent. The electrochemical study was performed by cyclic voltammetry. The SEM analysis of the films obtained at a -1.3 V constant polarization shows dendritic grains in the form of islet. The DRX spectra exhibit characteristic iron peaks according to the face centered cubic (Fcc) structure. These samples were annealed. At a temperature of 650 ° C, a single iron oxide phase was well formed, with the hematite structure. The SEM photos show a well-assembled columnar structure with formation of nano-wires at the surface of the deposit. The absorbance spectra reveal an absorption features in the ultraviolet range

1. Introduction

Hematite (α-Fe_2O_3) is a promising material for potential applications in spin electronics [1], catalysis [2], photocatalysis [3], in the fields of biology and biomedical [4], gas sensors and photodetectors energy-related applications. It has attracted a great attention because to its chemical stability in aqueous solutions for water splitting, the abundance in nature, a suitable band gap (Eg) of 2.14 eV and low fabrication cost [5]. In the literature, various α-Fe_2O_3 morphologies with different shapes have been reported, including α-Fe2O3 nanoparticles [6], nanowires [7], nanorods [8] and nanotubes [9]. Different approaches have been used for the synthesis of α-Fe_2O_3 nanostructures with higher surface to volume ratios, sol-gel process [10], thermal oxidation of iron foil [11] and solvothermal and hydrothermal [12]. α-Fe_2O_3 is also produced from electrochemical anodization [13] and electrodeposition [14- 15]. Moreover, it is reported that the films obtained with the electrochemical method possess relatively low band gap [16], which has a great interest in solar cell applications. Recently, Hanedar and co-workers [15] prepared grass-like nanostructured α-Fe_2O_3 photoelectrodes for the first time through a one-step new electrochemical method from an oxygenated aqueous solution of Fe^{3+}at room temperature without using surfactants, which are required in most applications, such as sensors and energy conversion and storage. Dalchiele et al. [16] used a different electrochemical approach based on a potential cycling and a potential pulsed electrodeposition procedure to fabricated α-Fe_2O_3 electrodeposits. Also, Fe_2O_3 is one of the most promising anode materials for the next generation lithium ion batteries (LIBs) to substitute graphite, demonstrated excellent reversible properties and high theoretical capacity (1007 mAh/g) [17]. The present investigation concerns iron oxide thin films formed by oxidation of iron film. The iron film is firstly deposited directly on the silicon surface by electrochemical method. This method allows to optimize deposition condition parameters such as bath composition and deposition time to get good iron films. Further thermal oxidation of iron films is performed at temperature 650 C°.

2. Experimental

The electrodeposition of Fe was carried out from an electrolyte bath containing aqueous sulfate solution with Ammonium Fluoride and Saccharin additives. The bath solution was composed of 0,4M H_3BO_3, 0,4 M NH_4Cl, 0.01M Saccharin and 0,1 M $FeSO_4$. The distance between anode and cathode was 0.5 cm. The cathode used was a silicon (n^+-Si (100)) a with a resistivity ρ=0.02-0.04 Ω.cm-^{1}. The platinum was used an anode electrode. The electrodeposition was performed using Biologic VMP 3 potentiostat with a constant potential at -1,3 V (vs. Ag/AgCl) for different deposition times. The morphologies of the Fe and α-Fe_2O_3 films were obtained by using scanning electron microscope (SEM). X-ray diffraction (XRD) (X' Pertpro PHILIPS with Cu K_ radiation, was used to study the structure of Fe and α-Fe_2O_3 films. The measurements of magnetic properties were made using a vibrating sample magnetometer (VSM) with the applied field parallel and perpendicular to the surface of sample. The optical properties of the deposits on the ITO-coated quartz plates were studied by recording the absorption spectra using a CARY 500-VARLAN UV-vis-NIR spectrometer at room temperature.

3. Results and discussion

3.1 Electrochemical study of Fe Films

The electrochemical behavior of Fe^{2+} on Si electrode in solution is further investigated by the cyclic voltammetry in 0.4 M H_3BO_3, 0,1M $FeSO_4$, as shown in Fig. 2 ((**a**): **solid line**)). The pH of solution is adjusted to 1.54

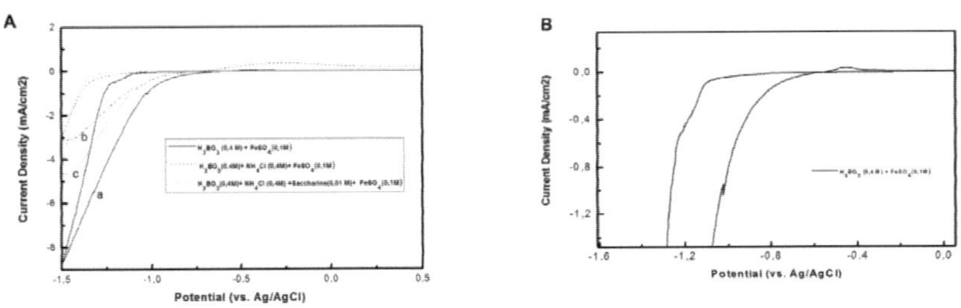

Figure 1.A) Cyclic voltammety of deposit solution: ((**a**): **solid line**) without additive; ((**b**): **dash line**) with additive NH_4Cl (0.4M); ((**c**): **dot line**) with additive NH_4Cl (0.4M), Saccharin (0.01M) pH=1,54, v = 10 mV/. B) the zoom of Fig. 2((**a**): **solid line**)).

The analysis of the (I-V) curve reveals clearly the oxidation-reduction process of the Fe^{+2}/ Fe ions (Fig. 1(a) and the zoom of Fig. 1(a)). It can be noted that the reduction reaction of Fe^{+2} ferric cations is predominant in the potential range [-1- -1.2 V (vs. Ag/AgCl)], characterized by a current plateau. Beyond -1.2 V (vs. Ag/AgCl), an increase in the fast cathodic current is observed, traducing the competition between the reduction reaction of the ferric ions and that of hydrogen. From the anodic peak intensity (the zoom of Fig. 1(a)), it can be deduced that a small amount of iron has been deposited on the Si electrode with an important release of hydrogen. Fig. 1((**b**): **dash line**) shows cyclic voltammetry of 0.1M $FeSO_4$, 0.4 M H_3BO_3 and 0.4 M NH_4Cl at pH values (1.54). In the presence of 0.4 M NH_4Cl in solution, an important cathodic and anodic currents are recorded corresponding to the reduction-oxidation of Fe^{2+}/Fe and $FeCl^+$/Fe with a displacement of the cathodic current towards the negative potentials, according to the following reactions.

- $Fe^{2+} + 2e^- \rightarrow Fe$ (1)

- $FeCl^+ + 2e^- \rightarrow Fe + Cl^-$ (2)

Fig. 2 illustrates the iron speciation pattern in the presence of the chloride ligand in the aqueous medium.

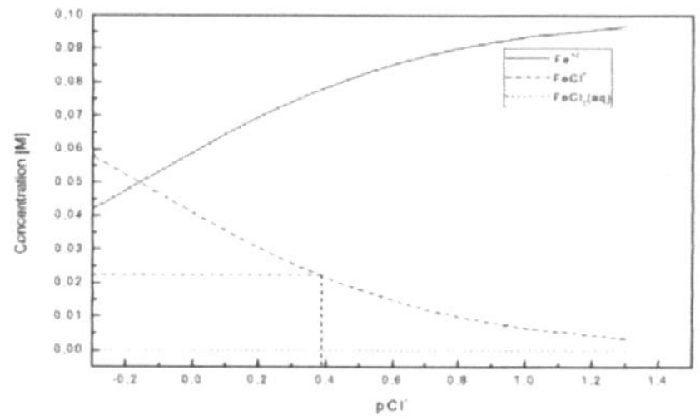

Figure 2. Distribution diagram of iron complexes in the chloride milieu, as a function of pCl-.

The total concentration (Fe_{Tot}) is equal to 0.1 M and the total chlorine concentration is 0.4 M (NH_4Cl). The total concentration of different (Fe (II)) complexed and uncomplexed electro-active species in the chloride medium is given by the iron mass conservation equation as follows:

- $[Fe(II)]_{Tot} = C = [Fe^{+2}] + [FeCl^+] + [FeCl_2(aq)]$ (3)

On the other hand we have:

- $\beta_1 = \frac{[FeCl^+]}{[Fe^{+2}][Cl^-]} = 10^{-0.6}$ [18] (4)
- $\beta_2 = \frac{[FeCl_2]}{[Fe^{+}][Cl^-]^2} = 10^{-8.17}$ [18] (5)

This is given:

- $[Fe^{+2}] = \frac{C}{1+\sum_{i=1}^{i=2}\beta_i[Cl^-]^i}$; $[FeCl^+] = \frac{C\beta_1[Cl^-]}{1+\sum_{i=1}^{i=2}\beta_i[Cl^-]^i}$; $[FeCl_2(aq)] = \frac{C\beta_2[Cl^-]^2}{1+\sum_{i=1}^{i=2}\beta_i[Cl^-]^i}$ (6)

According to the iron speciation diagram, Fe^{+2} ions are the most predominant species in the solution. The percentage of species present in the deposition bath for 0.39 pCl is given in Table. 1. It is found that the presence of the Cl^- ions in the electrolytic bath leads to the adsorption of the ion $FeCl^+$ on the silicon surface which will subsequently slow down the migration of the protons towards the silicon surface.

Table 1. Percentage of complex and uncomplexed electroactive species of iron in chloride medium for pCl = 0.39.

electro-active specie	Fe^{+2}	$FeCl^+$	$FeCl_2$
molar Percentage %	78	22	0

The influence of the simultaneous addition of 0.1 M saccharin and 0.4 M (NH_4Cl) in sulfate bath has been studied, the cyclic voltammetry is given in Fig. 1((c): **dot line).** In the presence of saccharin, the cathodic current density is increased and the iron dissolution peak increased also. This can be attributed to the adsorption of Fe^{+2} complexed with saccharin and Cl^- on the surface of the electrode.

3.2 Morphological observations

Fig. 3(a-b) and (c-d) show the SEM images of the electrodeposited Fe film at -1.3 V (vs. Ag/AgCl) for 1 min and 2 min, respectively. All Fe films reveal dendritic grains in the form of islet, with an average size of 100 nm. By increasing of deposition time, the thickness of Fe layers increases to 350 nm for 2min.

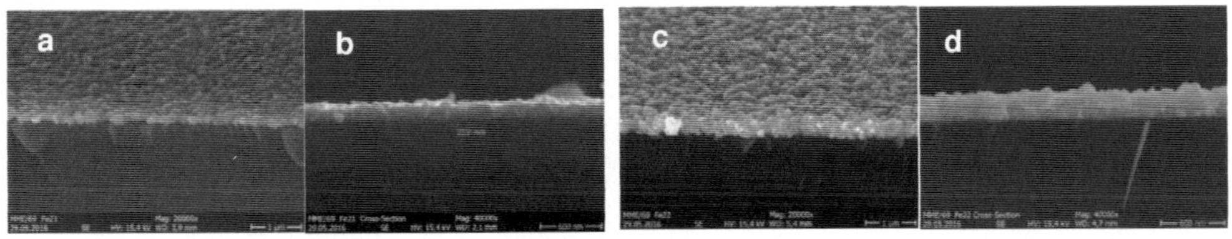

Figure 3. SEM images of iron deposition on silicon substrate at different deposition times for a potential of -1.3 V (vs. Ag/AgCl); (a and b) 1min; (c and d) 2min.

Fig. 4 shows the XRD patterns of the as-prepared Fe at -1.3 V(vs. Ag/AgCl) for 1, 2 and 3 min. It bring out the characteristic diffraction peaks which can be indexed to (110) plane of Fe (cfc) phase (JCPDS, card no.00-001-1267). The presence of weak characteristic diffraction peaks at 2Θ= 46,3° (004) and 56,4° (423), corresponding to the formation of $FeSi_2$ phase and a sharp diffraction peak at 2Θ= 61,98° (221), corresponding to the FeSi phase were observed. Fig .5 exhibits also the characteristic diffraction peaks, corresponding to Si substrate used as a working electrode which can be identified (JCPDS, card no. 01-080-0018).

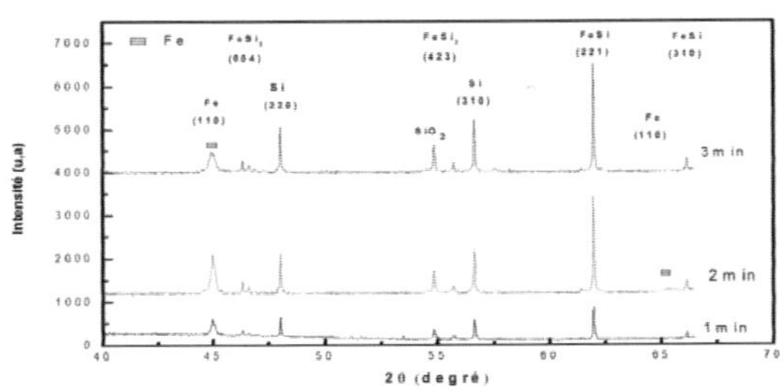

Figure 4. X-ray diffraction pattern of iron deposits obtained at -1.3V (vs. Ag/AgCl), for different deposition times.

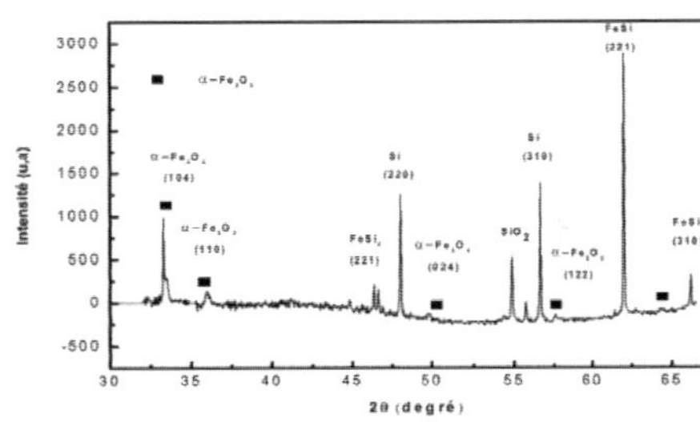

Figure 5. X-ray diffraction pattern of iron deposit obtained at -1.3 V (vs. Ag/AgCl), 2 min, after annealed at 600◦C for 10 h

4. Thermal oxidation of Fe Films

The electrodeposited Fe films were annealed in a furnace at 650◦C for 5 h30 in air environment. The formation of single oxide α-Fe_2O_3 nanostructures was confirmed by the XRD pattern displayed in Fig.5. All new diffraction peaks were attributed to the rhombohedral structure of α-Fe_2O_3 (JCPDS no. 33-0664).

4.1 Morphological observations of α-Fe_2O_3

The effect of Fe film electrodeposition time, after annealing at 650 °C for 5h 30 on morphology of the α-Fe_2O_3 was investigated (Fig.6). As shown in Fig 6(a), for 1min of electrodeposition, the formation of well-assembled columnar structure with the growth of nanowires on the hematite surface layer α-Fe_2O_3 is clear. When the electrodeposition time is increased to 2min, the SEM image Fig 6(b) reveals that the α-Fe_2O_3 films have a similar morphology as that to 1 min of electrodeposition. Fig. 6 indicates obviously that the thickness of the α-Fe_2O_3 depends on the electrodeposition time of Fe film and it is higher of these Fe films as a result to the diffusion of oxygen inside the Fe lattice during thermal oxidation, producing an elongation of lattice.

IOP Conf. Series: Journal of Physics: Conf. Series **987** (2018) 012040 doi:10.1088/1742-6596/987/1/012040

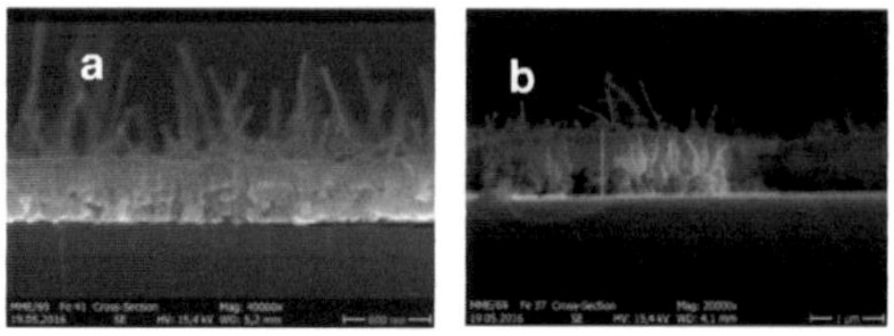

Figure 6. : SEM images in cross-section of thin layer of α-Fe_2O_3 after annealed at 650 °C for 5h 30, of iron films deposit (a and b) 1min; (c and d) 2min.

4.2 Optical properties

Fig. 7 shows the absorbance spectrums of the α-Fe_2O_3 thin layers produced by thermal oxidation at 650°C for 5 h30 of iron films for deposition time 1 and 2 min. The figure depicts the variation of the absorbance spectrum as a function of the hematite thickness layer obtained. An increase in the absorbance is observed (Fig. 7) when the thickness layer is increased.

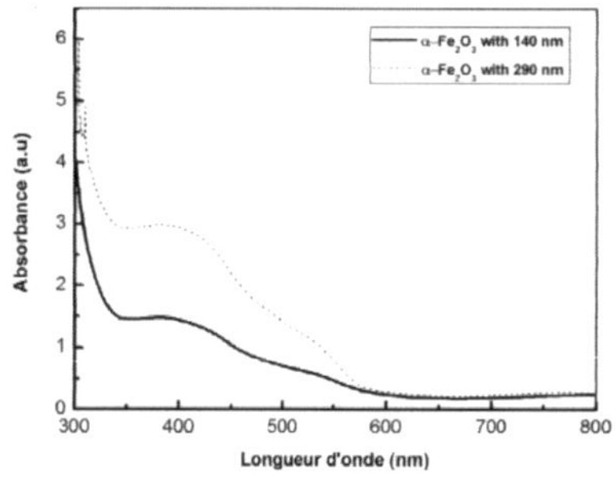

Figure 7. Variation of the absorbance spectrum as a function of the thickness α-Fe_2O_3 films

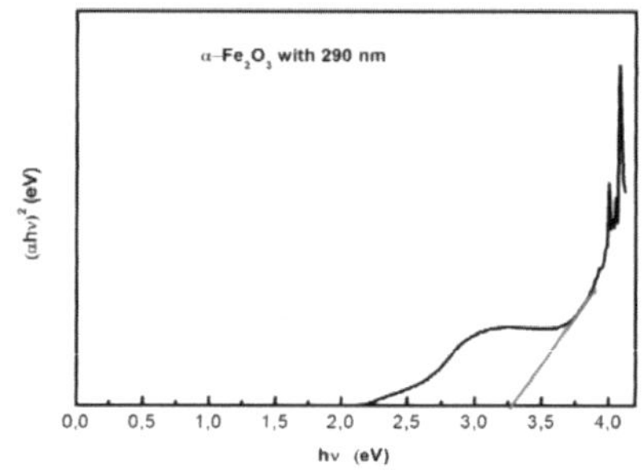

Figure 8. Estimated band gap of α-Fe_2O_3 films after annealed at 650 °C for 5h 30 of iron films 2min.

The plot of $(\alpha h\nu)^2$ versus photon energy (hν) for iron oxide thin film is presented on Fig 8. The optical band gap energy of the α-Fe_2O_3 is determinated by plotting the $(\alpha h\nu)^2$ as a function of (hν). α-Fe_2O_3 film possess a direct band gap transition. The optical band gap energy is found 3.2 eV for α-Fe_2O_3 thickness of 290 nm, which is higher to those reported values for α-Fe2O3 nanostructures [16, 17].

α-Fe_2O_3 photoanodes has been extensively studied by several groups for use in solar water splitting [17, 19]. The photoelectrochemical current densities of the α-Fe_2O_3 nanostructures electrodeposited onto ITO electrode were carried out by linearly sweeping the potential from 0.0 to 1.8 V (vs. Ag/AgCl) with a scan rate of 10 mV/s in 0.1 M KOH aqueous solution, in the dark and under illumination. The (J-V) curves under illumination shows a significant increase in the photocurrent at low bias due to the photogeneration of electrons and holes, Fig. 10.

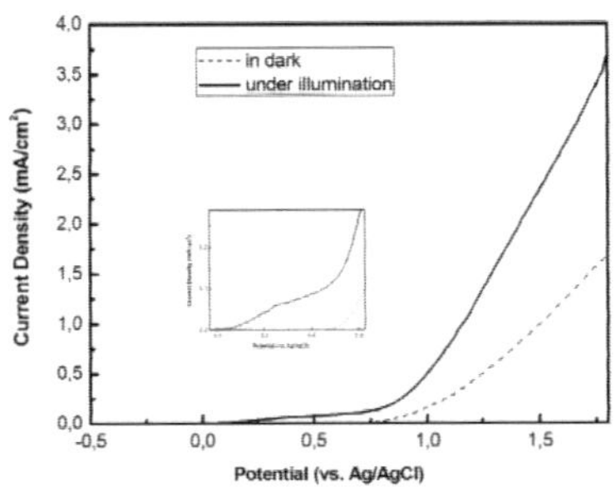

Figure 9. Photocurrent-voltage characteristics in 0.1 M KOH with AM 1.5 illumination (100 mW/cm^2) of the as-electrodeposited α-Fe_2O_3.

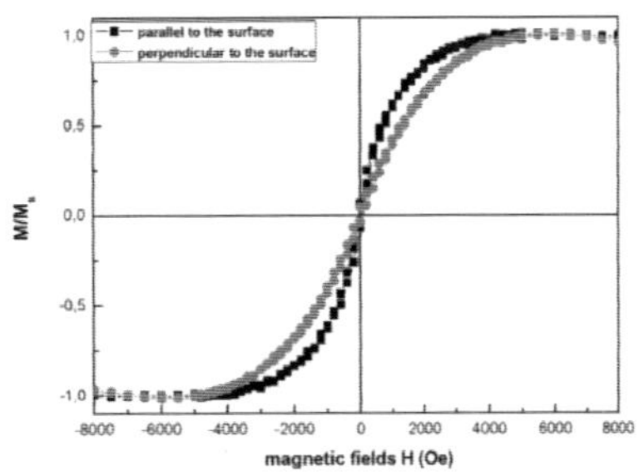

Figure 10. (M/M_s–H) loops of α-Fe_2O_3 film, after annealing at 650 °C for 5h 30, of iron film deposit 1min.

The photocurrent density of the columnar α-Fe_2O_3 was 0.5 mA/cm^2 at 1 V(vs. Ag/AgCl) and it reaches to approximately 3.5 mA/cm^2 at 1.8 V (vs. Ag/AgCl). By comparison with other works, the characteristic in the dark (J-V) demonstrates a high current density at less positive potential with respect to that obtained by Y. Hanedar et al [17] of grass-like nanostructured hematite and by R. Lui et al [19] of Au nanoparticles incorporated into α-Fe_2O_3 platelels by plasmonic. Under light illumination, the α-Fe_2O_3 columnar structure shows a remarkable photocurrent improvement, this result is comparable to that R. Lui et al [19]. It was found that a various design of morphology can be affected an enhancement of hematite photoelectrodes by exploiting a plasmonic [19].

4.3 Magnetic properties

Fig. 10 shows the magnetic hysteresis curves recorded at room temperature for α-Fe_2O_3 film, after annealed at 650◦C for 5 h30 of Fe sample (1min and -1.3 V (vs. Ag/AgCl)), where '//' and '⊥' represent the applied field H is parallel and perpendicular, respectively, to the surface of sample. The different shapes of magnetic hysteresis loops in parallel and perpendicular directions indicate that the arrays exhibit a small anisotropy of sample with squareness ratios (*Mr*/*Ms*) (//) about 0.06 and 0.04 for *Mr*/*Ms* (⊥). The results indicate that the coercivity *Hc* (//) is about 56 Oe, which is small than the coercivity *Hc* (⊥) 85 Oe. From the perpendicular and the parallel M-H curves, we can deduce that the α-Fe_2O_3 film has soft magnetic properties.

5. Conclusion

Columnar structure α-Fe_2O_3 thin films have been fabricated onto silicon substrate using a two-step electrochemical method and thermal oxidation of iron films. Optimized preparative parameters using cyclic voltammetry leads to iron films with dendritic grains in the form of islet. After thermal oxidation, the DRX result demonstrated that a single phase has been formed with the hematite structure. The UV-Visible characterization showed that the thin hematite film has been a direct gap of Eg = 3.2 eV for thickness 290 nm. Annealed film exhibits soft magnetic properties with a small ratio Mr/Ms of 0.06. The columnar structure α-Fe2O3 photoanode showed significantly photocatalytic performance.

References

[1] Janisch R, Gopal P, Spaldin 2005 *J. Phys. Condens Matter* **17** 657
[2] Li C, Shen Y, Jia M, Sheng S, Adebajo M O, Zhu 2008 *J. Catal. Commun* ***9*** *355*
[3] Huang M C, Chang W S, Lin J C, Chang Y H, Wu C C 2015 *J. Alloys. Comp* ***636*** *176-182*
[4] Kim J S, Yoon T J, Kim B G, Park S J, Kim H W, Lee K H, Park S B, Lee J K, Cho 2006. *J. Tox Sci,* ***89*** *338*
[5] Sivula K, Formal F L, Grätzel M 2011 *J. Chem Sus Chem* ***4*** *432-49*

[6] Zhang H, Wang W W, Li H F, Meng S L, Li D Q 2008 *J. Mater Lett 62 1230–3*
[7] Pascal C, Pascal J L, Favier F, Moubtassim M L, Payen C. 1999 *J. Mater Chem **11** 141–7.*
[8] Lian S, Wang E, Kang Z, Bai Y, Gao L, Jiang M, Hu C and Xu L 2004 *J. Solid State Commun **129** 485–90*
[9] Wen X G, Wang S H, Ding Y, Wang Z L, Yang S H. 2005 *J. Phys Chem **109** B 215–20*
[10] Watanabe A, Kozuka H 2003 *J. Phys Chem **107** B 12713-12720*
[11] Liao L, Zheng Z, Yan B, Zhang J X, Gong H, Li J C, Liu C, Shen Z X, Yu T 2008 *J. Phys Chem **112** C 10784-10788*
[12] Du M, Xu C, Sun J, Gao L 2012 *J. Electrochim Acta **80** 302-307*
[13] Wang L, Fei T, Lou Z., Zhang T 2011 *J.ACS Appl. Mater Interfaces **3** 4689-4694*
[14] Kulkarni S S, Lokhande C D 2003 *Mater Chem and Phys **82** 151-156*
[15] Hanedar Y, Demir U, Oznuluer T 2016 *J. Superlattices Microstruct 98 371-378*
[16] Schrebler R S, Altamirano H, Grez P, Herrera F V, Munoz E C, Ballesteros L A, Cordova R A, Gomez H, Dalchiele EA 2010 *J. Thin Solid Films 518:6844–6852*
[17] Park Y, Oh M, Park J S, Baek S H., Kim M, Kim S, Kim J H 2015 *J. Carbon **94** 9-17*
[18] Skoog D, Analisis Instrumental 1989 McGraw Hill México
[19] Liu R, Yang X, Anfuso C L, Huang Z and Han L 2001 *J. Rev. Adv Sci Eng **4** 1-9*

Acknowledgements: This work was financially supported by the DGRSDT/MESRS, Algeria.

IOP Conf. Series: Journal of Physics: Conf. Series **987** (2018) 012041 doi:10.1088/1742-6596/987/1/012041

Simple morphological control over functional diversity of SERS materials

A A Semenova[1] **and E A Goodilin**[1-4]

[1] Department of Materials Science, Lomonosov Moscow State University, 1 Leninskie gory, building 73, Moscow 119991, Russian Federation
[2] Kurnakov Institute of General and Inorganic Chemistry of the Russian Academy of Sciences, 31 Leninsky prospect, Moscow 119991, Russian Federation
[3] Department of Chemistry, Lomonosov Moscow State University, 1 Leninskie gory, building 3, Moscow 119991, Russian Federation
[4] Author to whom any correspondence should be addressed: goodilin@yandex.ru

Abstract. Nowadays, surface-enhanced Raman spectroscopy (SERS) becomes a promising universal low-cost and real-time tool in biomedical applications, medical screening or forensic analysis allowing for detection of different molecules below nanomolar concentrations. Silver nanoparticles and nanostructures have proven to be a common choice for SERS measurements due to a tunable plasmon resonance, high stability and facile fabrication methods. However, a proper design of silver-based nanomaterials for highly sensitive SERS applications still remains a challenge. In this work, effective and simple preparation methods of various silver nanostructures are proposed and systematically developed using aqueous diamminesilver (I) hydroxide as a precursor.

1. Introduction

The great number of effective and diverse ways of nanoparticle design is known nowadays [1, 2] thus allowing various practical applications from catalysis to the development of new innovative analytical methods based on the Surface-enhanced Raman spectroscopy (SERS). In the latter, a major part of all the studies is biomedically oriented and demands for a high chemical purity of the prepared nanoparticles with no residuals of reagents, surfactants and other byproducts which can be cell toxic or could harm analytically representative measurements because of worsening of a signal-to-noise ratio, a signal background shape, reproducibility and sensitivity of the SERS analysis. A possible solution of this important problem is based on the application of aqueous diamminsilver (I) hydroxide as the only reagent containing volatile components favouring reduction of silver ions into pure metallic silver nanoparticles under gentle conditions [3–5]. In this paper, we present our experimental results demonstrating a flexibility of this approach leading to a number of SERS-active nanostructured materials as predetermined by processing conditions.

2. Materials and methods

Silver nanoparticles and nanostructures were prepared using several novel routes based on transformation of diamminsilver (I) hydroxide via formation of silver (I) oxide and pure silver under different conditions such as hydrolytic heating [3], aerosol pyrolysis [4,5] or reduction by H_2O_2 [6].

Silver nitrate ($AgNO_3$), sodium hydroxide (NaOH), ammonium hydroxide (NH_4OH, 30%), polyvinylpyrrolidone (PVP), hydrogen peroxide (H_2O_2, 10%), Rhodamine 6G (Rh6G), cyclohexanethiol were of high purity grade. In all the experiments, ultra-pure water with resistivity ca. 18 MΩ·cm (Milli-Q, Millipore) was used.

2.1. Hydrolytic heating
In a typical preparation route [3], 0.5 M aqueous NaOH solution was added dropwise to 20 ml of freshly prepared 0.1 M aqueous $AgNO_3$ solution until complete precipitation of black-brown silver (I) oxide. Then this as-prepared precipitate was washed with water and dissolved in 5 ml of 10% aqueous ammonia solution. The obtained transparent silver complex solution was filtered through Millex-LCR syringe driven filter units (Millipore, 0.45 μm pores). The aliquot of the resulting solution (0.1, 0.5 or 2.5 ml) was added to 50 ml portions of boiling water. After 20, 30 and 60 min of the decomposition reaction the aliquots of pure silver sols were collected and immediately cooled down and stored in a refrigerator at about 5 °C overnight.

2.2. Aerosol pyrolysis
In a typical experiment, 0.1 M aqueous NaOH was added dropwise to freshly prepared 0.01 M aqueous $AgNO_3$ until complete precipitation of the black-brown silver (I) oxide. The as-prepared oxide was thoroughly washed with water and dissolved in 10% aqueous NH_4OH to have 0.01 M solution of the diamminesilver (I) hydroxide. The obtained silver complex solution was filtered through Millex-LCR syringe driven filter units with 0.45 μm pores. In the ultrasonic silver rain (USR) deposition process, this initial ammonia solution of silver (I) oxide was nebulized into mist and 1–5 micron droplets were streamed either onto hot glass substrates heated up to 250–300 °C without preheating the mist and resulted in formation of silver rings [4] or into a hot zone (550–950 °C) of a tubular furnace followed by the condensation of the decomposition products at about 60 °C and formation of more complex shapes of silver nanoparticles [5]. The entire run of nanostructured substrate preparation demanded typically 30–60 min only.

2.3. Silver octahedral particles
The synthesis of silver octahedral particles was based on the protocol proposed originally by J.Fang, et al. [6] and customized to provide porous octahedra with a morphologically predetermined large number of hot spots. The first stage included the formation of silver (I) oxide octahedral particles by adding 0.5 M NH_4OH solution dropwise into 50 ml 100 mM aqueous solution of $AgNO_3$ in the presence of 0.1 mM of PVP solution as a specific face-selective surfactant until the mixture became transparent. After that, 0.3 ml of 2M NaOH aqueous solution was added to the colorless solution under vigorous stirring. The brownish-black precipitate immediately formed. The obtained silver (I) oxide was washed with distilled water and ethanol several times using centrifugation (4000 rpm, 10 min, Sartorius Sigma 3-30 K) and finally redispersed in ethanol. The second stage allowed the synthesis of porous facetted silver octahedral pseudomorphs by the reduction of silver (I) oxide solution by H_2O_2 and, unlike the initial protocol, by additional etching with ammonia solution. For this purpose, 10 ml of 10% H_2O_2 was added to the obtained silver (I) oxide under stirring followed by the injection of 10 ml of 0.5 M ammonia solution. The synthesized silver particles were collected by centrifugation (4000 rpm, 10 min) and washed with distilled water and ethanol.

2.4. Characterization
The obtained nanomaterials were comprehensively characterized by a transmission electron microscopy (TEM) and electron diffraction (ED) using LEO912 AB OMEGA (Carl Zeiss) electron microscope. The pristine sample microstructure was studied using a scanning electron microscope (Carl Zeiss NVision 40) at 0.5–5 kV accelerating voltage. UV-vis absorption spectra were recorded using the UV-vis spectrophotometer Lambda 35 (Perkin-Elmer) with an attached diffuse reflectance accessory. The obtained substrates were examined using Rigaku D/MAX 2500 (Japan) with a rotating copper anode (CuKa irradiation, 5–80° 2Q range, 0.02° step). Diffraction maxima were indexed using the PDF-2 database. Raman and SERS experiments were performed using an InVia Raman

microscope (Renishaw, UK) equipped with a 20 mW 514 nm argon laser and power neutral density filter (10%). All the spectra were collected using a 50x objective lens and 10–120 s of acquisition time. A silicon wafer was used for calibration.

3. Results

Decomposition of aqueous diamminesilver (I) hydroxide (figure 1) allows a wide variation of given morphologies of thus prepared materials controlled by simple changes of concentrations and thermal treatment procedures (figure 2). Mild heating of bulk or dispersed aqueous solutions in the form of aerosol drops as well as shape-preserving preparation of pseudomorphs results in pure silver nanoparticles or hierarchic silver nanostructures with following morphologies: spheres, rings, porous spheroids, cuboids, octahedra and more complex structures with multiple hot spots. All the prepared nanostructures demonstrate their high efficiency in SERS measurements with enhancement factors of about 10^4 – 10^8 and were used in a form of SERS-active substrates for a study of different biomolecules and pollutants (figure 3).

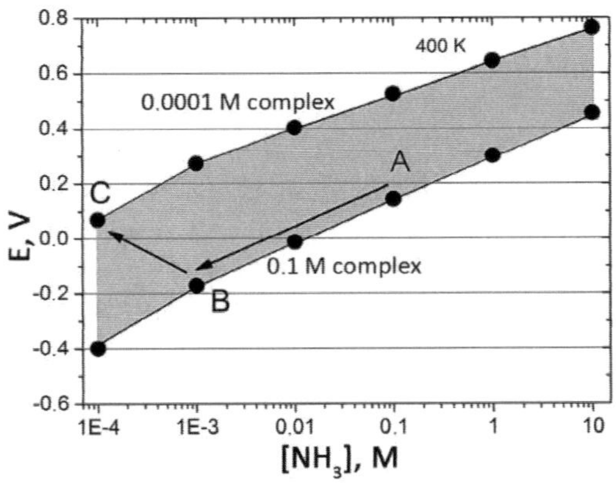

Figure 1. Calculated potentials for the reaction $Ag + 1/4\ O_2 + 2NH_3 + 1/2H_2O = [Ag(NH_3)_2]^+ + OH^-$ as a function of ammonia concentration in boiling water, the lower curve is equal to 0.1 M concentrations of $[Ag(NH_3)_2]^+$, the upper one is diluted, 10^{-4} M, solution of the complex.

4. Discussion

We consider a group of robust and effective processing methods of silver nanostructures supported by our previous works [3–5] which is based on gradual self-reduction of one of the most useful and well-known ammonia complex of silver, $[Ag(NH_3)_2]^+$, due to the lost of its stability upon the ammonia ligand elimination. This route plays with drastic stabilization of silver ions within the complex ($K_{stab}([Ag(NH_3)_2]^+) = 1.3 \cdot 10^7$, $Ag^+ + 2NH_3 = [Ag(NH_3)_2]^+$) and a huge variation of the redox potential of silver $E_o(Ag/Ag^+) = 0.8$ V in the presence of ammonia of different concentrations and also a possibility to precipitate the Ag_2O solid phase or dissolve it by ammonia ($Ag_2O + H_2O + 4NH_3 = 2[Ag(NH_3)_2]^+ + 2OH^-$). This method is characterized by the absence of side-anions in the redox system except OH^-, which is a natural part of all the aqueous solutions. Another part of the system is oxygen which can be dissolved in the solution as an oxidizing agent or evolved from the solution as a product of silver oxide decomposition, the standard potential of the pair O_2/OH^- is $E_o\ (O_2/OH^-) = 0.401$ V. That is why the shift of equilibrium in the overall process $Ag + 1/4\ O_2 + 2NH_3 + 1/2H_2O = [Ag(NH_3)_2]^+ + OH^-$ serves as a good control tool for generation of silver nanoparticles simply by changing the concentration of ammonia which could generate enough Ag^+ ions and cause their reduction to metallic silver otherwise Ag^+ is stabilized as $[Ag(NH_3)_2]^+$ and can not be reduced. The calculated data of the potentials are presented in figure 1. It is evident that even very small concentrations of ammonia makes the reaction to shift to the right side forming $[Ag(NH_3)_2]^+$, no silver nanoparticles could be formed, they will be dissolved due to stabilization of silver ions within the stable ammonia complex. However if the NH_3 ligand is lost for some reasons (heating and evolving from the solution), there is an area of negative potentials where diamminesilver (I) hydroxide is not stable thermodynamically and decomposes forming metallic silver which deposits as nanoparticles. The ABC pathway means a starting point of a high concentration of separately prepared $[Ag(NH_3)_2]^+$ in an excess of ammonia, then a sharp drop of the concentration of ammonia should happen, for example, due to the ligand evaporation with partial decomposition of the complex, and, afterwards, the process ends up at a very small concentration of $[Ag(NH_3)_2]^+$ and a higher concentration of bare silver ions.

The use of the aqueous diamminesilver (I) hydroxide to prepare silver nanostructures gives a large benefit owing to the simplicity, promptness and low temperature conditions of the production process (typically lower than 250 °C). In the case of pure silver hydrosols (figure 2, left – top), their formation occurs in bulk solutions and they represent 2–30 nm, mostly spherical, nanoparticles with a broad size distribution which is evidently connected with nearly equilibrium conditions of their nucleation and growth under supersaturation generated by the above described redox reactions due to a drop of ammonia concentration in hot water and its step-by-step lost.

Evaporation of water from aerosol droplets as microreactors together with ammonia ligands lost gives polyhedral silver nanoparticles in a cooling zone of a hot reactor at 50–70 °C (figure 2, right – top). Their etching with ammonia and water vapors results in shape-preserving transformation of silver (I) oxide single crystalline nanoparticles into aggregates of smaller silver nanoparticles of the same shapes. However, the difference in molar volumes of metallic silver and its oxide leads to shrinkage and also pore formation. As a result, porous polyhedral nanoparticles, pseudomorphs, are formed due to such topochemical transformations.

Silver octahedra preparation could be also performed in a more controllable manner (figure 2, left – bottom) although two conditions have to be obeyed: surfactants (PVP) should be added to confine crystallization of silver (I) oxide in alkaline media, also an appropriate reducing agent like ascorbic acid has to be used to provide fast redox transformation of silver (I) oxide into metallic silver since no elevated temperatures are used in such a procedure as compared to aerosol pyrolysis.

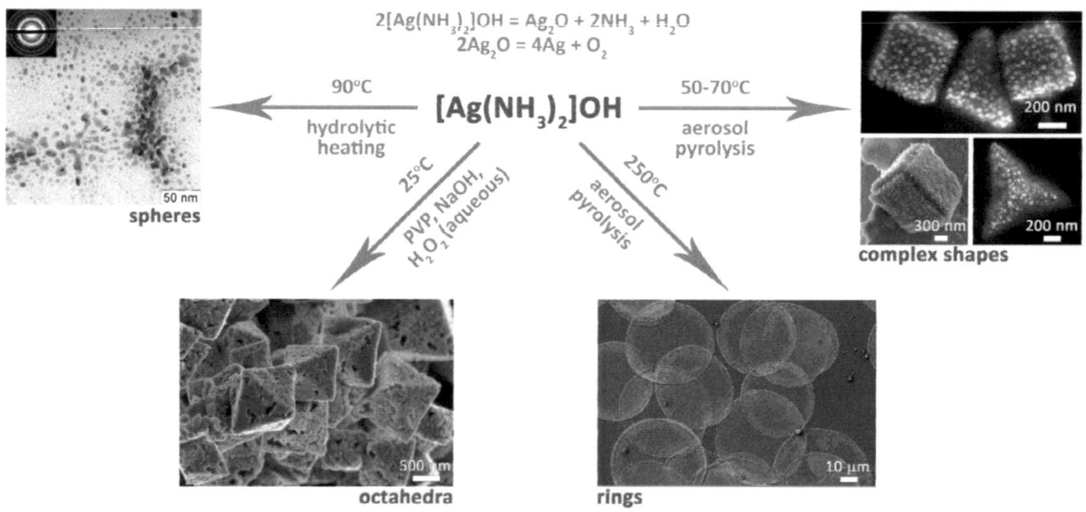

Figure 2. The diversity of silver nanostructures prepared from diamminesilver (I) hydroxide as an universal precursor of metallic silver.

The most nonequilibrium process among the depicted in figure 2 is aerosol spray deposition of hierarchic ring nanostructures onto substrates preheated up to 250 – 300 °C (figure 2, right – bottom). Such substrates after processing consist of overlapping pure silver rings of a complex morphology originating from decomposition of micron-sized droplets of ultrasonic mist of diamminesilver (I) hydroxide solution. Indeed, if a droplet of liquid dries on a solid surface, forming nanoparticles, the latter would be deposited in a ring-like fashion. During the drying process, droplet edges become pinned to the substrate, and capillary flow outward from the centre of the droplet brings suspended particles to the edge as evaporation proceeds. The particles are left highly concentrated along the original droplet edge. As soon as the formed nanoparticles are moved to the rim by the outward flow, their local attractions lead to the formation of loosely packed structures near the rim. The complexity of this "coffee ring" phenomenon yields unusual silver ring structures observed for the substrates. In our experiments, silver deposition gives intersecting circles of 30–100 microns in diameter. This value is several times larger than the expected size of 1–10 microns of the falling mist droplets therefore the liquid from the droplets spreads laterally over the substrate. Solvent evaporation increases the concentration of the silver complex and then metallic silver resides on rims of the spreading circles, producing walls of silver craters according to the reactions mentioned above. The thickness of walls typically lies in a range of 1–3 microns while the residual part is covered with silver nanoclusters gradually increasing in their sizes from 10–20 nm in the centers of circles to about 100 nm in the wall vicinities. The nanoclusters increase the role of capillary forces and, afterwards, more droplets stack onto the surface and boil; an increase of sputtering time leads to rough and porous layers. Most of the circles form groups or long chains until the surface becomes rather uniformly covered with stochastically intersecting silver rings. "Coffee ring" formation onto the substrates is accompanied by nanostructuring. When silver forms, large and smaller gaps appear in the region of rims and this gives random blocks ("bricks of the wall") of 200–500 nm. The bricks and the walls are composed of nanometer-sized interconnected silver particles because every act of droplet transformation on the substrate gives newborn silver nanoparticles of a very small size of several nanometers. Later, these nanoparticles grow, join each other and form channels, making the rims highly porous. Instant capillary transport of nutrient solution to the top of walls through the channels leads to the growth of 10–50 nm "sesame seed" grains of silver covering the bricks.

Formation of various nanoparticle morphologies is a highly suitable way to optimize materials for different desired SERS measurements. The analysis of enhancement coefficients among the nanomaterials demonstrated in figure 2 prove that the best SERS functions are revealed for silver rings and octahedral silver skeleton porous particles (figure 3).

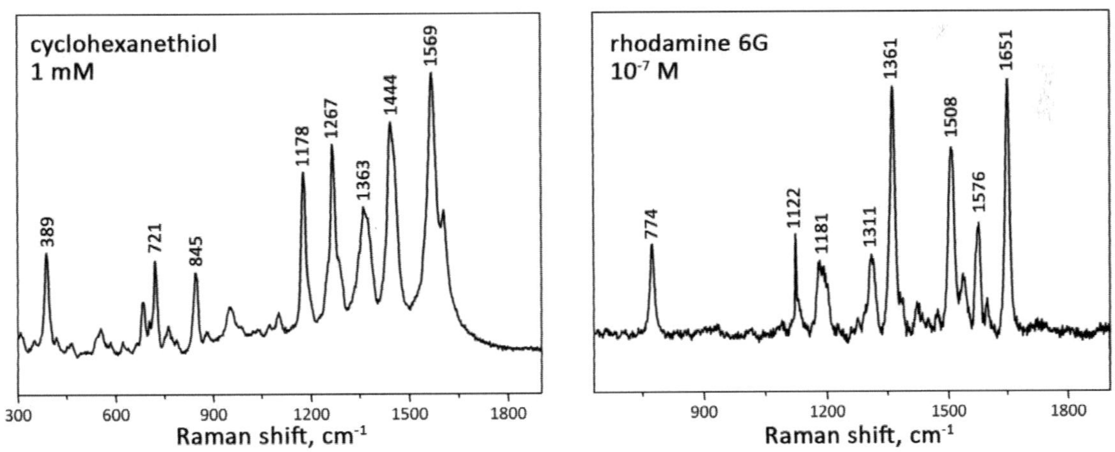

Figure 3. SERS spectra of a model dye rhodamine 6G and cyclohexanethiol recorded using silver ring substrates (right) and porous octahedral pseudomorphs (left), respectively.

This finding is not surprising since both the materials possess hierarchic structures allowing the development of numerous "hot spots" [1, 2] greatly enhancing the Raman spectral signal from molecular analytes. The spectra in figure 3 revealed the most important vibration modes of the analytes at submicromolar concentrations, such bright spectra with distinct and intense peaks make a strong proof that the suggested preparation method of SERS-active materials is effective and can be highly recommended for a practical use.

5. References

[1] Sharma B, Frontiera R R, Henry A-I, Ringe E and Van Duyne R P 2012 *Mater. Today* **15** 16–25
[2] Langer J, Novikov S M and Liz-Marzan L M 2015 *Nanotechnology* **26** 322001
[3] Semenova A A, Goodilin E A and Tretyakov Yu D 2011 *Mendeleev Commun.* **21** 312–4
[4] Semenova A A, et al 2012 *J. Mater. Chem.* **22** 24530–44
[5] Semenova A A, Ivanov V K, Savilov S V and Goodilin E A 2013 *CrystEngComm* **15** 7863–71
[6] Fang J, Liu S and Li Z 2011 Biomaterials **32** 4877–84

Acknowledgments

This research is supported by the Grant of the President of Russian Federation for the governmental support of young scientists (grant #MK-1146.2017.3). Authors thank A.E.Barantchikov and V.K.Ivanov for SEM analysis and fruitful discussions.

IOP Conf. Series: Journal of Physics: Conf. Series **987** (2018) 012042 doi:10.1088/1742-6596/987/1/012042

Characterization and Comparison of Photocatalytic Activity Silver Ion doped on TiO_2(TiO_2/Ag+) and Silver Ion doped on Black TiO_2(Black TiO_2/Ag^+)

Jin Yi Kim[1], Ho Hyung Sim[1], Sinae Song[2], Yeoung Ah Noh[2], Hong Woon Lee, Hee Taik Kim[1,2]*.

[1] Department of Fusion Chemical Engineering, Hanyang University, 55 Hanyangdaehak-ro, Sangnok-gu, Ansan Kyeonggi-do, Korea.

[2] Department of Advanced Materials Science and Engineering, Hanyang University, , 55 Hanyangdaehak-ro, Sangnok-gu, Ansan Kyeonggi-do, Korea.

khtaik@hanyang.ac.kr

Abstract. Titanium dioxide (TiO_2) is one of the representative ceramic materials containing photocatalyst, optic and antibacterial activity. The hydroxyl radical in TiO_2 applies to the intensive oxidizing agent, hence TiO_2 is suitable to use photocatalytic materials. Black $TiO_2$was prepared through reduction of amorphous TiO_2 conducting under H_2 which leads to color changes. Its black color is proven that absorbs 100% light across the whole-visible light, drawing enhancement of photocatalytic property. In this study, we aimed to compare the photocatalytic activity of silver ion doped on TiO_2(TiO_2/Ag^+) and silver ion doped on black TiO_2(black TiO_2/Ag^+) under visible light range. TiO_2/Ag+ was fabricated following steps. 1) TiO_2 was synthesized by a sol-gel method from Titanium tetraisopropoxide (TTIP). 2) Then $AgNO_3$ was added during an aging process step for silver ion doping on the surface of TiO_2. Moreover, Black TiO_2/Ag+ was obtained same as TiO_2/Ag^+ except for calcination under H_2. The samples were characterized X-ray diffraction (XRD), UV-visible reflectance (UV-vis DRS), and Methylene Blue degradation test. XRD analysis confirmed morphology of TiO_2. The band gap of black TiO_2/Ag^+ was confirmed (2.6 eV) through UV-vis DRS, which was lower than TiO_2/Ag^+(2.9 eV). The photocatalytic effect was conducted by methylene blue degradation test. It demonstrated that black TiO_2/Ag^+ had a photocatalytic effect under UV light also visible light.

1. Introduction

The Sol-gel synthesis method is one of the suitable methods for producing metal oxides having better purity and uniformity under specific conditions. In this synthesis method, necessary nanoparticles can be produced by a relatively simple method and synthesis can be performed even at a low temperature. Among the metal oxides obtained by this synthesis method, Titanium dioxide has optical effects, photocatalyst, antibacterial effect, etc. as one of the typical ceramic materials commonly used. It has a particularly excellent photocatalytic effect, but synthesizes superoxide anion and hydroxy radical which are activated by ultraviolet light and have high oxidizing power [1,3]. TiO_2 which is a semiconductor material having a strong oxidizing power has three forms of rutile, anatase and brookite, and generally has a band gap of 3.0 eV. In order to lower the band gap of TiO_2, which has a high oxidizing power and a relatively large bandgap energy, in this study, after synthesis by the same method, one is the general atmospheric atmosphere, And calcined in an H_2 gas atmosphere [2,3]. TiO_2 calcined in a hydrogen atmosphere draws black, unlike general TiO_2. Since black absorbs light, unlike white TiO_2 which is not good in the visible light region, it can be seen in the hope that it will absorb light in the visible light region to produce photocatalytic effect [4,5].

2. **Experimentals**

2.1. *Materials*

Titanium tetra-isopropoxide (TTIP) and Silver nitrate were purchased from Sigma-Aldrich (97%). Isopropyl alcohol was bought from Junsei Co. Ltd, Japan. Polyethylene glycol 600 (PEG 600) were procured from Daejung Chemical and Material Co Ltd, South Korea. The solution of 0.5 M Nitric acid (Daejung) was prepared by dissolving concentrated nitric acid in deionized (DI) water.

2.2. Synthesis of Ag-doped TiO_2

Ag-doped TiO_2 was synthesized using one-step sol-gel method process from TTIP. As following these steps: A basis solution was prepared by dispersing 2.0 g of PEG in 120 ml of Isopropyl alcohol. Then, 20 ml of TTIP was mixed with the basic solution using magnetic stirrer for 1 h, on 60 °C. The mixed solution was titrated with 0.5 M HNO_3 to pH 1,3,5 and 7 as well as the temperature was maintained. After titration, an Ag solution (0.2 M) that silver nitrate was dispersed in DI water was added to the titrated composite. And the composite was aged under constant stirring and heating for 6 h, keeping temperature. The aged composite was filtered by DI water and Ethanol for 3 times using an aspirator. The filtered composite was dried at 70 °C. Finally, the obtained powder was calcined at 550 °C for 2 h using tube furnace. An Ag-doped black TiO_2 was obtained from general Ag-doped TiO_2. The fabricating method significantly similar to the previous process except for final step. The substance which after drying was calcined at 550 °C for 2 h in the H_2 atmosphere. This powder was modified from white to black (Figure 1)

2.3. Photodegradation of Methylene blue

The photocatalytic activities of calcined Ag-doped TiO_2 and Ag-doped black TiO_2 was evaluated by measuring degradation rates of Methylene blue. A certain amount of sample was placed in a methylene blue aqueous solution (100 ml), and then light was irradiated with a halogen lamp. From the initial methylene blue solution, Abosrbance was measured using a UV-visible instrument at intervals of 30 minutes.

2.4. Characterization

The acquired material was characterized via a X-ray diffraction (XRD, Rigaku D/MAX-2500/PC) to decide crystallinity. The XRD was measured by using Cu Kα radiation in the 2θ range from 20° to 80 ° with a scanning rate of 4°/min. The Photocatalytic activity was confirmed through Methylene blue degradation. Lastly, it was able to calculate the band gap energy of synthesized nanoparticles from UV-vis diffuse reflectance spectroscopy (UV-vis DRS).

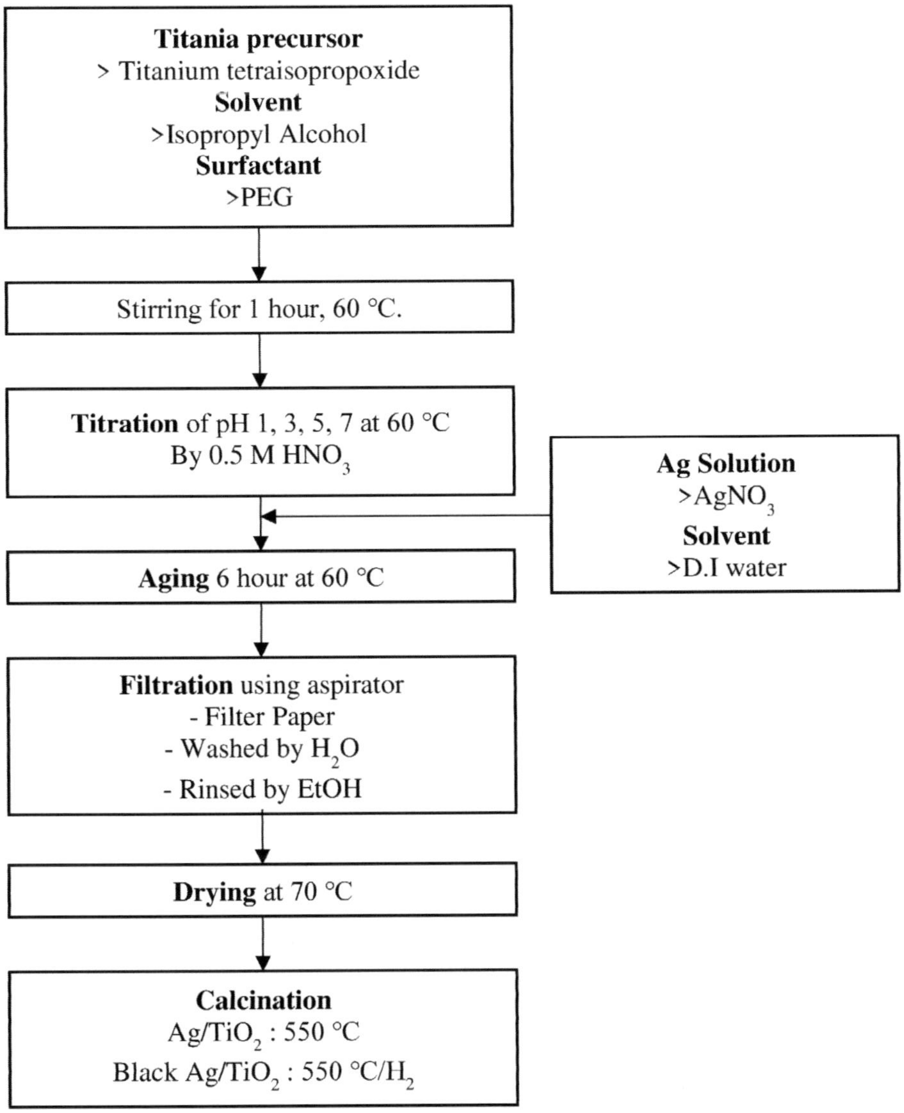

Figure 1. This flow chart showed synthesis method of two types Ag-TiO_2.

3. Results

3.1. Crystal structure and morpology

Through a series of experimental processes, White Ag-doped TiO_2 and black Ag-doped TiO_2 were synthesized. In Figure 2 you can see the nanopowder made from other colors. Nanoparticles synthesized under general air conditions showed white, and nanoparticles calcined in H_2 atmosphere gave black. As a result of measurement at these speeds of 4 ° per minute at a rate of 20 ° to 80 ° for structural analysis, the Anatase shape such as 550 ° C, whichever is lower, 1, 3, etc., the black Ag - TiO_2 is rutile day Anatase shape was shown. The higher the value of pH, the higher the percentage of rutile was shown. The crystallite size of particles through Scherrer equation is on average 13 - 15 nm for white TiO_2 and black TiO_2 about 31 - 35 nm in size. And also, it can be cofirmed through Figure 3

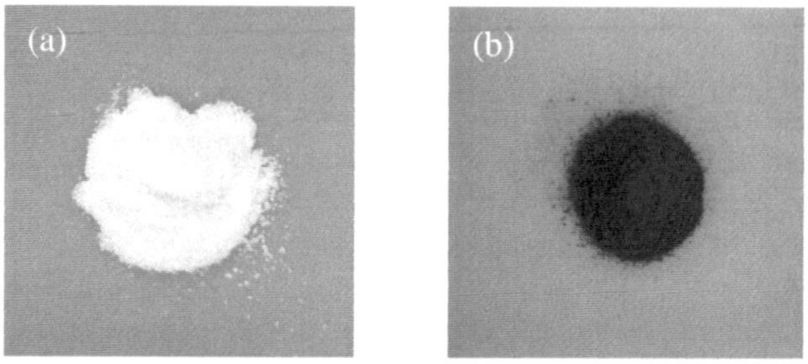

Figure 2. (a) white TiO_2/Ag^+, and (b) black TiO_2/Ag^+

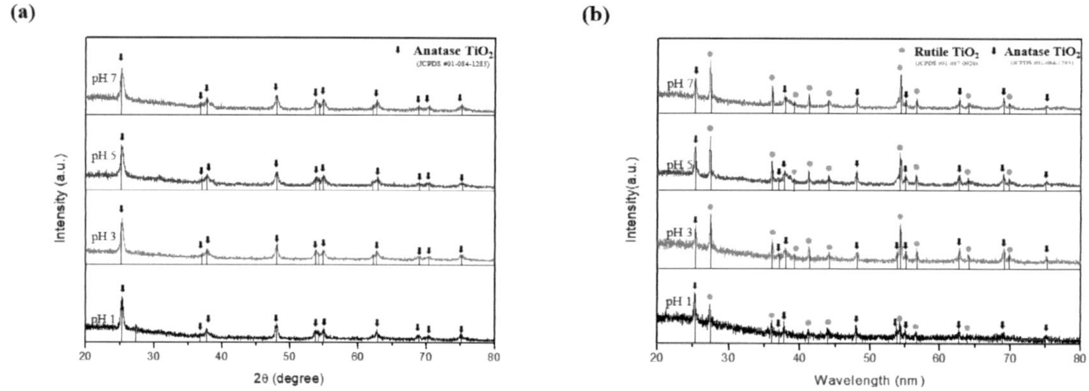

Figure 3. XRD analysis of Ag doped on TiO_2 ; (A) calcined at 550 °C, and (b) 550 °C in H_2 atmosphere

3.2. Photocatalytic studies

In order to investigate photocatalytic activity, methylene blue decomposition experiment was conducted. Place a specimen to investigate the photocatalytic effect in a solution diluted appropriately with methylene blue and irradiate the light through a light source. (Figure 4), white TiO_2 on anatase shows better decomposition reaction than black TiO_2. It was confirmed that black TiO_2/Ag^+ has a photocatalytic activity lower than that of general TiO_2/Ag^+ , but the photocatalytic activity also showed on the surface reduced black TiO_2/Ag^+. Figure 5 was showed degradation ratio from Figure 4, Although the reduction rate of (b) is gentle compared to (a), you can see that it is a shape that is decreasing compared with the initial stage. Photocatalytic activity can be confirmed also with this black Ag-doped TiO_2.

IOP Conf. Series: Journal of Physics: Conf. Series **987** (2018) 012042 doi:10.1088/1742-6596/987/1/012042

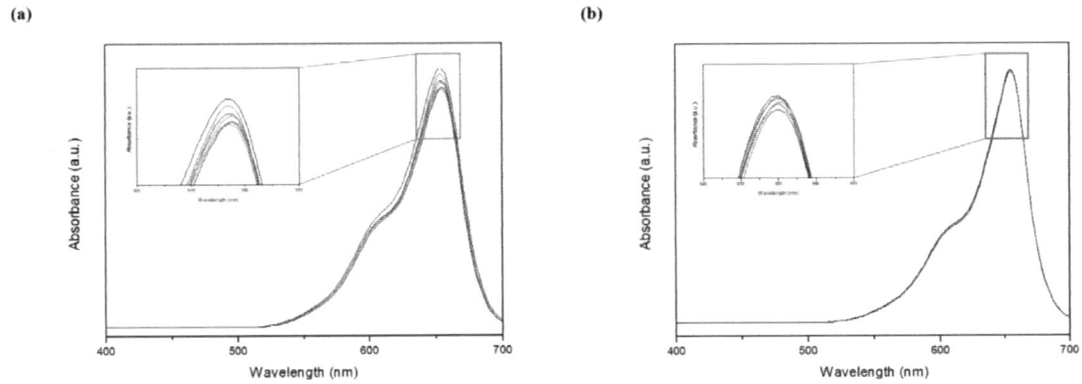

Figure 4. MB degradation analysis of (a) Ag doped TiO_2 calcined at 550 °C, and (b) calcined at 550 °C in H_2 atmosphere.

3.3. UV-visible diffuse reflectance spectreoscopy

The synthesized Ag-doped TiO_2 band gap was confirmed via UV-visible diffuse reflectance spectroscopy. Looking at Figure 5 (a), it can be confirmed that they represent pH 1, 3, 5, 7 in order and absorb a small amount of light even in the visible light region. Based on the results of Reflectance, the bandgap energy was calculated using the Kubelka-munk equation. (equation 1, 2).

$$F(R) = \frac{[(1-R)]^2}{2R} \text{ ------------------------------ } (1)$$

$$[F(R)hv]^{\frac{1}{n}} = A(hv - E_g) \text{ --------------------- } (2)$$

$\boldsymbol{R}$: *percentage of reflected light*
$\boldsymbol{h}$:*Plank constant* ($4.13566733 \times 10^{-15}$) *eV·s*)
$\boldsymbol{v}$:*Radiation frequency*
$\boldsymbol{E_g}$:*Bandgap*
$\boldsymbol{A}$:*Constant*
n : indirent bandgap = 2, direct bandgap=1/2
$\boldsymbol{\lambda}$:*Wavelength*
$\boldsymbol{c}$:*Light velocity* ($3.0 \times 10^8 m/s$)

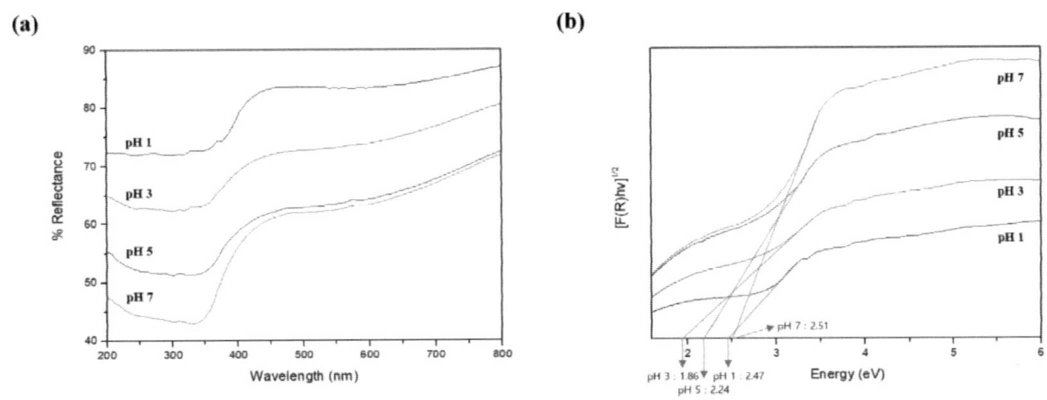

Figure 5. UV-vis DRS analysis of Ag doped TiO_2. : (a) calcined at 550 C, (b) calcined at 550 C in H_2 atmophere.

4. Conclusion

TiO_2/Ag^+ is one of the well-known ceramic materials for the photocatalytic activity. We synthesized TiO_2/Ag^+ with general TiO_2/Ag^+ (white TiO_2/Ag^+) and black TiO_2/Ag^+ to confirm the characteristics. Black TiO_2/Ag^+ shows the black part as the part of the table is reduced. In general, black TiO_2/Ag^+ has anatase and rutile mixed phase. In the degradation experiment of methylene blue, both photocatalytic activities of TiO_2/Ag^+ were shown. As a result, the white TiO_2/Ag^+ showed better photocatalytic characteristics, but the black TiO_2/Ag^+ absorbs light even in the visible light range.

Acknowledgments

This work was supported by the Human Resources Development program(No.20174030201830) of the Korea Institute of Energy Technology Evaluation and Planning(KETEP) grant funded by the Korea government Ministry of Trade, Industry and Energy.

5. References

[1] R. S. D, A. E, A. M and S. D, 2016, Optik - International Journal for Light and Electron Optics, **127**, 7143-54

[2] S. M, Amit kumar P, Thomas B, P. P. R, Misha H, T. M. L, P. R, V. P. N. N and C. P. G. V, 2012, *Journal of Fluorescence*, **22**, 1563-69

[3] Godlisten N. S, S. M. I, Sun Jeong J, Shin Jae K, S. M. H and Hee Taik K, 2015, *Appl. Surf. Sci.*, **351**, 1213-23

[4] Mingchun X, Youkun G, Elias Martinez M, Marinus K, Martin M, Yuemin W, Hicham I and Christof W, 2011, *Phys. Rev. Lett.*, **106**, 138302

[5] Wei Z, Wei L, Jian-Qiang W, Yang Q, Ying Y, Ying X, Kaifu Z, Lei W, Honggang F and Dongyuan Z, 2014, *J. Am. Chem. Soc.*, **136**, 9280-83

IOP Conf. Series: Journal of Physics: Conf. Series **987** (2018) 012043 doi:10.1088/1742-6596/987/1/012043

On low-dimensional models at NMR line shape analysis in nanomaterial systems

M M Kucherov[1] **and O V Falaleev**[2]

[1] Institute of Space and Information Technology, Siberian Federal University, 26B Kirenskogo St., 660074 Krasnoyarsk, Russia
[2] Krasnoyarsk Scientific Center, SB RAS, 50 Akademgorodok, 660036 Krasnoyarsk, Russia

E-mail: mkucherov@sfu-kras.ru

Abstract. We present a model of localized spin dynamics at room temperature for the low-dimensional solid-state spin system, which contains small ensembles of magnetic nuclei ($N \sim 40$). The standard spin Hamiltonian (XXZ model) is the sum of the Zeeman term in a strong external magnetic field and the magnetic dipole interaction secular term. The ^{19}F spins in a single crystal of fluorapatite [$Ca_5(PO_4)_3F$] have often been used to approximate a one-dimensional spin system. If the constant external field is parallel to the *c* axis, the 3D ^{19}F system may be treated as a collection of many identical spin chains. When considering the longitudinal part of the secular term, we suggest that transverse component of a spin in a certain site rotates in a constant local magnetic field. This field changes if the spin jumps to another site. On return, this spin continues to rotate in the former field. Then we expand the density matrix in a set of eigenoperators of the Zeeman Hamiltonian. A system of coupled differential equations for the expansion coefficients then solved by straightforward numerical methods, and the fluorine NMR line shapes of fluorapatite for different chain lengths are calculated.

1. Introduction

Our work is devoted to possibilities of one-dimensional models in the statistical theory of shape of the NMR absorption line. Recently, considerable progress has been made in studies on mechanisms responsible for the wings of a magnetic resonance line [1], but in absence of isolated groups of magnetic nuclei, a problem 'as old as NMR' of calculating the shape of the NMR absorption line is still one of the most difficult problems of solid state radiospectroscopy. The difficulties of the theory are compounded by the clear lack of experimental data: the absence of fine structure of the spectra makes them insufficiently critical in comparing theory and experiment. Therefore, considerable interest is associated with the experimental data obtained on the samples, in which the magnetic nuclei form low-dimensional lattice. Non-ergodic quasi-equilibria was also observed in the linear spin ½ chains [2].

We consider a system of *N* particles of spins ½ in a sample of volume *V*. The Hamiltonian for such a system in the rotating frame can be written as

$$H_R = H_{0,R} + \sum_{i \neq j}^{N} H^{(ij)} . \tag{1}$$

Here the double sum is taken over all N spins i and j forming a regular lattice. For like spins the unperturbed Zeeman Hamiltonian $H_{0,R}$ is given by

$$H_{0,\mathrm{R}} = \sum_{i=1}(\Omega_i - \omega_{\mathrm{rf}})S_i^z = \Delta\Omega\sum_{i=1} S_i^z \; . \tag{2}$$

For the case of like spins, one can eliminate $\Delta\Omega$ by performing the transformation into the interaction representation.

The shape of the absorption line plays an important role in NMR theory. However, in its calculation, in view of the absence of an explicit small parameter and the many-particle nature of the problem, it is necessary to make approximations based on the simplest estimates and physical intuition. One of such approaches [3–5] is based on the non-equivalence of interactions between longitudinal and transverse spin components in a truncated dipole Hamiltonian, obtained for $\xi = ½$ from a general Hamiltonian

$$H_{\mathrm{R}}^{(ij)} = H^{(ij)} = \gamma^2\hbar\frac{1-3\cos^2\theta_{ij}}{r_{ij}^3}\left(S_i^z S_j^z - \xi S_i^+ S_j^-\right), \tag{3}$$

where r_{ij} is the distance between the ith and jth spins, θ_{ij} is the orientation of the vector $\vec{r}_{ij}$, connecting the two spins with respect to the main magnetic field, $\xi = ½$ – magnetic anisotropy. Here we consider the high-field approximation and retain only the first term of the dipolar Hamiltonian. This separation follows from the axial symmetry of the Hamiltonian, by which the longitudinal component of the total spin moment of the system is conserved, and the transverse damped to zero. This leads to a difference several times between the dephasing time (T_2) of the transverse spin components and the slow variation of the longitudinal components (spectral diffusion). With increasing magnetic anisotropy ($\xi \to 0$), these differences are amplified.

A special case of the open chain is when we consider only the so-called "nearest-neighbour couplings"; there are $N - 1$ of them, all with $b = \dfrac{1-3\cos^2\theta_{i,i+1}}{r_{i,i+1}^3}$. The Hamiltonian above then simplifies to what we will refer to as our 1D NN model. The 1D NN case is known to be integrable (i.e. it has an exact analytical solution) via an algebraic method known as the Bethe Ansatz [6]. It is fair to say that in one dimension it is always important to consider boundaries, since physical systems only contain finite chains even in the absence of doping [7]. This is especially also true for artificially created spin chains using surface structures [8, 9], ion-traps [10], or ultra-cold gases [11–13] as quantum simulators, where measurements of spin correlations are in principle possible.

2. Theory and calculation

2.1. 1D dipolar model

Following direct-product formalism for calculating magnetic resonance signals [14], we seek the solution of the density matrix equation in the form

$$\rho(t) = \sum_{\{\varepsilon\}} g_{\{\varepsilon\}}(t) E_{\{\varepsilon\}} \, , \tag{4}$$

where the coefficients $g_{\{\varepsilon\}}(t)$ are functions of time, and the $E_{\{\varepsilon\}}$ are a set of basis operators with numbering scheme (ε), such that $H_{0,\mathrm{R}}E_\varepsilon = [H_{0,\mathrm{R}}, E_\varepsilon] = \mu\Delta\Omega E_\varepsilon$, where $H_{0,\mathrm{R}}$ is the Zeeman energy, μ is the difference between the number of operators S^+ and S^- in an eigenoperator $E_{\{\varepsilon\}}$, and $\Delta\Omega$ is the resonant frequency offset of spectral packet with respect to the oscillating rf magnetic field ω_{rf} . These eigenoperators are linearly independent and orthonormal.

$$\mathrm{Tr}\left(E_{\{\varepsilon\}}^\dagger E_{\{\varepsilon'\}}\right) = \delta_{\{\varepsilon\},\{\varepsilon'\}} \, , \tag{5}$$

For a system of N particles of spin ½, operators $E_{\{\varepsilon\}}$ can be written as a direct product of N spin operators corresponding to different particles

$$E_{\{\varepsilon\}} = \prod_{i=1}^{N} S_i^{(\varepsilon_i)}, \tag{6}$$

were $\varepsilon_i = \alpha, \beta, +,$ or $-$. S^+ and S^- are conventional raising and lowering spin operators and the polarization operators S^α and S^β. By using the above orthogonality property of the basis operators, one obtains a system of equations which couple the FID (free induction decay) components $g_{\{\varepsilon\}}(t)$. That is

$$\frac{\partial \rho_{\{\varepsilon\}}(t)}{\partial t} = -\mathrm{i}\sum_{i \neq j}^{N} C^{(ij)}_{\{\varepsilon\}\{\varepsilon'\}} F(\vec{r}_{ij})\, g_{\{\varepsilon'\}}(t). \tag{7}$$

The matrix $C^{(ij)}$ is defined by

$$C^{(ij)}_{\{\varepsilon\}\{\varepsilon'\}} F(\vec{r}_{ij}) = \mathrm{Tr}\left(E^\dagger_{\{\varepsilon\}}\left[H^{(ij)}, E_{\{\varepsilon'\}}\right]\right), \tag{8}$$

where the dagger † denotes the Hermitian conjugate. The explicit form of $C^{(ij)}$ clearly depends on the commutation relations between the basis operators and the Hamiltonian describing the interaction among the spin particles. In the case of the dipolar interaction, one obtains the following commutation properties for different types of pairwise combinations $S_i^{(\varepsilon_i)} S_j^{(\varepsilon_j)}$

$$\left[H^{(ij)}, S_i^+\right] = F(\vec{r}_{ij})\left(S_j^z S_i^+ + \xi S_i^z S_j^+\right),$$
$$\left[H^{(ij)}, S_i^+ S_j^z\right] = \tfrac{1}{2}\xi F(\vec{r}_{ij})\left(S_i^+ + \xi S_j^+\right), \text{ etc.} \tag{9}$$

where $F(\vec{r}_{ij}) \equiv \dfrac{1-3\cos^2\theta_{ij}}{r_{ij}^3}$. Let's move on to assessing the effect of localization on the spectra. Denoting local field on site i as b, we get so–called the spin-isochromatic components [14]:

$$E(b) = 2\sqrt{2}(2s+1)^{-N/2} N^{-1/2} \sum_i S^\alpha_{i-1} S^+_i S^\alpha_{i+1},$$
$$E(0) = 2(2s+1)^{-N/2} N^{-1/2} \sum_i \left(S^\alpha_{i-1} S^+_i S^\beta_{i+1} + S^\beta_{i-1} S^+_i S^\alpha_{i+1}\right), \tag{10}$$
$$E(-b) = 2\sqrt{2}(2s+1)^{-N/2} N^{-1/2} \sum_i S^\beta_{i-1} S^+_i S^\beta_{i+1},$$

as well as new components which are "boundary terms" that need to be considered for return to the previous site:

$$E(b/2) = 2\sqrt{2}(2s+1)^{-N/2} N^{-1/2} \sum_i \left(S^+_i S^\alpha_{i+1} S^z_{i+2} + S^z_{i-2} S^\alpha_{i-1} S^+_i\right). \tag{11}$$
$$E(-b/2) = 2\sqrt{2}(2s+1)^{-N/2} N^{-1/2} \sum_i \left(S^+_{i-1} S^\beta_i S^z_{i+1} + S^z_{i-1} S^\beta_i S^+_{i+1}\right).$$

Having applied an identity substitution

$$S^\alpha_{i-2} S^\alpha_{i-1} S^+_i + S^+_i S^\alpha_{i+1} S^\alpha_{i+2} = S^z_{i-2} S^\alpha_{i-1} S^+_i + S^+_i S^\alpha_{i+1} S^z_{i+2} + S^\alpha_{i-1} S^+_i S^\alpha_{i+1} + \tfrac{1}{2}\left(S^\alpha_{i-1} S^+_i S^\beta_{i+1} + S^\beta_{i-1} S^+_i S^\alpha_{i+1}\right), \tag{12}$$

we get from the equation of motion for the many-body density matrix,

$$-\mathrm{i}\left[H^{(ij)}, E(b)\right] \to b\left[\left(1+\tfrac{1}{2}\xi\right)E(b) + \tfrac{1}{2}\xi E(b/2) + \tfrac{1}{2\sqrt{2}}\xi E(0)\right], \tag{13}$$

where boundary term is an eigenoperator $E(b/2)$, linearly independent and orthonormal. From the solution of the density matrix equation, it follows

$$\rho(t) = g_1(t)E(b) + g_2(t)E(b/2) + g_3(t)E(0) + g_4(t)E(-b/2) + g_5(t)E(-b), \tag{14}$$

which implies the equations for the components $g_{\{\varepsilon\}}(t)$: $\frac{\partial \vec{g}(t)}{\partial t} = -\mathrm{i}b\mathbf{C}\vec{g}$, for short.

After the nonselective $(\pi/2)_x$ pulse the first no constant term of the density matrix $\rho(t)$ is given by

$$\rho_1(0) = -2^{-N}\frac{\hbar\Omega}{kT}\sum_{i=1}^{N} S_i^+ = 2^{-N/2} q\left[E(b) + \sqrt{2}E(0) + E(-b)\right]. \tag{15}$$

where $q = \frac{\hbar\Omega\sqrt{N/2}}{2kT}$. The vector of initial values from equation (14) is $\vec{g}(0) = \left(1, \sqrt{2}, 1, 0, 0\right)$ that yields 5 possible spectral components. The amplitudes of individual spectral lines were determined by eigenvectors and eigenvalues of matrix **C**.

Finally, the separation Δ between line shape components in fluorapatite was found 3.50 Oe (the experimental value 3.64 ± 0.20 Oe [15]). This estimation demonstrates the strong influence of the localization on the NMR absorption line shape. Fluorapatite spin dynamics involves the orthonormal states such as, for example, $E(b)$ and $E(b/2)$, and constructs the actual eigenvectors of spin system as a linear superposition of the eigenvectors representing the two structures. As both structures have different energy, they are not equal contributors to the overall structure – the superposition is a weighted average, or a 0.959:0.267 linear combination of $E(b)$ and $E(b/2)$. When more contributing structures are included, the eigenvectors become more accurate and more excited states can be derived from different combinations of the contributing structures.

The final results, frequencies and intensities (squared amplitudes), are given in Table I.

Table 1. Calculated NMR spectrum for the linear three-spin chain model (relative to the Larmor frequency)[1].

Frequency (Oe)	Intensity
± 3.50	0.294
± 1.36	0.135
0	0.143

As expected, the next five-spin model transitions are grouped about the three-spin model transitions and after a Gaussian convolution, the line shapes from the two models would not differ significantly.

2.2. 3D dipolar model

The present work is essentially a computer simulation with identical, rigid spins. The first results justified our expectations [16]. They were obtained for 3D lattice structure of the single crystal CaF_2. The following expression was obtained for the shape of the NMR line, which is the Fourier transform of the FID components $g_{\{\varepsilon\}}(t)$:

$$g(\omega) = \mathrm{Re}\frac{\Gamma_x(\mathrm{i}\omega)}{1-\left(\xi^2+2\xi\right)\left(\mathrm{i}\omega\Gamma_x(\mathrm{i}\omega)-1\right)-\Gamma_x^2(\mathrm{i}\omega)\xi^2\Delta_z^2(1+\xi)^2}, \tag{16}$$

[1] For the largest ratio between the strongest intra- and cross- chain couplings.

where $\Delta_z^2 = \frac{1}{4}\sum_j b_{ij}^2$, and $\Gamma_x(i\omega)$ is connected by the Laplace transform with $\Gamma_x(t)$, correlation function of the total magnetic moment of the system. The difference was from contributions of cross-correlation functions in the expression for $F_x(t)$, autocorrelation function for the spin in one of the nodes, because we neglected the possibility of returning to the original node in the 3D CaF_2 spin system, which is strict for large-dimensional lattices if dimension $d \to \infty$. Under these assumptions, the following equation was obtained for $\Gamma_x(i\omega)$:

$$\Gamma_x(i\omega) = \int_0^\infty \exp\left\{-i\omega t - t\xi^2\Delta_z^2\Gamma_x(i\omega) - \frac{\Delta_z^2 t^2}{2}\right\} dt\,, \tag{17}$$

This result differs from the result of [3], in which not all the terms retained in the above approximation are considered. The fourth moment of the absorption line by formula (16) is less than the total value by 14%, and by the results of [3] without renormalization, by 19%. The value of the fourth moment for 1D fluorapatite is less than the total by only 10%, which confirms the applicability of our approximation.

3. Discussion

In the calculation of the spectra of correlation functions, we will follow an approximation proposed in [3–5]; in other words, we consider the longitudinal part of interaction in full while the transversal part is considered partly. Although the basic assumption about the constancy of fields is strict in the limit $\xi \to 0$, formula (16) qualitatively correctly describes the change in the form of the spectrum with a change in the magnetic anisotropy parameter ξ: the narrowing from Gaussian at 0 to the δ function at $\xi = -1$, and broadening for $\xi > 0$ too. We found a numerical solution of equation (17) for various values of the parameter ξ and calculated the atlas of the spectra using formula (17). For a truncated dipole Hamiltonian, a line with a flat top is obtained, somewhat flatter than in the experiment. The best agreement with experiment has spectra at lower values $|\xi|$. Similar expansion was obtained in the calculations of other authors [3, 5]. It was explained because of the difference in the action of close and distant spins on the spin. The inclusion of the processes discarded above, connected with the change in the longitudinal components of the spins, will act in the same direction.

In the formulas (16), (17) it is easy to include the heteronuclear interaction and inhomogeneous broadening of the Gaussian form. The mean square of the corresponding field Δ_H^2 is added to Δ_Z^2 in the last term of the exponent in (17), and in the formula (16) in the denominator it is necessary to replace in the second term ξ by $\xi_1 = \xi\Delta_z^2/(\Delta_z^2 + \Delta_H^2)$, and in the last term $\xi^2\Delta_z^2(1+\xi)^2$ by $\xi_1^2(\Delta_z^2 + \Delta_H^2) + \xi^2\Delta_z^2(2\xi_1 + \xi_1^2)$. The calculation performed with the obtained formulas makes it possible to trace the approximation of the shape of the NMR line to the Gaussian line in such systems as the parameter increases.

In conclusion, we considered model of localized states in fluorine spin system of fluorapatite in comparison with 3D case. We have considered low-dimensional spin system and found strong dependence the absorption NMR line shape on spin localization.

References

[1] Zobov V E and Kucherov M M 2017 *JETP* **124** 151
[2] Madi Z L, Bruschweiler R, and Ernst R R 1998 *J. Chem. Phys.* **109** 10603
[3] Sauermann G and Wiegand M 1981 *Physica B* **103** 309
[4] Kucherov M M 1978 *NMR in crystals* ed A G Lundin (Krasnoyarsk: IP SB RAS) p 47
[5] Karnaukh G E, Lundin A A, Provotorov B N, and Summanen K T 1986 *Sov. Phys. JETP* **64** 1324
[6] Giamarchi T 2004 *Quantum physics in one dimension (International series of monographs on physics)* (Oxford)
[7] Takigawa M, Motoyama N, Eisaki H, and Uchida S 1997 *Phys. Rev. B* **55** 14129
[8] Chen X *et al.* 2008 *Phys. Rev. Lett.* **101** 197208

[9] Hirjibehedin C F, Lutz C P, Heinrich A J 2006 *Science* **312** 1021
[10] Porras D and Cirac J I 2004 *Phys. Rev. Lett.* **92** 207901
[11] Hilker T A, Salomon G, Grusdt F, Omran A, Boll M, Demler E, Bloch I, Gross C 2017 *Science* **357** 484
[12] Vogler A, Labouvie R, Barontini G, Eggert S, Guarrera V, and Ott H 2014 *Phys. Rev. Lett.* **113** 215301
[13] Mazurenko A, Chiu C S, Ji G, Parsons M F, Kanasz-Nagy M, Schmidt R, Grusdt F, Demler E, Greif D, and Greiner M 2017 *Nature* **545** 462
[14] Nevzorov A A and Freed J H 2000 *J. Chem. Phys.* **112** 1425
[15] Van der Lugt W and Caspers W J 1964 *Physica (Utr.)* **30** 1058
[16] Zobov V E and Kucherov M M 1990 *Modern NMR and EPR methods in solid state chemistry* ed L N Gavrilova (Chernogolovka: IPCP RAS) p 103

IOP Conf. Series: Journal of Physics: Conf. Series **987** (2018) 012044 doi:10.1088/1742-6596/987/1/012044

Adherence of amino acids functionalized iron oxide nanoparticles on bacterial models *E. Coli* and *B. subtilis*

W. Trujillo[1], J. Zarria[2], J. Pino[2], L. Menacho[1], M. Coca[2] and A. Bustamante[1]

[1] Laboratorio de Cerámicos y Nanomateriales, Facultad de Ciencias Físicas, Universidad Nacional Mayor de San Marcos, Lima, Perú
[2] Laboratorio de Reproducción y Biología de Desarrollo, Facultad de Ciencias Biológicas, Universidad Nacional Mayor de San Marcos, Lima, Perú

E-mail: jacquelyne.zarria@unmsm.edu.pe

Abstract. Magnetic iron oxides nanoparticles (NPs) functionalized with lysine (Lys) and arginine (Arg) was obtained by following chemical co-precipitation route in basic medium. The synthesis was performed by mixing ferrous chloride ($FeCl_2$•$4H_2O$), ferric chloride ($FeCl_3$•$6H_2O$) and the specific amino acid in a molar ratio of 1: 2: 0.5, respectively. High pH sample was washed several times with distilled water to reach a pH similar to distilled water (Ph=7) after the synthesis process, part of the NPs obtained was dried. Of the measurements of XRD and MS was obtained that the samples are magnetic nanoparticles of maghemite of about 9 nm in diameter. Of the F-TIR and zeta potential measures was obtained that the amino acids Lys and Arg were correctly functionalized at magnetic nanoparticles, referred to herein as M@Lys and M@Arg. In order to demonstrate the capture and adhesion of the nanoparticles to the bacteria, scanning electron microscopy (SEM) was performed. The obtained visualization of both bacteria shows that they are coated by the magnetic particles. In addition, M@Lys (B. sutilis) were cultured to verify the inhibition of growth measured by colony forming units (CFU), the concentrations of M@Lys were $1.75x10^2$ g/mL and $0.875x10^2$ g/mL. After the confrontation obtained efficiencies of 75.63% and 98.75% respectively for the third dilution. While for the fourth dilution were 90% and 98.57% respectively were obtained for each concentration of nanoparticles. Hinting that a high efficiency of bacterial capture at very low concentrations of NPs, which gives us a tool to capture nanobiotechnology bacteria in liquid cultures with application to capture them in wastewater. Based on our results we concluded that NPS functionalized with the amino acids Lys and Arg adhere to the bacteria efficiently in low concentrations.

1. Introduction

The traditional chemicals disinfectants such as sodium hypochlorite, ozone, antibiotics, among others commonly used for water disinfection are left behind due to the more than 600 by-products derived from these, many of them carcinogenic ([1] [2]). In addition, the use of antibiotic-based disinfectants generates resistance in the microorganism which implies the inversion and synthesis of new disinfectants with more potent or conjugated antibiotics which has been proven to have a side effect and an unfavorable impact on the surrounding biota ([3] [4]). According to the current regulation in Peru of the *Estándares Nacionales de Calidad Ambiental para el Agua* (D.S.002.2008 ECA AGUA), category 1: Population and recreational. It shows parameters and concentrations of permissible components in the water that reaches the population, including the category of coliforms where we

find thermotolerant coliforms, total coliforms, fecal enterococci, Escherichia Coli, parasitic forms, Giardia duodenalis, salmonella and Vibrio cholerae. In the aforementioned list we observed the presence of Gram-positives (V. cholerae) and Gram-negatives (E. coli, fecal enterococcus) bacteria.
In this context, the functionalized magnetic nanoparticles (NPs) are a very interesting alternative because they have shown a great application potential in water treatment processes, especially for the bacteria elimination ([5]). In this study, magnetite NPs (Fe_3O_4) functionalized with two amino acids (arginine and lysine) denominated as Mgh@Arg and Mgh@Lys respectively were synthesized. In general, this functionalization will be referred as M@AA. The synthesis and functionalization of M@AA will be performed by the chemical route of co-precipitation ([6] [7]).
The M@AA application tests were performed on the bacteria: *Escherichia Coli* (E. Coli) and *Bacillus Subtilis* (B. Subtilis), which have shown affinity for these nanomaterials functionalized with amino acids ([5]). This is a result of its organic composition of the cell wall. Both bacteria have been used as a models of Gram-negative and Gram-positive bacteria.
The results obtained from the synthesis, functionalization and application of these functionalized nanomaterials, with the objective of being applied in the future in the capture of bacteria for the disinfection of water and giving a great scientific contribution in the basic research, lays the bases for further studies in the use and application of nanomaterials for an environmentally friendly industry. Also, the high efficiency of these processes reach scales of high impact not only environmental, but also social, such as for the reuse and cleaning of treated water used for irrigation, enhance the purification of water for human consumption and avoid the loss of this vital resource ([2] [8] [9] [10]).

2. Materials and Methods

Synthesis and characterization of the functionalized magnetic nanoparticles

The magnetic nanoparticles were synthesized in the Laboratory of Ceramics and Nanomaterials in the Faculty of Physical Sciences of the UNMSM, following the chemical method of co-precipitation synthesis in basic medium from ferric and ferrous salts. The obtained samples were characterized by X-ray diffraction (XRD), Mössbauer spectroscopy (MS), Z potential (ζ) and infrared spectroscopy (FTIR), which were performed at the Faculty of Physical Sciences of the UNMSM, National University of Trujillo and Texas A&M University. Finally, a stock of functionalized magnetite nanoparticles was obtained at concentrations of Mgh@Arg (15.1 mg/ml) and Mgh@Lys (17.5 mg/ml), of which three concentrations were tested for Mgh@Lys of 1.75 and 0.875 mg/lm in bacterial culture.

Culture for Gram-positive bacteria (B. subtilis)

The culture of Gram-positive bacteria was prepared in a volume of 100ml of growth medium with 15 g/L tryptone, 5 g/L yeast extract and 5 g/L NaCl. *Bacillus subtilis* of the NRRL strain was used as the biological model of positive bacteria.
Both cultures were homogenized at 200 rpm in a temperature-controlled homogenizer. Thereafter, E. Coli will be allowed to grow for a period of 16 hours at 37 °C while *B. subtilis* will be cultured for 32 hours at 30 °C.

Bacterial count

A volume of 10ml will be taken into conical tubes and centrifuged for 10 minutes at 4000 g at 4 °C. The obtained pellet will be isolated and washed three times with sterile saline (0.9% NaCl at pH 7) to remove all medium. After this, it is resuspended in saline solution and a stock of bacteria is separated and cultured in growth nutrient agar. The quantification was performed by counting of CFU/ml. This procedure was repeated after the confrontation with the nanoparticles.

Bacterial capture by nanoparticles

The functionalized magnetic nanoparticles will be homogenized in physiological saline solution for the dispersion and conjugation of the bacteria, taking care of the pH values using HCI and NaOH. The bacterial capture will be performed in the shaker tempered at 200 rpm for 20 minutes.

Quantification of post-exposure bacteria to functionalized nanoparticles

In the same way in which the bacteria were quantified to verify their growth and viability, the bacteria were centrifuged, diluted and cultured in order to know the amount of CFU/ml and to verify the bacterial growth.

Observation of bacteria captured by nanoparticles by Scanning Electron Microscopy

During the bacterial capture procedure, 1 ml of solution was taken, which will be gradually fixed with 2% Glutaraldehyde (diluted with phosphate buffer). After the gradual exchange of saline solution by the fixative solution, the sample will be left for a period of 12h for the correct fixation of the bacteria so these do not lose their form for the visualization under the electron microscope of the contact of the bacterium with functionalized nanoparticles.

After 12h, the bacteria were washed with phosphate buffer pH 7, and after that the dehydration in the alcohol battery from 20 °, 30 °, 40 °, 50 °, 60 °, 60 °, 70 °, 80 °, 90 °, 100 ° and finally by acetone, each of these steps was performed for 10 minutes each. The last two steps were repeated in duplicate.

After dehydrating the sample, it went through a process of desiccation at critical point where the sample already fixed and dehydrated loses the water completely, taking advantage of the critical point of the same 5 °C and the exchange by CO_2. This procedure was carried out in the Laboratory of Specialized Equipment (LSE) of the Faculty of Biological Sciences of the UNMSM, in the desiccator at Critical Point EMS.

After the desiccation process was carried out at the critical point, the sample was completely dry. It was adhered to a metal support with self-adhesive conductive carbon tapes. Once the bacteria captured by the nanos on the surface of the support entered a process of coating with plasma gold, this process will be carried out with the LSE Metalizer of FBS-UNMSM, brand SPI supplies.

In the same laboratory was carried out the visualization of the bacteria captured by the nanoparticles in the Scanning Electron Microscope of the FEI Mark model Inspect S50, in high vacuum mode with the ETD detector to identify the respective ultrastructure.

3. Results and Discussions

Characterization:

X-ray diffraction and Mössbauer spectroscopy

X-ray diffraction (XRD) measurements were performed on a Bruker diffuser, model D8-Focus, with LYNSEIS detector, using copper Kalfa radiation with an average wavelength of 1.54184 Å. The sweep angle 2θ was 8 to 80 sexagesimal degrees with a pitch of $\Delta\theta = 0.01°$ and time of 2s per data, as shown in figure 1.

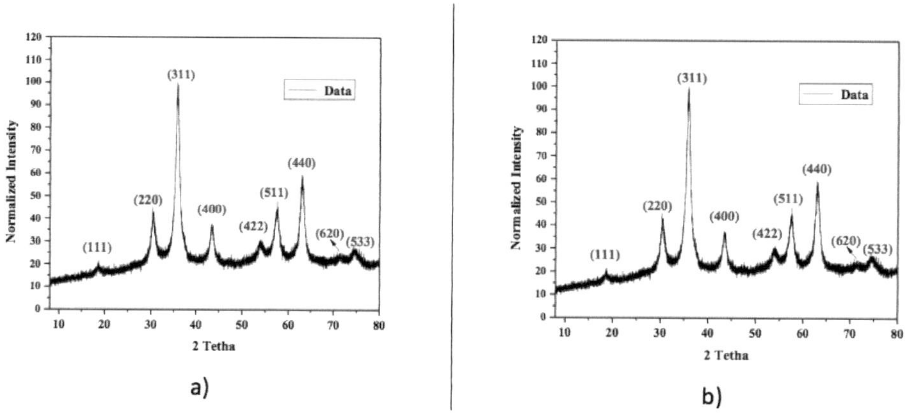

Figure 1. X-ray diffractogram of Maghemite with arginine (a) and lysine (b)

The analysis of X-ray diffractograms indicates the presence of maghemite (γ-Fe_2O_3) which has indexed the major peaks using PDF N° 39-1346 with widened lines in the samples covered with arginine (M@Arg) and with lysine (M@Lys). On the other hand, by performing a more detailed analysis on the peak (511) (Fig. 2), we see that in the samples there is a distribution of two sizes of crystallites: one of 8.5 nm and 19.2 nm in the Mgh@Arg and another one of 6.4 nm and 16.1nm for the Mgh@Lys, which was calculated using the Debye-Scherrer equation.
The fitted to the peak (511) located at $2\theta = 57.43°$ of the Mh@Arg used two Gaussians and with a $\chi 2$ of 1.346, with the following parameters to the first line of $Y_{01} = 21.88663$ for the base, $X_{c1} = 57.35115°$ central position, the area $A_1 = 12.17818$ and the line width at half of the maximum height was $w_1 = 1.85978°$ (red line) and for the second peak with $Y_{02} = 21.88663$, $X_{c2} = 57.47709°$, $A_2 = 9.79644$ and $w2 = 0.82529$ (green line). While the parameters for the Mgh@Lys, the adjustment had a $\chi 2$ of 1.556, whose parameters are for the first line $Y_{01} = 23.64279$, $X_{c1} = 57.30808°$, $A_1 = 15.20828$ and $w_1 = 2.47269°$ (red line) and $Y_{02} = 23.64279$, $X_{c2} = 57.46954$ °, $A_2 = 14.13105$ and $w_2 = 0.98138°$ for the second line (green line).

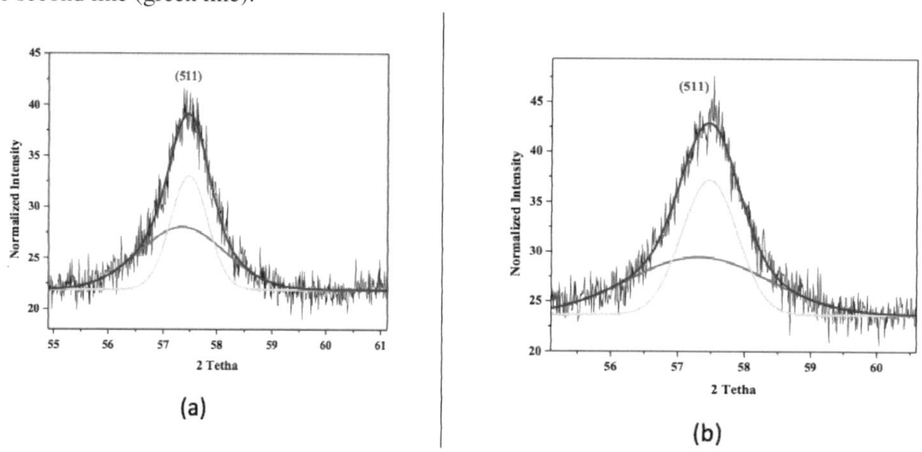

Figure 2. Fitted with 2 Gaussian of the peak (511) of the Maghemite with arginine (a) and lysine (b)

The Mössbauer Spectroscopy (MS) measurements at room temperature were performed with a transducer with sinusoidal signal, a source of Co^{57} and 1024 channels (Fig. 3). The fitted were made with the software NORMOS and for the sample Mgh@Arg is shown a sextet with widened line widths typical of a magnetic field distribution (B_{hf}) due to the small particle size less than 100Å (10 nm), corresponding to a superparamagnetic relaxation. The distribution has a mean field B_{hf} = 46.7T and an isomer shift (IS) of 0.333 mm/s characteristic of Fe^{3+} of the maghemite, as shown in the upper part of figure 3a and, to the right, its distribution of hyperfine magnetic fields. For the sample Mgh@Lys was adjusted with a distribution of hyperfine magnetic fields with a mean field B_{hf}= 43.16T and with IS of 0.387 mm/s characteristic of Fe^{3+} as seen in the lower left of figure 3a and. to the right, its distribution of B_{hf} fields; the sextet corresponds to a magnetic field of 47.21T with IS of 0.325 mm/s, and a quadrupolar perturbation ε = -0.002 mm/s. When one compares the two Mössbauer spectra, we observe that the Mgh@Lys spectrum is more collapsed than the Mgh@Arg due not only to the small size of the particles, but also to the effect of the covering that affects the magnetic dipole interaction.

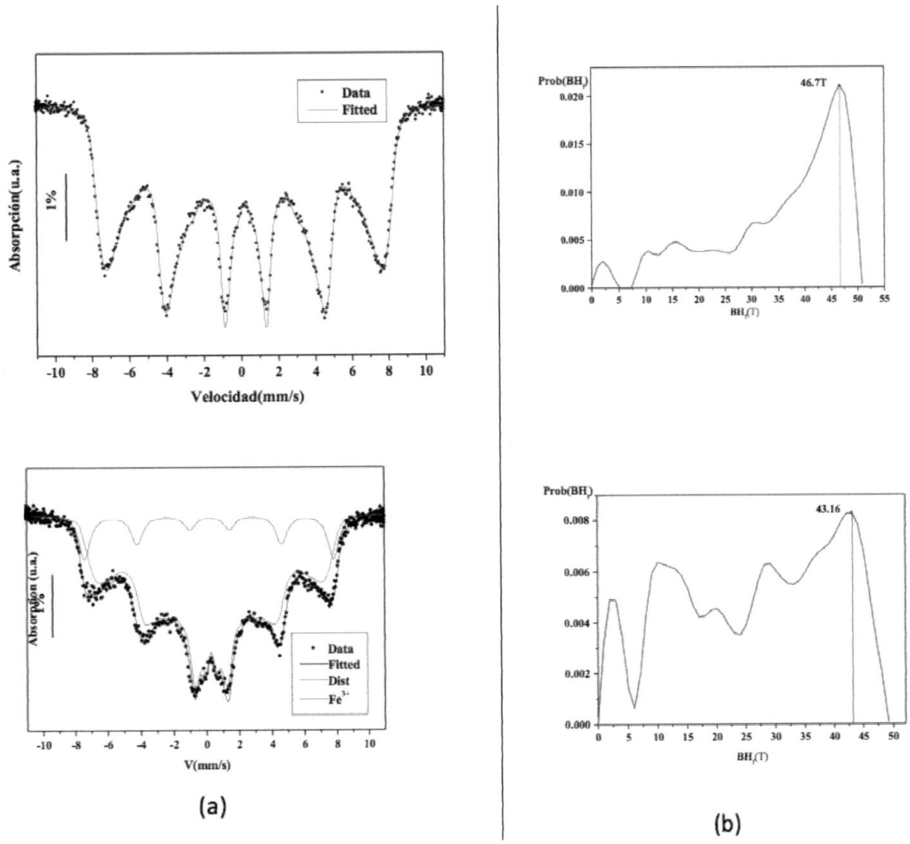

Figure 3. ME at RT from the Mgh@Arg in the the upper left, with Mgh@Lys at the bottom and to the right, their respective distributions of hyperfine magnetic fields.

Culture of Gram-positive bacteria

Centrifuge and obtain the bacterial pellet, which was resuspended in 10 ml of saline solution. *Bacillus subtilis* from the NRRL strain was washed three times with saline. Cultures obtained from the dilutions obtained averaged 9.5 x 10^6 CFU/ml.

Clash of bacteria with functionalized nanoparticles

When cultivating the third dilution under sterile conditions, *Bacillus subtilis* of the NRRL strain obtained a post challenge growth of 2,3 x 10^6 CFU ml for the concentration of M@Lys 1.75 mg/ml; 1.2x10^5 CFU/ml for 0.875 mg/ml, resulting 75.63 in efficiency and 98.75% efficiency in bacterial capture and inhibition of growth as seen in Fig. 4.

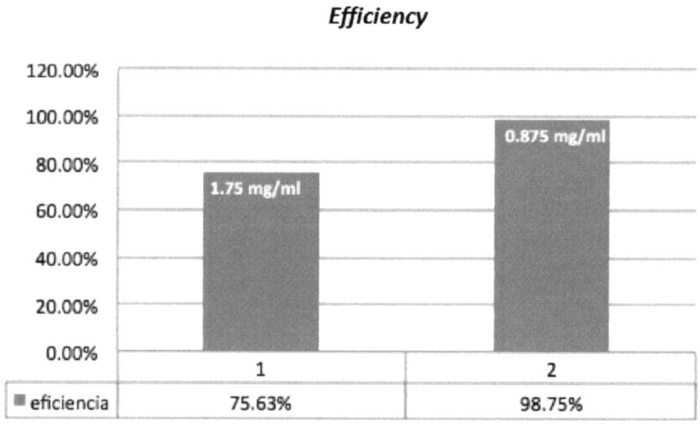

Figure 4. The efficiency of *B. subtilis* growth inhibition at two concentrations of M@Lys is observed in the bar graphs, at concentrations of 1.75 and 0.875 mg/ml. For the third dilution before cultivation.

Cultivation of the fourth dilution under sterile conditions to *Bacillus subtilis* of the NRRL strain resulted in post-challenge growth of 9,5 x 10^5 CFU/ml for the concentration of M@Lys 1.75 mg/ml; 1.4 x 10^5 CFU/ml for 0.875 mg/ml, resulting 90% in efficiency and 98.57% efficiency in the bacterial capture and growth inhibition as seen in Fig. 5.

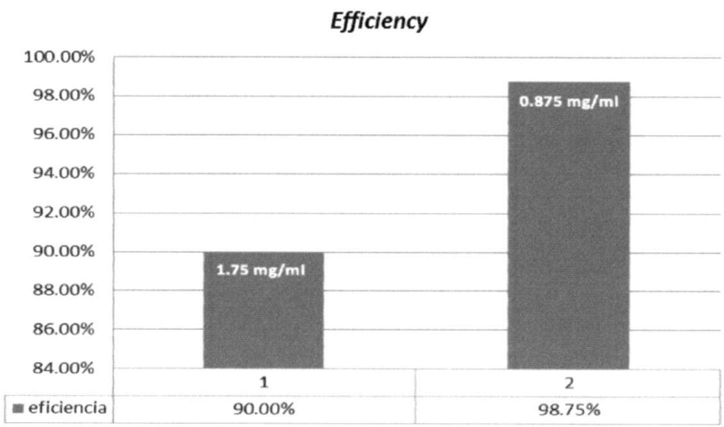

Figure 5. The growth inhibition efficiency of *B. subtilis* is shown in the bar graph for two concentrations of NP@Lys at concentrations of 1.75 and 0.875 mg/ml. For the fourth dilution before cultivation.

Evidence of bacterial capture by Scanning Electron Microscopy

After the fixation, dehydration, drying and metallization of the samples, micrographs of E. coli were obtained, where it was successfully observed—as shown in Fig. 6 (a, b, c and d) where the bacterial population is visualized—.Many of them were accumulated by centrifugation, surrounded by the M@Arg nanoparticles.

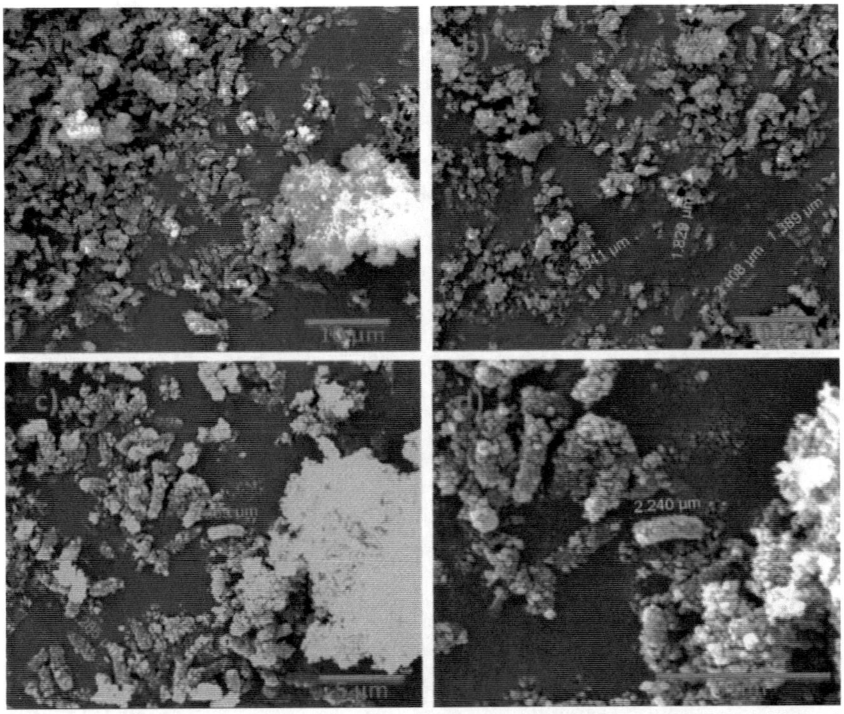

Figure 6. Four microphotographs are observed a) 6000X at 10 µm scale, b) 6000X at 10 µm in which 4 bacteria (E. coli) with an average size of 1.8 µm are observed, c) 12000X at 5 µm of scale, d) 24000X at 5 µm scale where a bacteria of 2,240 µm is observed. All bacteria totally captured by M @ Arg nanoparticles. The microphotographs were obtained at voltages between 20-30 kV, 2.62e-5 Torr pressure and with the ETD detector of the FEI-Inspect S50 Microscope.

On the other hand, *B. subtilis* was visualized in the same way by the Scanning microscopy. However, its visualization was faster due to its size as we can see in Fig. 7 and 8. The individuals of *B. subtilis* captured by the M@Lys nanoparticles can be seen.

IOP Conf. Series: Journal of Physics: Conf. Series **987** (2018) 012044 doi:10.1088/1742-6596/987/1/012044

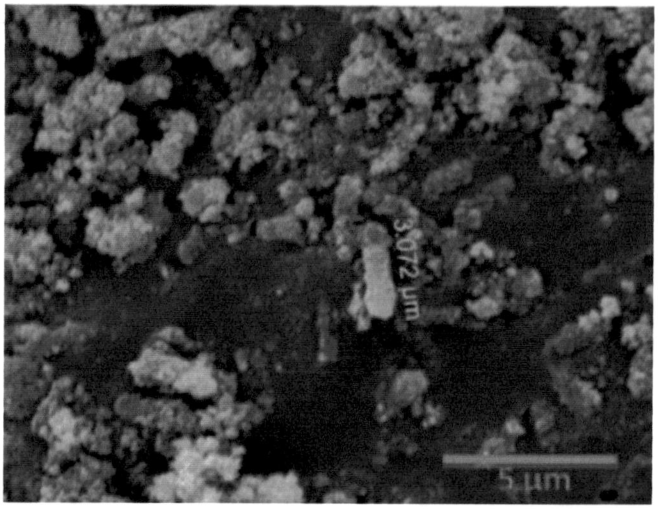

Figure 7. **The *B. subtilis* bacillus is observed at the center of the 5 µm scale image with an approximate size of 3,072 µm, captured by the M@Lys nanoparticles. The photomicrograph was obtained at 15000 magnification, 12.5kV, 1.37e-5 Torr pressure and with the ETD detector of the FEI-Inspect S50 microscope.**

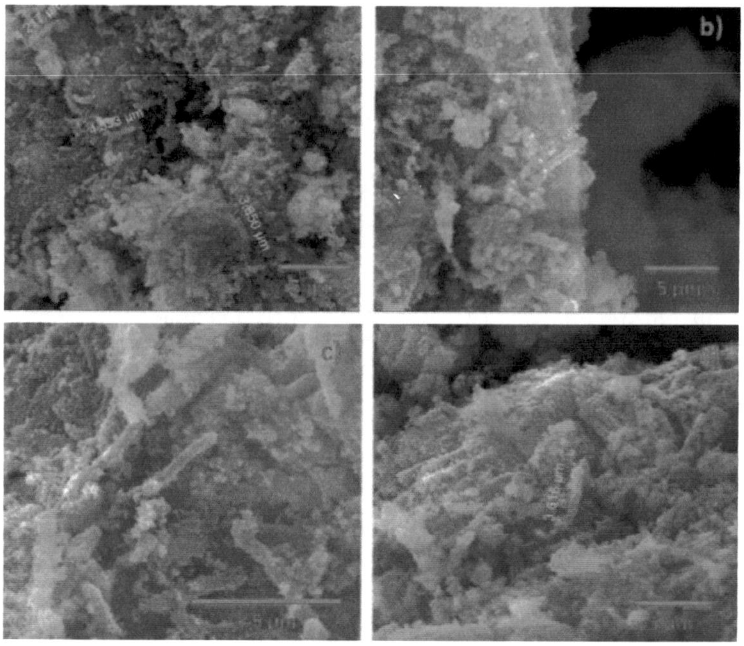

Figure 8. **There are four microphotographs: a) 6000X at 10 µm of scale, three measurements of *B. subtilis* with an average size of 4.15 µm are observed, b) 12000X at 5 µm where *B. subtilis* is observed with a size of 3,706 µm, c) 24000X at 5 µm scale, d) 12000X at 5 µm scale where a 3,615 µm bacterium is observed, all the bacilli captured by the M @ Lys nanoparticles. The microphotographs were obtained at voltages between 20-30 kV, 4.15e-5 Torr of pressure and with the ETD detector of the FEI-Inspect S50 Microscope.**

 doi:10.1088/1742-6596/987/1/012044

4. Conclusions

The M@Lys NPs showed more efficience in the fourth dilution tan the third dilution, this could be for bacterial saturation in the solution, also the afinitive between M@Lys and gram positive bacteria.
The results indicate that at very low concentrations of M@Lys and M@Arg, these can perform an efficient bacterial capture, which opens doors to the reduction of bacterial proliferation in liquid media—with low budget and easy laboratory manipulation—, in the particular case of the application to the bacterial capture in sewage.

5. References

[1] Krasner S, Weinberg H, Richardson S, Pastor S, Chinn R, Sclimenti M, Onstad G, and Thruston A 2006 Occurrence of a new generation of disinfection by products. *Environ. Sci. Technol.* **40** 7175-85

[2] Li Q, Mahendra S, Lyon D, Brunet L, Liga M, Li D and Álvarez P 2008 Antimicrobial nanomateriales for water disinfection and microbial control: potencial applications and implications *Water Res.* **42** 4591-4602

[3] Mahmoudi M and Serpooshan V 2012 Silver-coated Engineered magnetic nanoparticles are promising for the Success in the figth against antibacterial resistance threat *Acs Nano* **6** 2656-64

[4] Wu M, Deokar A, Liao J, Shih P and Ling Y 2013 Graphene-based photothermal agent for rapid and effective killing of bacteria *Acs Nano* **7** 1281-90

[5] Yinjia J, Fei L, Chao S, Meipin T and Yanglang H 2014 Efficient bacterial capture with amino acid modified magnetic nanoparticles *Water Research* **50** 124-34

[6] Herrera W 2013 Preparación y caracterización de nanopartículas de magnetita funcionalizados con ácido láurico, oleico y etilendiamino tetraacético para aplicaciones biomédicas y remediación ambiental. Tesis de Licenciatura, Facultad de Ciencias Físicas. Universidad Nacional Mayor de San Marcos.

[7] Jolivet J, Chanéaca C and Tronc E 2004 Iron oxide chemistry. From molecular clusters to extended solid networks *Chemical Communications* **10** 481-83

[8] Rastogi SK, Jabal JMF, Zhang H, Gibson CM, Haler KJ, Qiang Y, Aston D and Branen A 2011 Antibody@Silica Coated Iron Oxide Nanoparticles: Synthesis, Capture of E. Coli and Sers Titration of Biomolecules with Antibacterial Silver Colloid *J Nanomedic Nanotechnol* **2** 121

[9] Laurent S, Forge D, Port M, Roch A, Robic C, Elst L and Muller N 2008 Magnetic Iron Oxide Nanoparticles: Synthesis, Stabilization, Vectorization, Physicochemical Characterizations, and Biological Applications *Chem. Rev.* **108** 2064-2110

[10] Jin Y, Deng J, Yu J, Yang C, Tong M and Hou Y 2015 Fe_5C_2 nanoparticles: a reusable bactericidal material with photothermal efects under near-infrared irradiation *Journal of Materials Chemistry B* **3** 3993-4000

Institute of Physics
Dirac House, Temple Back
Bristol BS1 6BE UK

CURRAN ASSOCIATES INC.
proceedings
.com

ISSN: 1742-6588
ISBN 978-1-5108-6123-7